MATHEMATICAL SURVEYS
NUMBER VI

INTRODUCTION TO THE THEORY OF ALGEBRAIC FUNCTIONS OF ONE VARIABLE

BY

CLAUDE CHEVALLEY

PUBLISHED BY THE
AMERICAN MATHEMATICAL SOCIETY

1951

First printing, 1951
Second printing, 1963
Third printing, 1967
Fourth printing, 1971

International Standard Book Number 0-8218-1506-7
Library of Congress Catalog Number 51-4714

To

SYLVIE CHEVALLEY

INTRODUCTION

An algebraic function y of a complex variable x is a function which satisfies an equation of the form $F(x, y) = 0$, where F is a polynomial with complex coefficients; i.e., y is a root of an algebraic equation whose coefficients are rational functions of x. This very definition exhibits a strong similarity between the notions of algebraic function and algebraic number, the rational functions of x playing a role similar to that played by the rational numbers. On the other hand, the equation $F(x, y) = 0$ may be construed to represent a curve in a plane in which x and y are the coordinates, and this establishes an intimate link between the theory of algebraic functions of one variable and algebraic geometry.

Whoever wants to give an exposition of the theory of algebraic functions of one variable is more or less bound to lay more emphasis either on the algebraico-arithmetic aspect of this branch of mathematics or on its geometric aspect. Both points of view are acceptable and have been in fact held by various mathematicians. The algebraic attitude was first distinctly asserted in the paper *Theorie der algebraischen Funktionen einer Veränderlichen*, by R. Dedekind and H. Weber (Journ. für Math., 92, 1882, pp. 181–290), and inspires the book *Theorie der algebraischen Funktionen einer Variabeln*, by Hensel and Landsberg (Leipzig, 1902). The geometric approach was followed by Max Noether, Clebsch, Gordan, and, after them, by the geometers of the Italian school (cf. in particular the book *Lezione di Geometria algebrica*, by F. Severi, Padova, 1908). Whichever method is adopted, the main results to be established are of course essentially the same; but this common material is made to reflect a different light when treated by differently minded mathematicians. Familiar as we are with the idea that the pair "observed fact—observer" is probably a more real being than the inert fact or theorem by itself, we shall not neglect the diversity of these various angles under which a theory may be photographed. Such a neglect should be particularly avoided in the case of the theory of algebraic functions, as either mode of approach seems liable to provoke strong emotional reactions in mathematical minds, ranging from devout enthusiasm to unconditional rejection. However, this does not mean that the ideal should consist in a mixture or synthesis of the two attitudes in the writing of any one book: the only result of trying to obtain two interesting photographs of the same object on the same plate is a blurred and dull image. Thus, without attacking in any way the validity *per se* of the geometric approach, we have not tried to hide our partiality to the algebraic attitude, which has been ours in writing this book.

The main difference between the present treatment of the theory and the one to be found in Dedekind-Weber or in Hensel-Landsberg lies in the fact that the constants of the fields of algebraic functions to be considered are not necessarily the complex numbers, but the elements of a completely arbitrary field. There are several reasons which make such a generalization necessary. First, the analogy between algebraic functions and algebraic numbers becomes even closer if one considers algebraic functions over finite fields of constants. In that case, on the one hand class field theory has been extended to the case of fields of functions, and, on the other hand, the transcendental theory (zeta function, L-series) may also be generalized (cf. the paper of F. K. Schmidt, *Analytische Zahlentheorie in Körpern der Charakteristik p*, Math. Zeits., 33, 1931). Moreover, A. Weil has succeeded in proving the Riemann hypothesis for fields of algebraic functions over finite fields, thereby throwing an entirely new light on the classical, i.e., number theoretic, case (cf. the book of A. Weil, *Sur les courbes algébriques et les variétés qui s'en déduisent*, Paris, Hermann, 1948; this book contains an exposition of the theory from a geometric point of view, although this point of view is rather different from that of the Italian geometers). Secondly, if S is an algebraic surface, and R the field of rational functions on S, then R is a field of algebraic functions of one variable over $K\langle x\rangle$, where K is the basic field and x a non constant element of R. E. Picard, among others, has very successfully used the method of investigation of S which amounts to studying the relationship between R and various fields of the form $K\langle x\rangle$ (cf. E. Picard and G. Simart, *Théorie des fonctions algébriques de deux variables independantes*, Paris, Gauthier-Villars, 1897). Now, even when K is the field of complex numbers, $K\langle x\rangle$ is not algebraically closed, which makes it necessary to have a theory of fields of algebraic functions of one variable over fields which are not algebraically closed.

The theory of algebraic functions of one variable over non algebraically closed fields of arbitrary characteristic has been first developed by H. Hasse, who defined for these fields the notion of a differential (H. Hasse, *Theorie der Differentiale in algebraischen Funktionenkörpern mit volkommenen Konstantenkörper*, Journ. für Math., 172, 1934, pp. 55–64), and by F. K. Schmidt, who proved the Riemann-Roch theorem (F. K. Schmidt, *Zur arithmetischen Theorie der algebraischen Funktionen*, I, Math. Zeits., 41, 1936, p. 415). In this book, we have used the definition of differentials and the proof of the Riemann-Roch theorem which were given by A. Weil (A. Weil, *Zur algebraischen Theorie der algebraischen Funktionen*, Journ. für Math., 179, 1938, pp. 129–133).

As for contents, we have included only the elementary part of the theory, leaving out the more advanced parts such as class field theory or the theory of correspondences. However, we have been guided by the desire of furnishing a suitable base of knowledge for the study of these more advanced chapters. This is why we have placed much emphasis on the theory of extensions of fields of algebraic functions of one variable, and in particular of those extensions which are obtained by adjoining new constants, which may even be transcendental over the field of constants of the original field of functions. That the consideration of such extensions is desirable is evidenced by the paper of M. Deuring, *Arithmetische Theorie der Korrespondenzen algebraischen Funktionenkörper*, Journ. für Math., 177, 1937. The theory of differentials of the second kind has been given only in the case where the field under consideration is of characteristic 0. The reason for this restriction is that it is not yet clear what the "good" definition of the notion should be in the general case: should one demand only that the residues be all zero, or should one insist that the differential may be approximated as closely as one wants at any given place by exact differentials (or a suitable generalization of these)? Here is a net of problems which, it seems, would deserve some original research. The last chapter of the book is concerned with the theory of fields of algebraic functions of one variable over the field of complex numbers and their Riemann surfaces. The scissor and glue method of approach to the idea of a Riemann surface has been replaced by a more abstract definition, inspired by the one given by H. Weyl in his book on Riemann surfaces, which does not necessitate the artificial selecting of a particular generation of the field by means of an independant variable and a function of this variable. We have also avoided the cumbersome decomposition of the Riemann surface into triangles, this by making use of the singular homology theory, as developed by S. Eilenberg.

I have been greatly helped in the writing of this book by frequent conversations with E. Artin and O. Goldman; I wish to thank both of them sincerely for their valuable contribution in the form of advice and suggestions.

TABLE OF CONTENTS

NOTATIONS FREQUENTLY USED

Con $_{R/S}$: conorm from R to S (IV, 7).
Cosp $_{R/S}$: cotrace from R to S (for repartitions, IV, 7; for differentials, VI, 2 and VI, 6).
$d(\mathfrak{a})$: degree of a divisor $\mathfrak{a}$ (I, 7).
$\mathfrak{d}(x)$: divisor of an element x (I, 8).
$\mathfrak{d}(\omega)$: divisor of a differential ω (II, 6).
$\delta(\mathfrak{a})$: dimension of the space of differentials which are multiple of a divisor $\mathfrak{a}$ (II, 5).
∂: boundary (VII, 3).
$H_n(X, Y)$: n-dimensional homology group of X modulo Y (VII, 3).
$i(\gamma, \gamma')$: intersection numbers of the 1-chains γ and γ' (VII, 6).
$j(\omega, \omega')$: (VII, 5).
$K\langle\cdots\rangle$: field obtained by adjunction to the field K of the element or elements or set of elements whose symbols are between the sign $\langle$ and the sign $\rangle$; special meaning for fields of algebraic functions of one variable defined in V, 4.
$l(\mathfrak{a})$: dimension of the space of elements which are multiples of a divisor $\mathfrak{a}$ (II, 1).
$\nu_{\mathfrak{p}}$: order function at a place $\mathfrak{p}$ (for elements, I, 5 and III, 1; for repartitions, II, 4; for differentials, II, 6).
$N_{S/R}$: Norm from S to R (IV, 7).
$N^{\mathfrak{P}}_{S/R}$: (IV, 5).
$\omega^{\mathfrak{p}}$: $\mathfrak{p}$-component of a differential ω (II, 7).
res $_{\mathfrak{p}}\omega$: residue of a differential ω at a place $\mathfrak{p}$ (III, 5).
$\mathrm{Sp}_{S/R}$: Trace from S to R (for repartitions, IV, 7; for differentials, VI, 2).
$\mathrm{Sp}^{\mathfrak{P}}_{S/R}$: (IV, 5).
$|\gamma|$: set of points of a chain γ (VII, 3).

CHAPTER I

PLACES AND DIVISORS

§1. Fields of algebraic functions of one variable

Let K be a field. By a *field of algebraic functions of one variable over K* we mean a field R containing K as a subfield and which satisfies the following condition: R contains an element x which is transcendental over K, and R is algebraic of finite degree over $K\langle x\rangle$.

The element x is of course not uniquely determined. If x' is any element of R which is transcendental over K, then R is algebraic of finite degree over $K\langle x'\rangle$. In fact, the degree of transcendency of R over K being one, R is algebraic over $K\langle x'\rangle$. In particular, x is algebraic over $K\langle x'\rangle$, and $K\langle x, x'\rangle$ is of finite degree over $K\langle x'\rangle$. Since R is of finite degree over $K\langle x\rangle$, it is *a fortiori* of finite degree over $K\langle x, x'\rangle$, which proves that it is of finite degree over $K\langle x'\rangle$.

Those elements of R which are algebraic over K are called *constants*. They form a certain subfield K' of R, the *field of constants*. The field R is also a field of algebraic functions of one variable over K'. In fact, any element x of R which is transcendental over K is also transcendental over K', and R, which is algebraic of finite degree over $K\langle x\rangle$, is also algebraic of finite degree over $K'\langle x\rangle$.

It is important to keep in mind that, when discussing properties of a field R of algebraic functions of one variable, we shall consider in fact not properties of the field R alone but properties of the pair formed by K and R. For instance, let Z be any field, and set $R = Z\langle x, y, z\rangle$, where x and y are algebraically independent over Z and z is algebraic over $Z\langle x, y\rangle$. Set $K_1 = Z\langle x\rangle$, $K_2 = Z\langle y\rangle$. Then R is a field of algebraic functions of one variable over either one of the fields K_1 or K_2; but its properties as a field of algebraic functions of one variable over K_1 may be quite different from its properties as a field of algebraic functions of one variable over K_2.

However, when considering a field R of algebraic functions of one variable over a field K, the field of constants of R will appear more and more to be the essential object instead of K itself, which will gradually fade into the background.

§2. Places

Let R be a field and K a subfield of R. By a *V-ring* in R (over K) is meant a subring $\mathfrak{o}$ of R which satisfies the following conditions:

1. $\mathfrak{o}$ contains K;
2. $\mathfrak{o}$ is not identical with R;
3. If x is an element of R not in $\mathfrak{o}$, then x^{-1} is in $\mathfrak{o}$.

Let $\mathfrak{o}$ be a V-ring. Those elements in $\mathfrak{o}$ which are not units in $\mathfrak{o}$ (we call them "non-units") form an ideal $\mathfrak{p}$ in $\mathfrak{o}$. In fact, if x is a non-unit and $z \in \mathfrak{o}$, then xz

is a non-unit, because, if xz had an inverse u in $\mathfrak{o}$, zu would be in $\mathfrak{o}$ and inverse of x. Now, let x and y be non-units in $\mathfrak{o}$. If either x or y is 0, $x - y$ is clearly a non-unit. If x and y are both $\neq 0$, one at least of the pair of inverse elements x/y and y/x is in $\mathfrak{o}$. If $x/y \in \mathfrak{o}$, then $x - y = y(x/y - 1)$ is a non-unit; if $y/x \in \mathfrak{o}$, then $x - y = x(1 - y/x)$ is a non-unit. Thus the non-units of $\mathfrak{o}$ form an ideal $\mathfrak{p}$. Any ideal in $\mathfrak{o}$ containing $\mathfrak{p}$ but $\neq \mathfrak{p}$ contains a unit and therefore coincides with $\mathfrak{o}$.

Now, let R be a field of algebraic functions of one variable over a field K. By a *place* in R we mean a subset $\mathfrak{p}$ of R which is the ideal of non-units of some V-ring $\mathfrak{o}$ of R (over K). This V-ring is uniquely determined when $\mathfrak{p}$ is given. In fact, it is the set of all $x \in R$ such that $x\mathfrak{p} \subset \mathfrak{p}$ (we mean by $x\mathfrak{p}$ the set of products of x by elements of $\mathfrak{p}$). To show this, we observe first that any $x \in \mathfrak{o}$ has the required property; on the other hand, if $x \notin \mathfrak{o}$, then x^{-1} is in $\mathfrak{o}$ and is not a unit, whence $x^{-1} \in \mathfrak{p}$ and $1 \in x\mathfrak{p}$, $x\mathfrak{p} \not\subset \mathfrak{p}$. The ring $\mathfrak{o}$ is called the *ring of the place* $\mathfrak{p}$. The elements of $\mathfrak{o}$ are said to be *integral* at the place $\mathfrak{p}$.

Since every element of $\mathfrak{o}$ not in $\mathfrak{p}$ is a unit in $\mathfrak{o}$, we see immediately that the residue ring $\mathfrak{o}/\mathfrak{p}$ is a field. This field is called the *residue field* of the place $\mathfrak{p}$.

The ring $\mathfrak{o}$ *is integrally closed in* R, i.e., every element x of R which satisfies an equation of the form $x^n + \sum_{i=1}^{n} a_i x^{n-i} = 0$, with $a_1, \cdots, a_n$ in $\mathfrak{o}$, itself belongs to $\mathfrak{o}$. For, were this not the case, then x^{-1} would belong to $\mathfrak{p}$, and we would have $1 = -\sum_{i=1}^{n} a_i (x^{-1})^i \in \mathfrak{p}$, which is impossible. It follows in particular that the ring of any place contains the field of constants K' of R. This shows that the notion of place in R is the same whether we consider R as a field of algebraic functions over K or over K'. On the other hand, we have $K' \cap \mathfrak{p} = \{0\}$, which proves that the natural homomorphism of $\mathfrak{o}$ onto the residue field $\Sigma = \mathfrak{o}/\mathfrak{p}$ of $\mathfrak{p}$ maps K' isomorphically upon a subfield of Σ. We shall allow ourselves, whenever convenient, to speak of Σ as of an overfield of K'; this amounts to not distinguishing between the elements of K' and their residue classes modulo $\mathfrak{p}$.

§3. Places of the field $K\langle x\rangle$

Let us consider the special case where $R = K\langle x\rangle$, with, of course, x transcendental over K. Let $f = f(x)$ be an irreducible polynomial in x with coefficients in K. Any element u of R may be written in the form $u = g/h$, with g and h in the ring $K[x]$. Let $\mathfrak{o}_f$ be the set of elements u of the form g/h, with h not divisible by f. Since f is irreducible, it cannot divide a product of polynomials without dividing one of them. Thus the formulas

$$\frac{g_1}{h_1} - \frac{g_2}{h_2} = \frac{g_1h_2 - g_2h_1}{h_1h_2}, \qquad \frac{g_1}{h_1}\frac{g_2}{h_2} = \frac{g_1g_2}{h_1h_2}$$

show that $\mathfrak{o}_f$ is a subring of R. It is clear that this subring contains K. Moreover, $1/f$ is not in $\mathfrak{o}_f$, because, if we write $1/f = g/h$, where g and h are polynomials in x, $h = gf$ is divisible by f; this shows that $\mathfrak{o}_f \neq R$. Now, let u be any element of R not in $\mathfrak{o}_f$. We may write u in the form g/h, where g and h are polynomials in x without common factor. Since $u \notin \mathfrak{o}_f$, f divides h, and therefore f does not divide g, whence $u^{-1} = h/g \in \mathfrak{o}_f$. Thus $\mathfrak{o}_f$ is a V-ring; we shall denote by $\mathfrak{p}_f$ the corresponding place. It is clear that $\mathfrak{p}_f$ consists of all elements of the form fg/h, where g and h are polynomials in x, and h is not divisible by f.

Thus, to every irreducible polynomial f in x with coefficients in K we have associated a place $\mathfrak{p}_f$ of $K\langle x\rangle$. If f and f' are essentially distinct irreducible polynomials (i.e., f'/f not in K), the places $\mathfrak{p}_f$ and $\mathfrak{p}_{f'}$ are distinct because f^{-1} belongs to $\mathfrak{o}_{f'}$ but not to $\mathfrak{o}_f$.

Now, observe that, if we set $x' = x^{-1}$, we have $K\langle x'\rangle = K\langle x\rangle$. It follows that to every irreducible polynomial in x' with coefficients in K, there is associated a place of $K\langle x\rangle$. This applies in particular to the irreducible polynomial x'; we shall denote by $\mathfrak{p}_{1/x}$ the place defined by x' and by $\mathfrak{o}_{1/x}$ the ring of this place. The place $\mathfrak{p}_{1/x}$ is distinct from all the places $\mathfrak{p}_f$ defined above, because, if f is any irreducible polynomial in x, we have $x \in \mathfrak{o}_f$, while x clearly does not belong to $\mathfrak{o}_{1/x}$.

Now we assert that the places $\mathfrak{p}_f$ (for all irreducible polynomials f in x with coefficients in K) and $\mathfrak{p}_{1/x}$ exhaust all the places of R. Let $\mathfrak{p}$ be any place in R, and let $\mathfrak{o}$ be its ring. Assume first that $x \in \mathfrak{o}$. Since $\mathfrak{o}$ is a ring and contains K, it follows that $\mathfrak{o}$ contains the entire ring $K[x]$. Since $\mathfrak{p}$ is obviously a prime ideal in $\mathfrak{o}$, $\mathfrak{p} \cap K[x]$ is a prime ideal in $K[x]$. Thus $\mathfrak{p} \cap K[x]$ either is the zero ideal or is formed of all multiples of some irreducible polynomial f. The first case is impossible because every element $\neq 0$ of $K[x]$ would then be a unit in $\mathfrak{o}$, from which it follows immediately that every element of R would belong to $\mathfrak{o}$. Thus $\mathfrak{p} \cap K[x]$ consists of the multiples of some irreducible f. If g and h are in $K[x]$, and h is not divisible by f, then h is not in $\mathfrak{p}$ and is therefore a unit in $\mathfrak{o}$, whence $gh^{-1} \in \mathfrak{o}$, which proves that $\mathfrak{o}$ contains $\mathfrak{o}_f$. Let u be an element of R not in $\mathfrak{o}_f$; then we may write $u = g/h$, where g and h are in $K[x]$, have no common factor, and h is divisible by f. If u were in $\mathfrak{o}$, the same would be true of $h^{-1} = g^{-1}u$; but this is impossible because h, being in $\mathfrak{p}$, is not a unit in $\mathfrak{o}$. We have therefore proved that, if $x \in \mathfrak{o}$, then $\mathfrak{p}$ is one of the places $\mathfrak{p}_f$. If x is not in $\mathfrak{o}$, then $x' = x^{-1}$ is, and we see that $\mathfrak{p} \cap K[x']$ consists of all elements of $K[x']$ which are divisible (in $K[x']$) by some irreducible polynomial $f'(x')$ in x' with coefficients in K. Since x' is not a unit in $\mathfrak{o}$, it is in $\mathfrak{p} \cap K[x']$ and is therefore divisible by $f'(x')$ in $K[x']$. Thus we may assume that $f' = x'$; $\mathfrak{p}$ is then the place $\mathfrak{p}_{1/x}$.

It is clear that, if f is an irreducible polynomial in x with coefficients in K, then $\mathfrak{p}_f$ is the principal ideal generated in $\mathfrak{o}_f$ by f: $\mathfrak{p}_f = f\mathfrak{o}_f$. Similarly, $\mathfrak{p}_{1/x}$ is $(1/x)\mathfrak{o}_{1/x}$.

Let $\mathfrak{p}$ be any place of $R = K\langle x\rangle$. Denote by $\mathfrak{o}$ the ring of $\mathfrak{p}$ and by t a generator of $\mathfrak{p}$ (i.e., $\mathfrak{p} = t\mathfrak{o}$). Then no element $\neq 0$ of $\mathfrak{o}$ can belong to $t^n\mathfrak{o}$ for every $n > 0$. In fact, assume first that $\mathfrak{o} = \mathfrak{o}_f$ for some irreducible polynomial f in x. Let $u = g/h$ be an element of $\mathfrak{o}$ which belongs to $t^n\mathfrak{o}$ for every n (g and h are in $K[x]$, and h is not divisible by f). Since $\mathfrak{p} = t\mathfrak{o} = f\mathfrak{o}$, we see easily that $t^n\mathfrak{o} = f^n\mathfrak{o}$. We have by assumption, for each n, an equality of the form $g/h = f^n g_n/h_n$ with g_n and h_n in $K[x]$ and h_n not divisible by f. Thus $gh_n = f^n g_n h$; since f is irreducible and does not divide h_n, it follows easily that f^n divides g. This being true for every n, we have $g = 0$, whence $u = 0$. A similar argument applies if $\mathfrak{o} = \mathfrak{o}_{1/x}$.

Now, abandon for a moment the assumption that R is of the form $K\langle x\rangle$. Assuming only that R is a field of algebraic functions of one variable over K, let $\mathfrak{p}$ be a place of R which satisfies the following condition:

The ring $\mathfrak{o}$ *of* $\mathfrak{p}$ *contains an element* t *such that* $\mathfrak{p} = t\mathfrak{o}$ *and* $\bigcap_{n=1}^{\infty} t^n\mathfrak{o} = \{0\}$. (We shall see later that every place of R satisfies this condition.) If $x \in R$, there exists at least one integer n (which may be negative) such that $x \in t^n\mathfrak{o}$. In fact, if $x \in \mathfrak{o}$, we may take $n = 0$. If not, then x^{-1} is in $\mathfrak{o}$ and is $\neq 0$; therefore there exists an $m > 0$ such that $x^{-1} \in t^m\mathfrak{o}$, $x^{-1} \notin t^{m+1}\mathfrak{o}$, which means that $t^{-m}x^{-1}$ is in $\mathfrak{o}$ but not in $t\mathfrak{o} = \mathfrak{p}$; i.e., $t^{-m}x^{-1}$ is a unit in $\mathfrak{o}$, and $x = t^{-m}(t^{-m}x^{-1})^{-1}$ is in $t^{-m}\mathfrak{o}$. If $x \neq 0$, there is by assumption a largest integer n such that $x \in t^n\mathfrak{o}$; denote by $\nu_\mathfrak{p}(x)$ this integer. If x and y are elements $\neq 0$ in $\mathfrak{o}$, then

$$\nu_\mathfrak{p}(x) + \nu_\mathfrak{p}(y) = \nu_\mathfrak{p}(xy) \tag{1}$$

and, if $x + y \neq 0$,

$$\nu_\mathfrak{p}(x + y) \geqq \min\{\nu_\mathfrak{p}(x), \nu_\mathfrak{p}(y)\}. \tag{2}$$

In fact, xy clearly belongs to $t^{\nu_\mathfrak{p}(x)+\nu_\mathfrak{p}(y)}\mathfrak{o}$, whence $\nu_\mathfrak{p}(xy) \geqq \nu_\mathfrak{p}(x) + \nu_\mathfrak{p}(y)$. In particular, $0 = \nu_\mathfrak{p}(1) \geqq \nu_\mathfrak{p}(x) + \nu_\mathfrak{p}(x^{-1})$, i.e., $\nu_\mathfrak{p}(x^{-1}) \leqq -\nu_\mathfrak{p}(x)$. Now, if we write $x = t^{\nu_\mathfrak{p}(x)}u$, u belongs to $\mathfrak{o}$ but not to $t\mathfrak{o} = \mathfrak{p}$, i.e., u is a unit in $\mathfrak{o}$ and $x^{-1} = t^{-\nu_\mathfrak{p}(x)}u^{-1}$, whence $\nu_\mathfrak{p}(x^{-1}) \geqq -\nu_\mathfrak{p}(x)$, and therefore $\nu_\mathfrak{p}(x^{-1}) = -\nu_\mathfrak{p}(x)$. We conclude that $\nu_\mathfrak{p}(y) = \nu_\mathfrak{p}(xyx^{-1}) \geqq \nu_\mathfrak{p}(xy) - \nu_\mathfrak{p}(x)$, and, comparing with the inequality obtained above, $\nu_\mathfrak{p}(xy) = \nu_\mathfrak{p}(x) + \nu_\mathfrak{p}(y)$. On the other hand, if $\mu = \min\{\nu_\mathfrak{p}(x), \nu_\mathfrak{p}(y)\}$, we have $x \in t^\mu\mathfrak{o}$, $y \in t^\mu\mathfrak{o}$, whence $x + y \in t^\mu\mathfrak{o}$ and therefore $\nu_\mathfrak{p}(x + y) \geqq \mu$.

To complete the definition of the function $\nu_\mathfrak{p}$ (which has not been defined for 0), we make the convention to write $\nu_\mathfrak{p}(0) = \infty$, where ∞ is a symbol with which we compute according to the following rules: $\infty > n$ for every integer n; $\infty \geqq \infty$; $\infty + n = \infty$ for every integer n; $\infty + \infty = \infty$. Taking these conventions into account, the formulas (1) and (2) are valid in every case.

It should be observed that the equality $\nu_\mathfrak{p}(x + y) = \min\{\nu_\mathfrak{p}(x), \nu_\mathfrak{p}(y)\}$ holds whenever $\nu_\mathfrak{p}(x) \neq \nu_\mathfrak{p}(y)$. In fact, assume that $\nu_\mathfrak{p}(x) < \nu_\mathfrak{p}(y)$. We have $\nu_\mathfrak{p}(x) = \nu_\mathfrak{p}(x + y - y) \geqq \min\{\nu_\mathfrak{p}(x + y), \nu_\mathfrak{p}(-y)\}$; but it is easily seen that $\nu_\mathfrak{p}(-y) = \nu_\mathfrak{p}(-1) + \nu_\mathfrak{p}(y) = \nu_\mathfrak{p}(y)$; thus it is impossible that $\nu_\mathfrak{p}(x + y) > \nu_\mathfrak{p}(x)$.

More generally, we see easily by induction on m that, if $x_1, \cdots, x_m$ are any elements of R, then

$$\nu_\mathfrak{p}(x_1 + \cdots + x_m) \geqq \min\{\nu_\mathfrak{p}(x_1), \cdots, \nu_\mathfrak{p}(x_m)\}$$

and that the equality prevails if there is only one index i such that $\nu_\mathfrak{p}(x_i) = \min\{\nu_\mathfrak{p}(x_1), \cdots, \nu_\mathfrak{p}(x_m)\}$.

The definition of the function $\nu_\mathfrak{p}$ involves the choice of an element t such that $\mathfrak{p} = t\mathfrak{o}$; but actually the function $\nu_\mathfrak{p}$ depends only on the place $\mathfrak{p}$. In fact, assume that t' is any element of $\mathfrak{p}$ such that $\mathfrak{p} = t'\mathfrak{o}$. Then $t = t'u$ with $u \in \mathfrak{o}$, and, since $t' \in t\mathfrak{o}$, $u^{-1} \in \mathfrak{o}$. It follows immediately that $t^n\mathfrak{o} \subset t'^n\mathfrak{o}$ and $t'^n\mathfrak{o} \subset t^n\mathfrak{o}$ for every n, which proves our assertion. The function $\nu_\mathfrak{p}$ is called the *order function* at the place $\mathfrak{p}$; if $x \in R$, $\nu_\mathfrak{p}(x)$ is called the *order* of x at $\mathfrak{p}$. The knowledge of the order function at a place $\mathfrak{p}$ determines this place completely, because the ring $\mathfrak{o}$ of the place consists of all elements x for which $\nu_{\mathfrak{p}}(x) \geqq 0$. The elements of $\mathfrak{p}$ are the elements whose orders are >0, and the units of $\mathfrak{o}$ are the elements of order 0. The elements t for which $\mathfrak{p} = t\mathfrak{o}$ are the elements of order 1; they are also called *uniformizing variables* at $\mathfrak{p}$.

Now, let us return to the case where R is of the form $K\langle x\rangle$, with the supplementary assumption that K is the field of complex numbers. If $a \in K$, $x - a$ is an irreducible polynomial and every irreducible polynomial in x with coefficients in K is of the form $\lambda(x - a)$, $\lambda \in K$, $\lambda \neq 0$. Denote by $\mathfrak{p}_a$ the place which corresponds to the irreducible polynomial $x - a$. The polynomials h in x which are divisible by $x - a$ are those for which $h(a) = 0$. Thus the elements of the ring $\mathfrak{o}_a$ of $\mathfrak{p}_a$ are the rational fractions which do not admit a as a pole, while the elements of $\mathfrak{p}_a$ are the rational fractions which admit a as a zero. Furthermore, if ν_a is the order function at $\mathfrak{p}_a$, any rational fraction $u \neq 0$ may be written in the form $(x - a)^{\nu_a(u)}v$, where v admits a neither as a zero nor as a pole. It follows that, if $\nu_a(u) > 0$, then u admits a as a zero of order $\nu_a(u)$, while, if $\nu_a(u) < 0$, u admits a as a pole of order $-\nu_a(u)$.

Generalizing this terminology, we set up the following definitions (where R is a field of algebraic functions of one variable over a field K):

Let $\mathfrak{p}$ be a place of R. If an element $x \in R$ belongs to $\mathfrak{p}$, then we say that $\mathfrak{p}$ is a *zero of* x; if $x^{-1} \in \mathfrak{p}$, then we say that $\mathfrak{p}$ is a *pole of* x. Furthermore, if there exists an order function $\nu_{\mathfrak{p}}$ at $\mathfrak{p}$, and $\nu_{\mathfrak{p}}(x) > 0$, then we say that $\mathfrak{p}$ is a *zero of order* $\nu_{\mathfrak{p}}(x)$ *of* x, while, if $\nu_{\mathfrak{p}}(x) < 0$, we say that $\mathfrak{p}$ is a *pole of order* $-\nu_{\mathfrak{p}}(x)$ *of* x.

Consider in particular the case where $R = K\langle x\rangle$ and $\mathfrak{p} = \mathfrak{p}_{1/x}$. Let $g = a_0x^n + a_1x^{n-1} + \cdots + a_n$ be a polynomial of degree n in x with coefficients in K. If $x' = x^{-1}$, we can write

$$g = x^{-n}(a_0 + a_1x' + \cdots + a_nx'^n).$$

Now, $a_0 + a_1x' + \cdots + a_nx'^n$ belongs to the ring $\mathfrak{o}$ of $\mathfrak{p}$ and is not in $\mathfrak{p}$ because $a_0 \neq 0$. It follows that a polynomial of degree n in x admits $\mathfrak{p}_{1/x}$ as a pole of order n. If $u = g/h \in R$ (where g and h are polynomials in x), the order of u at $\mathfrak{p}$ is clearly the difference between the degrees of h and g.

Still assuming that $R = K\langle x\rangle$, let f be any irreducible polynomial in x with coefficients in K; we propose to investigate the residue field Σ of the place $\mathfrak{p}_f$, i.e., the ring $\mathfrak{o}_f/\mathfrak{p}_f$. Set $\mathfrak{q} = \mathfrak{p}_f \cap K[x]$; then those residue classes modulo $\mathfrak{p}_f$ which are represented by elements of $K[x]$ form a subring Σ_1 of Σ which is clearly isomorphic with $K[x]/\mathfrak{q}$. But $\mathfrak{q}$ is the set of multiples of f in $K[x]$; it follows that $K[x]/\mathfrak{q}$ is a field which can be obtained from K by adjunction of an element ζ which satisfies the equation $f(\zeta) = 0$. Now, every element u of $\mathfrak{o}_f$ can be written in the form g/h, with g and h in $K[x]$ and h not in $\mathfrak{q}$. Let $\bar{g}$, $\bar{h}$, and $\bar{u}$ be the residue classes of g, h, and u respectively; then $\bar{g} = \bar{u}\bar{h}$, $\bar{h} \neq 0$, whence $\bar{u} = \bar{g}\bar{h}^{-1}$. But $\bar{g}$ and $\bar{h}$ are in Σ_1 which is a field; therefore $\bar{u} \in \Sigma_1$ and $\Sigma = \Sigma_1$. Thus, the residue field Σ of $\mathfrak{p}_f$ can be obtained from K by adjunction of an element ζ such that $f(\zeta) = 0$. In particular, we see that Σ *is algebraic over* K, *of finite degree equal to the degree of the polynomial* f. In the case where $f = x$, we have $\Sigma = K$; this shows that the field of constants of $K\langle x\rangle$, which contains K and is contained in Σ, must be K itself. Replacing the consideration of x by that of $1/x$, we see that the residue field of $\mathfrak{p}_{1/x}$ is K.

If K is algebraically closed, every irreducible polynomial in x with coefficients in K is of degree 1, and therefore the residue field of any place coincides with K. Let then $\mathfrak{p}_a$ be the place which corresponds to $x - a$, and let u be an element of $K\langle x\rangle$ which does not have a as a pole. Write $u = g/h$, with g and h in $K[x]$ and $h(a) \neq 0$. We have

$$u - u(a) = \frac{g(x)h(a) - h(x)g(a)}{h(a)h(x)}$$

and $g(x)h(a) - h(x)g(a)$ is divisible by $x - a$. Thus, the value $u(a)$ taken by u at a is also the residue class of u modulo $\mathfrak{p}_a$.

More generally, let R be a field of algebraic functions of one variable over an arbitrary field K, and let $\mathfrak{p}$ be a place of R. Let x be an element of R for which $\mathfrak{p}$ is not a pole. Then the residue class of x modulo $\mathfrak{p}$ (which is an element of the residue field of $\mathfrak{p}$) will be called the *value taken by x at $\mathfrak{p}$*. It should be observed that, if K is not algebraically closed, the value taken by x at $\mathfrak{p}$ is not in general an element of K. The value taken by x at $\mathfrak{p}$ is denoted by $x(\mathfrak{p})$; it is clear that, if neither x nor y has $\mathfrak{p}$ as a pole, then $(x + y)(\mathfrak{p}) = x(\mathfrak{p}) + y(\mathfrak{p})$, $(xy)(\mathfrak{p}) = x(\mathfrak{p})y(\mathfrak{p})$. The elements which admit $\mathfrak{p}$ as a zero are those which take the value 0 at $\mathfrak{p}$.

It is often convenient to say that an element of R which has $\mathfrak{p}$ as a pole takes the value ∞ at $\mathfrak{p}$; ∞ is here a symbol which has no intrinsic connection with the symbol ∞ which was used to complete the definition of the order function at a place.

§4. Existence of places

We shall prove in this section a theorem which implies that any field of algebraic functions of one variable admits infinitely many places.

Theorem 1. *Let R be a field of algebraic functions of one variable over a field K. Assume that we are given a subring $\mathfrak{o}$ of R containing K and an ideal $\mathfrak{p}$ in $\mathfrak{o}$ not containing* 1 *but* $\neq \{0\}$. *Then there exists a place $\mathfrak{P}$ of R whose ring $\mathfrak{O}$ contains $\mathfrak{o}$ and which is such that $\mathfrak{p} \subset \mathfrak{P} \cap \mathfrak{o}$.*

If $\mathfrak{o}'$ is any subring of R containing $\mathfrak{o}$, we shall denote by $\mathfrak{p}\mathfrak{o}'$ the ideal generated in $\mathfrak{o}'$ by the elements of $\mathfrak{p}$. We denote by $\mathcal{F}$ the family of all subrings $\mathfrak{o}'$ of R containing $\mathfrak{o}$ which are such that $\mathfrak{p}\mathfrak{o}' \neq \mathfrak{o}'$; in particular, $\mathfrak{o}$ itself belongs to $\mathcal{F}$. We shall prove that $\mathcal{F}$ contains at least one maximal element (i.e., there is a ring in $\mathcal{F}$ which is not properly contained in any other ring of the family $\mathcal{F}$), and that any such maximal element is a V-ring.

To prove the first assertion, it is sufficient, in virtue of Zorn's Lemma, to prove that the family $\mathcal{F}$ is inductive, i.e., that if $\mathcal{F}'$ is a non empty subfamily of $\mathcal{F}$ such that, of any two rings in $\mathcal{F}'$, one contains the other, then $\mathcal{F}$ contains a ring which contains all rings of the family $\mathcal{F}'$. To do this, denote by $\mathfrak{o}_1$ the set-theoretic union of all rings of $\mathcal{F}'$. If x and y are in $\mathfrak{o}_1$, then $x \in \mathfrak{o}'$, $y \in \mathfrak{o}''$, where $\mathfrak{o}'$ and $\mathfrak{o}''$ are in $\mathcal{F}'$; one of the rings $\mathfrak{o}'$, $\mathfrak{o}''$ contains the other. If for instance $\mathfrak{o}'$ contains $\mathfrak{o}''$, then x and y are both in $\mathfrak{o}'$, whence $x - y \in \mathfrak{o}'$, $xy \in \mathfrak{o}'$, which shows that $x - y$ and xy are in $\mathfrak{o}_1$. The same conclusion subsists if $\mathfrak{o}' \subset \mathfrak{o}''$; therefore, $\mathfrak{o}_1$ is a ring. Since every ring belonging to $\mathcal{F}$ contains $\mathfrak{o}$, $\mathfrak{o}_1$ contains $\mathfrak{o}$. We assert that $\mathfrak{p}\mathfrak{o}_1 \neq \mathfrak{o}_1$. In fact, were this not the case, then we could represent 1 in the form $1 = x_1y_1 + \cdots + x_hy_h$, with $x_i \in \mathfrak{p}$, $y_i \in \mathfrak{o}_1$ $(1 \leqq i \leqq h)$. Each y_i would belong to some ring $\mathfrak{o}^{(i)} \in \mathcal{F}'$. For each pair (i, j), one of the rings $\mathfrak{o}^{(i)}$, $\mathfrak{o}^{(j)}$ would contain the other; since there are only a finite number of rings $\mathfrak{o}^{(i)}$, it follows easily that they would all be contained in one of them, say in $\mathfrak{o}^{(k)}$. But then we would have $1 = \sum_{i=1}^{h} x_iy_i \in \mathfrak{p}\mathfrak{o}^{(k)}$, whence $\mathfrak{p}\mathfrak{o}^{(k)} = \mathfrak{o}^{(k)}$, which is impossible since $\mathfrak{o}^{(k)} \in \mathcal{F}'$. Thus we have $\mathfrak{p}\mathfrak{o}_1 \neq \mathfrak{o}_1$, whence $\mathfrak{o}_1 \in \mathcal{F}'$, which proves that $\mathcal{F}$ is inductive.

Let $\mathfrak{O}$ be a maximal ring in $\mathcal{F}$; we shall prove that $\mathfrak{O}$ is a V-ring. First we show that any element in $\mathfrak{O}$ which is $\equiv 1 \pmod{\mathfrak{p}\mathfrak{O}}$ has an inverse in $\mathfrak{O}$. Let Q be the set of these elements; it is clear that products of elements in Q are in Q. Let $\mathfrak{O}'$ be the set of elements of the form xq^{-1}, with $x \in \mathfrak{O}$, $q \in Q$; then the formulas

$$\frac{x'}{q'} - \frac{x}{q} = \frac{qx' - q'x}{qq'}, \qquad \frac{x}{q}\frac{x'}{q'} = \frac{xx'}{qq'}$$

show that $\mathfrak{O}'$ is a ring. Since $1 \in Q$, $\mathfrak{O}'$ contains $\mathfrak{O}$. We assert that $1 \notin \mathfrak{p}\mathfrak{O}'$. In fact, assume for a moment that

$$1 = \sum_{i=1}^{h} x_i \frac{y_i}{q_i} \qquad x_i \in \mathfrak{p}, \qquad y_i \in \mathfrak{O}, \qquad q_i \in Q \qquad (1 \leqq i \leqq h).$$

Set $q = q_1 \cdots q_h$; then $q \in Q$, whence $q = 1 + \sum_{i=1}^{h'} x_i'y_i'$, with $x_i' \in \mathfrak{p}$, $y_i' \in \mathfrak{O}$, and we have

$$1 = \sum_{i=1}^{h} x_i \left(\prod_{j \neq i} q_j\right) y_i - \sum_{i=1}^{h'} x_i'y_i' \in \mathfrak{p}\mathfrak{O},$$

which is impossible. Thus, $\mathfrak{O}'$ belongs to $\mathcal{F}$; since $\mathfrak{O}$ is maximal, we have $\mathfrak{O}' = \mathfrak{O}$, whence, if $q \in Q$, $q^{-1} \in \mathfrak{O}$. Now, let u be any element of R not in $\mathfrak{O}$; then $\mathfrak{O}[u] \neq \mathfrak{O}$, whence $\mathfrak{O}[u] \notin \mathcal{F}$ and $\mathfrak{p}\mathfrak{O}[u] = \mathfrak{O}[u]$. It follows that we may represent 1 in the form $1 = \sum_{i=0}^{n} x_iu^i$, with $x_i \in \mathfrak{p}\mathfrak{O}$ $(1 \leqq i \leqq n)$. Since $1 - x_0 \in Q$, we may also write $1 = \sum_{i=1}^{n} x_i'u^i$ with $x_i' = x_i(1 - x_0)^{-1} \in \mathfrak{p}\mathfrak{O}$. We may furthermore assume that, among all representations of 1 in this form, we have selected the one with the lowest n; i.e., it is impossible to represent 1 in the form $1 = \sum_{i=1}^{n'} x_i''u^i$ with $n' < n$, $x_i'' \in \mathfrak{p}\mathfrak{O}$ $(1 \leqq i \leqq n')$. Now, assume for a moment that $u^{-1} \notin \mathfrak{O}$. Then we see in the same way that we can represent 1 in the form $\sum_{i=1}^{m} y_i'u^{-i}$ with $y_i' \in \mathfrak{p}\mathfrak{O}$ $(1 \leqq i \leqq m)$; furthermore, we assume that, among all representations of 1 in this second form, we have selected the one with the smallest possible m. If $n \geqq m$, we may write $u^n = \sum_{i=1}^{m} y_i'u^{n-i}$, whence

$$1 = \sum_{i=1}^{n-1} x_i'u^i + x_n'\left(\sum_{i=1}^{m} y_i'u^{n-i}\right),$$

which is impossible in virtue of our choice of n. Exchanging the roles played by u and u^{-1}, we see in the same way that the assumption that $n \leqq m$ likewise leads to an impossibility. Thus the assumption that $u^{-1} \notin \mathfrak{O}$ leads to a contradiction, and we have $u^{-1} \in \mathfrak{O}$. Since $\mathfrak{p}\mathfrak{O} \neq \mathfrak{O}$, we have $\mathfrak{O} \neq R$; therefore $\mathfrak{O}$ is a V-ring.

Let $\mathfrak{P}$ be the ideal of non-units of $\mathfrak{O}$. Then $\mathfrak{P}$ is a place of R; since $\mathfrak{p}\mathfrak{O} \neq \mathfrak{O}$, an element of $\mathfrak{p}$ cannot be a unit in $\mathfrak{O}$, whence $\mathfrak{p} \subset \mathfrak{P} \cap \mathfrak{o}$. Theorem 1 is thereby proved.

Remark 1. Neither the definition of a V-ring in R nor the proof of Theorem 1 makes any use of the fact that R is a field of algebraic functions of one variable. It follows that our proof of Theorem 1 yields a result which is valid for any pair of fields (K, R) such that K is a subfield of R.

Remark 2. When R is a field of algebraic functions of one variable over K, and when $\mathfrak{p}$ is a prime ideal in $\mathfrak{o}$, it can be proved that the intersection $\mathfrak{P} \cap \mathfrak{o}$ is necessarily equal to $\mathfrak{p}$; however, we shall not have to make use of this more refined result.

Corollary 1. *Let R be a field of algebraic functions of one variable over a field K, and let $x_1, \cdots, x_r$ be elements of R which are not all in K. Let $\mathfrak{a}$ be the set of polynomials $F(X_1, \cdots, X_r)$ in r letters with coefficients in K such that $F(x_1, \cdots, x_r) = 0$. Let $\xi_1, \cdots, \xi_r$ be elements of K such that $F(\xi_1, \cdots, \xi_r) = 0$ for all $F \in \mathfrak{a}$. Then there exists a place of R which is a common zero of $x_1 - \xi_1, \cdots, x_r - \xi_r$.*

Set $\mathfrak{o} \cdot = K[x_1, \cdots, x_r]$; an element y of $\mathfrak{o}$ can be represented in the form $P(x_1, \cdots, x_r)$ where P is a polynomial in coefficients in K, and, if $P(x_1, \cdots, x_r) = P'(x_1, \cdots, x_r)$, then $P' - P$ is in $\mathfrak{a}$. It follows that $P(\xi_1, \cdots, \xi_r)$ has the same value for all polynomials P such that $y = P(x_1, \cdots, x_r)$. Let $\mathfrak{p}$ be the set of elements y in $\mathfrak{o}$ for which this value is 0. Then $\mathfrak{p}$ is clearly an ideal in $\mathfrak{o}$; it is $\neq \mathfrak{o}$ because it has only 0 in common with K. If $\mathfrak{P}$ is a place of R such that $\mathfrak{p} \subset \mathfrak{P}$, then $\mathfrak{P}$ is a common zero of $x_1 - \xi_1, \cdots, x_r - \xi_r$.

Corollary 2. *Let R be a field of algebraic functions of one variable over a field K, and let x, y be elements of R, not both constants. Let F be an irreducible polynomial with coefficients in K such that $F(x, y) = 0$. If ξ, η are elements of K such that $F(\xi, \eta) = 0$, then there exists a place of R which is a common zero of $x - \xi$ and $y - \eta$.*

Assume for instance that x is not constant; then y is algebraic over $K\langle x\rangle$. Let $Y^n + \rho_1(x)Y^{n-1} + \cdots + \rho_r(x)$ be a polynomial in Y with coefficients in $K\langle x\rangle$, irreducible in $K\langle x\rangle[Y]$, which admits y as a zero. We may write $\rho_i(x) = P_i(x)/Q_i(x)$, where P_i and Q_i are polynomials with coefficients in K, relatively prime to each other. Let Q be a least common multiple of $Q_1, \cdots, Q_n$; then it is well known that $F_1(X, Y) = Q(X)Y^n + \sum_{i=1}^{n} (Q(X)/Q_i(X))P_i(X)Y^{n-i}$ is irreducible in $K[X, Y]$. Let F' be any polynomial in $K[X, Y]$ such that $F'(x, y) = 0$; then $F'(x, Y)$ is divisible by $F_1(x, Y)$ in $K\langle x\rangle[Y]$; from this and from the irreducibility of F_1 it follows that F' is divisible by F_1. In particular, we have $F = \alpha F_1$, $\alpha \in K$, whence $F_1(\xi, \eta) = 0$ and $F'(\xi, \eta) = 0$. Corollary 2 therefore follows from Corollary 1.

COROLLARY 3. *Let R be a field of algebraic functions of one variable over a field K, and let x be a non constant element of R. Then x admits at least one zero and one pole in R.*

The existence of a zero follows immediately from Corollary 2 (take $F = X$, $y = 0$). Since x^{-1} is not constant, it has at least one zero, and x has at least one pole.

COROLLARY 4. *A field of algebraic functions of one variable admits infinitely many places.*

Let R be the field in question, and let K be the field of constants in R. Let x be a non constant element of R; then $K[x]$ is the ring of polynomials in x with coefficients in K. This ring contains infinitely many essentially different irreducible polynomials. In fact, if K is infinite, we may take the polynomials $x - a$, $a \in K$; if K is finite, it is well known that, for every integer $n > 0$, there is an irreducible polynomial of degree n. If f is any irreducible element of $K[x]$, the ideal generated by f in $K[x]$ is not the unit ideal. Therefore, it follows from Theorem 1 that there exists a place $\mathfrak{p}_f$ of R which is a zero of f and whose ring contains $K[x]$. If f and g are essentially distinct irreducible polynomials in x, they are relatively prime to each other, and there exist elements u and v of $K[x]$ such that $uf + vg = 1$; it follows immediately that $\mathfrak{p}_f \neq \mathfrak{p}_g$.

§5. THE ORDER FUNCTION. THE DEGREE OF A PLACE

Let $\mathfrak{p}$ be a place of a field R of algebraic functions of one variable over a field K. We propose to prove that $\mathfrak{p}$ satisfies the condition stated in §3, which guarantees the existence of an order function; we shall also prove that the residue field Σ of $\mathfrak{p}$ is algebraic of finite degree over K.

Let x be an element $\neq 0$ of $\mathfrak{p}$. Assume that we have a finite sequence $(t_1, \cdots, t_{e+1})$ of elements of the ring $\mathfrak{o}$ of $\mathfrak{p}$ which satisfy the following conditions: $t_1 = x$; if $1 \leqq i \leqq e$, t_i/t_{i+1} is in $\mathfrak{p}$; $t_{e+1} = 1$. Let also $(u_1, \cdots, u_d)$ be a finite sequence of elements of $\mathfrak{o}$ whose residue classes $\bar{u}_1, \cdots, \bar{u}_d$ modulo $\mathfrak{p}$ are linearly independent over K. We propose to prove that the ed elements t_iu_j ($2 \leqq i \leqq e + 1$, $1 \leqq j \leqq d$) of R are linearly independent over the field $K\langle x\rangle$. Assume for a moment that there is a linear relation of the form

$$\sum_{i=2}^{e+1} \sum_{j=1}^{d} \rho_{ij}(x) t_i u_j = 0$$

with $\rho_{ij}(x) \in K\langle x\rangle$, the ρ_{ij}'s being not all zero. Multiplying this relation by a common denominator of all rational fractions $\rho_{ij}(x)$, we obtain a relation of the same form where the ρ_{ij}'s are polynomials. Dividing if necessary this relation by some power of x, we obtain a relation

$$(1) \qquad \sum_{i=2}^{e+1} \sum_{j=1}^{d} f_{ij}(x) t_i u_j = 0$$

where the $f_{ij}(x)$ are polynomials which are not all divisible by x. Set $f_{ij}(0) = a_{ij}$;

then $f_{ij}(x) - a_{ij} = xg_{ij}(x) \in x\mathfrak{o}$ (where g_{ij} is a polynomial). Let k be the index $\geqq 2$ which is determined by the following conditions: there exists an l $(1 \leqq l \leqq d)$ such that $a_{kl} \neq 0$, but, if $k < i \leqq e + 1$, then $a_{ij} = 0$ for $1 \leqq j \leqq d$. The equality (1) gives

$$\sum_{i=2}^{k} \sum_{j=1}^{d} a_{ij}\, t_i\, u_j = xw$$

with $w = -\sum_{i=2}^{e+1} \sum_{j=1}^{d} g_{ij}(x) t_i u_j \in \mathfrak{o}$. We write this in the form

$$\sum_{j=1}^{d} a_{kj}\, u_j = w \frac{x}{t_k} - \sum_{i=2}^{k-1} \sum_{j=1}^{d} a_{ij} \frac{t_i}{t_k} u_j .$$

Since $k \geqq 2$, x/t_k belongs to $\mathfrak{p}$; if $i < k$, t_i/t_k belongs to $\mathfrak{p}$. It follows that the right side of our last formula represents an element of $\mathfrak{p}$, whence $\sum_{j=1}^{d} a_{kj}\bar{u}_j = 0$. But this is impossible since $a_{kl} \neq 0$ and $\bar{u}_1, \cdots, \bar{u}_d$ are linearly independent over K. Thus our assertion that the elements $t_i u_j$ $(2 \leqq i \leqq e + 1, 1 \leqq j \leqq d)$ are linearly independent over $K\langle x\rangle$ is proved.

Since $x \in \mathfrak{p}$, x is not constant (cf. §2); the field R is therefore finite algebraic over $K\langle x\rangle$. Set $n = [R: K\langle x\rangle]$; then we see that $de \leqq n$.

Taking first $e = 1$, $t_1 = x$, $t_2 = 1$, we conclude that $d \leqq n$. Therefore Σ cannot contain more than n elements linearly independent over K. This proves that Σ is algebraic of finite degree over K.

Next, taking $d = 1$, $u_1 = 1$, we see that the number of terms in a sequence $(t_1, \cdots, t_{e+1})$ which satisfies the conditions specified above is bounded by $n + 1$. Among all sequences which satisfy these conditions, take one, say $(t_1, \cdots, t_{e+1})$ with the largest possible number of terms, and set $t = t_e$. Let z be any element of $\mathfrak{p}$; we shall prove that $z \in t\mathfrak{o}$. Were this not the case, z/t would not be in $\mathfrak{o}$, and therefore t/z would be a non-unit in $\mathfrak{o}$, i.e. t/z would be in $\mathfrak{p}$. But then $(t_1, \cdots, t_e, t_e/z, t_{e+1})$ would be a sequence of $e + 2$ terms satisfying our requirements, which is impossible. This proves that $\mathfrak{p} = t\mathfrak{o}$ is a principal ideal. Let y be an element $\neq 0$ of $\mathfrak{p}$; if $y \in t^m\mathfrak{o}$, set $u_1 = y$, $u_{i+1} = y/t^i$ $(1 \leqq i \leqq m - 1)$, $u_{m+1} = 1$. Then we have $u_i/u_{i+1} \in \mathfrak{p}$ $(1 \leqq i \leqq m)$; therefore, it follows from what we have proved above that $m \leqq [R: K\langle y\rangle]$. This proves that no element $\neq 0$ can belong to $t^m\mathfrak{o}$ for every m. Thus we have now proved

THEOREM 2. *Let $\mathfrak{p}$ be a place of a field of algebraic functions of one variable over a field K. Then there exists an order function at $\mathfrak{p}$, and the residue field of $\mathfrak{p}$ is algebraic of finite degree over K.*

COROLLARY. *Let R be a field of algebraic functions of one variable over a field K. Then the field of constants of R is of finite degree over K.*

This follows immediately from the fact that the field of constants is contained in the residue field of any place of R.

We see also that the residue field of a place $\mathfrak{p}$ is of finite degree over the field of constants. This degree is called the *degree of the place* $\mathfrak{p}$.

REMARK. Throughout this book, we shall use (without defining it anew) the notation $\nu_{\mathfrak{p}}$ to represent the order function at a place $\mathfrak{p}$.

§6. The theorem of independence

Consider the field R of rational functions of a complex variable with complex coefficients. If a is any complex number, there exists one and only one place $\mathfrak{p}_a$ at which x takes the value a. If $u \in R$, we may represent u in a neighbourhood of a by its Laurent series

$$u = \sum_{k=r}^{\infty} c_k(x - a)^k,$$

the c_k's being complex numbers. If $c_r \neq 0$, then $(x - a)^{-r}u$ has no zero or pole at $\mathfrak{p}_a$, from which it follows that the order of u at $\mathfrak{p}_a$ is r. Now, if we require more information on the behavior of u in the neighbourhood of a, we may give the values of a certain number of the coefficients c_k appearing at the beginning of our series, say of those for which $r \leqq k < m$, where m is an integer $>r$. If we set $v = \sum_{k=r}^{m-1} c_k(x - a)^k$, the requirement that u should have the given coefficients $c_r, \cdots, c_{m-1}$ may be expressed by the condition that $\nu_{\mathfrak{p}_a}(u - v) \geqq m$. We shall then say that u behaves as v up to the order $m - 1$ at $\mathfrak{p}_a$. Now, we consider the following problem. Let there be given a finite number of distinct complex numbers $a_1, \cdots, a_h$ and, for each a_i, a certain rational function v_i and an integer m_i; does there exist a rational function u which satisfies simultaneously the h conditions $\nu_i(u - v_i) \geqq m_i$, where ν_i is the order function at $\mathfrak{p}_{a_i}$? In other words, is there any necessary relation between the modes of behavior of a rational function at several distinct places (behavior meaning of course behavior up to a certain order)? We shall prove in the general case of an arbitrary field of algebraic functions of one variable that there is no such necessary relation. More precisely, we shall prove

THEOREM 3. *Let R be a field of algebraic functions of one variable, and let $\mathfrak{p}_1, \cdots, \mathfrak{p}_h$ be h distinct places of R. Assign to each $\mathfrak{p}_i$ some element v_i of R and some integer m_i. Then there exists an element u of R which satisfies the h conditions $\nu_i(u - v_i) \geqq m_i$ $(1 \leqq i \leqq h)$ (where ν_i represents the order function at $\mathfrak{p}_i$).*

We proceed by induction on h. There is nothing to prove if $h = 1$. Assume that $h > 1$ and that Theorem 3 holds for systems of $h - 1$ places. Then, given $h - 1$ integers $e_1, \cdots, e_{h-1}$, there always exists an element u of R such that $\nu_i(u) = e_i$ $(1 \leqq i \leqq h - 1)$. For, there exist elements v_i such that $\nu_i(v_i) = e_i$ $(1 \leqq i \leqq h - 1)$, and there exists an element u such that $\nu_i(u - v_i) \geqq e_i + 1$. Since $u = (u - v_i) + v_i$, we then have $\nu_i(u) = \nu_i(v_i) = e_i$ (cf. §3).

We shall prove that it is impossible that there should exist rational numbers $\rho_1, \cdots, \rho_{h-1}$ such that $\nu_h(z) = \sum_{i=1}^{h-1} \rho_i \nu_i(z)$ for every $z \neq 0$. Assume for a moment that such numbers exist, and consider first the case where at least one of them is <0. There exist elements z and z' of R such that

$$\nu_i(z) = 1, \qquad \nu_i(z') = 0 \quad \text{if} \quad \rho_i \geqq 0,$$

$$\nu_i(z) = 0, \qquad \nu_i(z') = 1 \quad \text{if} \quad \rho_i < 0.$$

Then, clearly, $\nu_h(z) \geqq 0$, $\nu_h(z') < 0$. On the other hand, since $\nu_i(z) \neq \nu_i(z')$, we have $\nu_i(z + z') = \min\{\nu_i(z), \nu_i(z')\} = 0$ $(1 \leqq i \leqq h - 1)$, whence $\nu_h(z + z') = 0$; this is however impossible, since $\nu_h(z') < \nu_h(z)$, whence $\nu_h(z + z') = \nu_h(z') < 0$. Assume now that the numbers ρ_i are all $\geqq 0$. They can obviously not all be 0; assume that $\rho_1 > 0$. If $\rho_2, \cdots, \rho_{h-1}$ are all $= 0$, then the conditions $\nu_h(z) > 0$ and $\nu_1(z) > 0$ are equivalent to each other, whence $\mathfrak{p}_h = \mathfrak{p}_1$, which is not the case. If one at least of $\rho_2, \cdots, \rho_{h-1}$ is $\neq 0$, then we write our equality in the form $\nu_1(z) = \rho_1^{-1}\nu_h(z) - \sum_{i=2}^{h-1} \rho_1^{-1}\rho_i\nu_i(z)$, and we are brought back to the case studied before. Our assertion is thereby proved.

It follows that there does not exist any system of h rational numbers $\rho_1, \cdots, \rho_h$ not all zero such that $\sum_{i=1}^h \rho_i\nu_i(z) = 0$ for all $z \neq 0$. From this we can conclude that there exist h elements $z_1, \cdots, z_h$ of R, all $\neq 0$, such that $\det(\nu_i(z_j))_{i,j} \neq 0$. To show this, we shall construct these elements step by step. For z_1 we take any element $\neq 0$ such that $\nu_1(z_1) \neq 0$. Then the set of rational linear combinations of the functions $\nu_1, \cdots, \nu_h$ which vanish at z_1 is of dimension $h - 1$ over the field Q of rational numbers. Assume that we have already determined $z_1, \cdots, z_k$ (where $k < h$) in such a way that no more than $h - k$ linearly independent rational linear combinations of $\nu_1, \cdots, \nu_h$ vanish simultaneously at $z_1, \cdots, z_k$. We take one of these, which is $\neq 0$, say $\sum_{i=1}^h \rho_i\nu_i$, and we determine $z_{k+1} \neq 0$ in such a way that $\sum_{i=1}^h \rho_i\nu_i(z_{k+1}) \neq 0$. Then no more than $h - (k + 1)$ linearly independent rational linear combinations of $\nu_1, \cdots, \nu_h$ vanish at $z_1, \cdots, z_{k+1}$. At the end of our construction, we have h elements $\neq 0$, $z_1, \cdots, z_h$ of R, such that no non trivial rational linear combination of $\nu_1, \cdots, \nu_h$ vanishes at $z_1, \cdots, z_h$. The numbers $\nu_i(z_j)$ being rational, it follows that $\det(\nu_i(z_j))_{i,j} \neq 0$.

The equations $\sum_{j=1}^h \sigma_{j,k}\nu_i(z_j) = -1$ for $i = k$, $\sum_{j=1}^h \sigma_{j,k}\nu_i(z_j) = +1$ for $i \neq k$ (where k is any index from 1 to h) have a solution $(\sigma_{1,k}, \cdots, \sigma_{h,k})$ in rational numbers. Let d be a positive integer such that the numbers $d\sigma_{i,k}$ $(1 \leqq i \leqq h, 1 \leqq k \leqq h)$ are all integral. If we set $\zeta_k = \prod_{j=1}^h z_j^{d\sigma_{j,k}}$, we have $\nu_k(\zeta_k) = -d$, $\nu_i(\zeta_k) = +d$ for $i \neq k$. Set $t_k = (1 + \zeta_k^{-1})^{-1}$. If $i \neq k$, we have $\nu_i(\zeta_k^{-1}) < 0 = \nu_i(1)$, whence $\nu_i(1 + \zeta_k^{-1}) = \nu_i(\zeta_k^{-1}) = -d$, and $\nu_i(t_k) = d$. On the other hand, $t_k - 1 = -\zeta_k^{-1}(1 + \zeta_k^{-1})^{-1}$ and $\nu_k(\zeta_k^{-1}) = d$, whence $\nu_k(t_k - 1) = d$. Now, we observe that the number d may be selected as large as we please. We choose it in such a way that $d + \nu_i(v_j) \geqq \max\{m_1, \cdots, m_h\}$ $(1 \leqq i, j \leqq h)$ and we set $u = \sum_{i=1}^h t_iv_i$. Then $u - v_i = (t_i - 1)v_i + \sum_{k \neq i} t_kv_k$; we have $\nu_i((t_i - 1)v_i) = d + \nu_i(v_i) \geqq m_i$, and, if $k \neq i$, $\nu_i(t_kv_k) = d + \nu_i(v_k) \geqq m_i$, from which it follows that $\nu_i(u - v_i) \geqq m_i$ $(1 \leqq i \leqq h)$. Theorem 3 is thereby proved.

In the course of the proof of the above theorem, we have established the following

Corollary. *Let $\mathfrak{p}_1, \cdots, \mathfrak{p}_h$ be h distinct places of a field R of algebraic functions of one variable. Assign to each $\mathfrak{p}_i$ some integer m_i. Then there exists an element u*

of R which satisfies the h conditions $\nu_i(u) = m_i$ $(1 \leqq i \leqq h)$ (where ν_i is the order function at $\mathfrak{p}_i$).

§7. Divisors

Let R be a field of algebraic functions of one variable over a field K. If we assign to every place $\mathfrak{p}$ of R an integer $e(\mathfrak{p})$ in such a way that $e(\mathfrak{p}) \neq 0$ for only a finite number of places $\mathfrak{p}$, we obtain what is called a *divisor* in R; this divisor is denoted symbolically by $\prod_{\mathfrak{p}} \mathfrak{p}^{e(\mathfrak{p})}$. The integer $e(\mathfrak{p})$ is called the *exponent* with which $\mathfrak{p}$ enters in the divisor; if this exponent is 0, we also say that $\mathfrak{p}$ does not enter (or occur) in the divisor.

Divisors are multiplied with each other according to the formal rule

$$(\textstyle\prod_{\mathfrak{p}} \mathfrak{p}^{e(\mathfrak{p})})(\prod_{\mathfrak{p}} \mathfrak{p}^{e'(\mathfrak{p})}) = \prod_{\mathfrak{p}} \mathfrak{p}^{e(\mathfrak{p})+e'(\mathfrak{p})}.$$

It is clear that they form an abelian group. The unit element of this group is $\prod_{\mathfrak{p}} \mathfrak{p}^0$, which is called the *unit divisor.*

If $\mathfrak{p}$ is a place, we shall identify $\mathfrak{p}$ with the divisor which assigns 1 to $\mathfrak{p}$ and 0 to every other place. Thus, if $\mathfrak{p}_1, \cdots, \mathfrak{p}_n$ are places and $e_1, \cdots, e_n$ integers, $\mathfrak{p}_1^{e_1} \cdots \mathfrak{p}_n^{e_n}$ represents a divisor; if $\mathfrak{p}_1, \cdots, \mathfrak{p}_n$ are distinct, $\mathfrak{p}_i$ enters in this divisor with the exponent e_i.

A divisor $\prod_{\mathfrak{p}} \mathfrak{p}^{e(\mathfrak{p})}$ is called *integral* if all exponents $e(\mathfrak{p})$ are $\geqq 0$. Products of integral divisors are clearly integral. If $\mathfrak{a}$ and $\mathfrak{b}$ are divisors, we say that $\mathfrak{b}$ is a *multiple* of $\mathfrak{a}$, or that $\mathfrak{a}$ *divides* $\mathfrak{b}$, if $\mathfrak{b}\mathfrak{a}^{-1}$ is an integral divisor. It is clear that, if $\mathfrak{a}$ divides $\mathfrak{b}$ and $\mathfrak{b}$ divides $\mathfrak{c}$, then $\mathfrak{a}$ divides $\mathfrak{c}$.

If $\mathfrak{a} = \prod_{\mathfrak{p}} \mathfrak{p}^{e(\mathfrak{p})}$ and $\mathfrak{b} = \prod_{\mathfrak{p}} \mathfrak{p}^{f(\mathfrak{p})}$ are divisors, there always exist divisors which divide both $\mathfrak{a}$ and $\mathfrak{b}$, and they are all the divisors which divide $\mathfrak{c} = \prod_{\mathfrak{p}} \mathfrak{p}^{g(\mathfrak{p})}$, where $g(\mathfrak{p}) = \min\{e(\mathfrak{p}), f(\mathfrak{p})\}$. For this reason, $\mathfrak{c}$ is called the *highest common divisor* of $\mathfrak{a}$ and $\mathfrak{b}$. The highest common divisor of two integral divisors is integral. Similarly, the divisors which are multiples of $\mathfrak{a}$ and of $\mathfrak{b}$ are all the multiples of $\mathfrak{m} = \prod_{\mathfrak{p}} \mathfrak{p}^{h(\mathfrak{p})}$, where $h(\mathfrak{p}) = \max\{e(\mathfrak{p}), f(\mathfrak{p})\}$. The divisor $\mathfrak{m}$ is called the *least common multiple* of $\mathfrak{a}$ and $\mathfrak{b}$. It is easily seen that the least common multiple of $\mathfrak{a}^{-1}$ and $\mathfrak{b}^{-1}$ is the inverse of the highest common divisor of $\mathfrak{a}$ and $\mathfrak{b}$.

Denote by $d(\mathfrak{p})$ the degree of a place $\mathfrak{p}$. If $\mathfrak{a} = \prod_{\mathfrak{p}} \mathfrak{p}^{e(\mathfrak{p})}$ is any divisor, the number $\sum_{\mathfrak{p}} e(\mathfrak{p})d(\mathfrak{p})$ (this sum has a meaning since only a finite number of exponents $e(\mathfrak{p})$ are $\neq 0$) is called the *degree* of $\mathfrak{a}$ and is denoted by $d(\mathfrak{a})$ (we see immediately that, if $\mathfrak{a} = \mathfrak{p}$ is a place, then the degree of $\mathfrak{a}$ as a divisor is equal to the degree of the place $\mathfrak{p}$). If $\mathfrak{a}$ and $\mathfrak{b}$ are divisors, we have

$$d(\mathfrak{a}\mathfrak{b}) = d(\mathfrak{a}) + d(\mathfrak{b}); \qquad d(\mathfrak{a}^{-1}) = -d(\mathfrak{a}).$$

The degree of an integral divisor is $\geqq 0$; therefore, if a divisor $\mathfrak{a}$ divides a divisor $\mathfrak{b}$, we have $d(\mathfrak{a}) \leqq d(\mathfrak{b})$.

Let $\mathfrak{a} = \prod_{\mathfrak{p}} \mathfrak{p}^{e(\mathfrak{p})}$ be a divisor, and let x, y be elements of R. Then we say that x is *congruent to y modulo $\mathfrak{a}$*, and we write

$$x \equiv y \pmod{\mathfrak{a}}$$

if the condition $\nu_{\mathfrak{p}}(x - y) \geqq e(\mathfrak{p})$ is satisfied for every place $\mathfrak{p}$. Let x, y, x', y' be elements of R and let λ, λ' be elements of the field of constants K' of R; then it is clear that the congruences $x \equiv y \pmod{\mathfrak{a}}$ and $x' \equiv y' \pmod{\mathfrak{a}}$ imply $\lambda x + \lambda' x' \equiv \lambda y + \lambda' y' \pmod{\mathfrak{a}}$. In particular, the elements of R which are $\equiv 0$ modulo $\mathfrak{a}$ form a vector space $\mathfrak{D}$ over K', and the condition for x and y to be congruent to each other modulo $\mathfrak{a}$ is equivalent to the condition that x and y should be in the same coset modulo $\mathfrak{D}$.

If $\mathfrak{a} = \mathfrak{p}_1^{e_1} \cdots \mathfrak{p}_m^{e_m} \mathfrak{q}_1^{f_1} \cdots \mathfrak{q}_n^{f_n}$ is a divisor (where $\mathfrak{p}_1, \cdots, \mathfrak{p}_m, \mathfrak{q}_1, \cdots, \mathfrak{q}_n$ are distinct places and $e_1 > 0, \cdots, e_m > 0, f_1 < 0, \cdots, f_n < 0$), the condition for an element x of R to be $\equiv 0 \pmod{\mathfrak{a}}$ can be analyzed in the following way: x should have $\mathfrak{p}_i$ as a zero of order $\geqq e_i$ $(1 \leqq i \leqq m)$; x should not have any pole outside the set $\{\mathfrak{q}_1, \cdots, \mathfrak{q}_n\}$, and, if it has a pole at $\mathfrak{q}_j$, the order of this pole should be $\leqq -f_j$. We see that this implies infinitely many conditions on x, one for each place.

It will be convenient to introduce a new type of congruence which involves only conditions bearing on the behavior of the element under consideration at a finite number of places. Let S be any finite set of places. Let x and y be elements of R and let $\mathfrak{a} = \prod_{\mathfrak{p}} \mathfrak{p}^{e(\mathfrak{p})}$ be a divisor. Then we shall write

$$x \equiv_S y \pmod{\mathfrak{a}}$$

when the conditions $\nu_{\mathfrak{p}}(x - y) \geqq e(\mathfrak{p})$ are satisfied for all places $\mathfrak{p} \in S$. It is clear that the elements x which are $\equiv_S 0 \pmod{\mathfrak{a}}$ form a vector space over K'; we shall denote this space by $\mathfrak{K}(\mathfrak{a}, S)$. The condition $x \equiv_S y \pmod{\mathfrak{a}}$ is equivalent to $x - y \in \mathfrak{K}(\mathfrak{a}, S)$.

LEMMA 1. *Let* $\mathfrak{a} = \prod_{\mathfrak{p}} \mathfrak{p}^{e(\mathfrak{p})}$ *and* $\mathfrak{b} = \prod_{\mathfrak{p}} \mathfrak{p}^{f(\mathfrak{p})}$ *be divisors such that* $\mathfrak{b}$ *is multiple of* $\mathfrak{a}$, *and let* S *be a finite set of places. Then* $\mathfrak{K}(\mathfrak{b}, S)$ *is contained in* $\mathfrak{K}(\mathfrak{a}, S)$, *and the vector space* $\mathfrak{K}(\mathfrak{a}, S)/\mathfrak{K}(\mathfrak{b}, S)$ *(over the field of constants) is of finite dimension equal to* $\sum_{\mathfrak{p} \in S} (f(\mathfrak{p}) - e(\mathfrak{p}))d(\mathfrak{p})$.

The inclusion $\mathfrak{K}(\mathfrak{b}, S) \subset \mathfrak{K}(\mathfrak{a}, S)$ is obvious. Set $\mathfrak{a}_S = \prod_{\mathfrak{p}} \mathfrak{p}^{e'(\mathfrak{p})}$, where $e'(\mathfrak{p})$ is $e(\mathfrak{p})$ when $\mathfrak{p} \in S$ and 0 otherwise, and define similarly $\mathfrak{b}_S$. Then $\mathfrak{K}(\mathfrak{b}, S) = \mathfrak{K}(\mathfrak{b}_S, S)$, $\mathfrak{K}(\mathfrak{a}, S) = \mathfrak{K}(\mathfrak{a}_S, S)$ and $\mathfrak{b}_S$ is a multiple of $\mathfrak{a}_S$. On the other hand we have $\sum_{\mathfrak{p} \in S} (f(\mathfrak{p}) - e(\mathfrak{p}))d(\mathfrak{p}) = d(\mathfrak{b}_S) - d(\mathfrak{a}_S)$. Now, we can find a finite sequence $(\mathfrak{a}_0, \mathfrak{a}_1, \cdots, \mathfrak{a}_h)$ of divisors which satisfy the following conditions: $\mathfrak{a}_0 = \mathfrak{a}_S$; if $1 \leqq i \leqq h$, $\mathfrak{a}_i$ is of the form $\mathfrak{a}_{i-1}\mathfrak{p}$, where $\mathfrak{p}$ is a place belonging to S; $\mathfrak{a}_h = \mathfrak{b}_S$. We have $\mathfrak{K}(\mathfrak{a}_S, S) = \mathfrak{K}(\mathfrak{a}_0, S) \supset \mathfrak{K}(\mathfrak{a}_1, S) \supset \cdots \supset \mathfrak{K}(\mathfrak{a}_h, S) = \mathfrak{K}(\mathfrak{b}_S, S)$. If we can prove that each $\mathfrak{K}(\mathfrak{a}_{i-1}, S)/\mathfrak{K}(\mathfrak{a}_i, S)$ is of finite dimension equal to $d(\mathfrak{a}_i) - d(\mathfrak{a}_{i-1})$, it will follow that $\mathfrak{K}(\mathfrak{a}_S, S)/\mathfrak{K}(\mathfrak{b}_S, S)$ is of finite dimension equal to $\sum_{i=1}^{h} (d(\mathfrak{a}_i) - d(\mathfrak{a}_{i-1})) = d(\mathfrak{b}_S) - d(\mathfrak{a}_S)$. Thus we see that it will be sufficient to prove Lemma 1 in the case where $\mathfrak{b} = \mathfrak{b}_S = \mathfrak{a}_S\mathfrak{p} = \mathfrak{a}\mathfrak{p}$, with $\mathfrak{p} \in S$. Let $\mathfrak{p}_1, \cdots, \mathfrak{p}_n$ be the distinct places belonging to S, with, say, $\mathfrak{p}_1 = \mathfrak{p}$. Making use of the Corollary to Theorem 3, §6, we determine an element $u \in R$ such that $\nu_{\mathfrak{p}_i}(u) = e(\mathfrak{p}_i)$ $(1 \leqq i \leqq n)$. Let d be the degree of $\mathfrak{p}_1$. We wish to prove that $\mathfrak{K}(\mathfrak{a}, S)/\mathfrak{K}(\mathfrak{b}, S)$ is of dimension d over the field of constants K' of R. The residue field of $\mathfrak{p}$ has a base $\{\bar{\omega}_1, \cdots, \bar{\omega}_d\}$ with respect to K' which is composed of d elements. Let ω_k' be an element of the ring of $\mathfrak{p}$ which belongs to the residue class $\bar{\omega}_k$ modulo $\mathfrak{p}$ ($1 \leqq k \leqq d$). By Theorem 3, §6, we can find elements $\omega_1, \cdots, \omega_d$ of R such that $\nu_\mathfrak{p}(\omega_k - \omega_k') \geqq 1$, $\nu_{\mathfrak{p}_i}(\omega_k) \geqq 0$ ($2 \leqq i \leqq n$). It is clear that ω_k still belongs to the residue class $\bar{\omega}_k$. Any element of the residue field of $\mathfrak{p}$ can be expressed in the form $\sum_{k=1}^d \lambda_k \bar{\omega}_k$ ($\lambda_k \in K'$, $1 \leqq k \leqq d$); therefore, every element z of the ring of $\mathfrak{p}$ may be expressed in the form $z = \sum_{k=1}^d \lambda_k \omega_k + z'$, with $\lambda_k \in K'$ ($1 \leqq k \leqq d$) and $z' \in \mathfrak{p}$.

The elements $u\omega_k$ ($1 \leqq k \leqq d$) belong to $\mathfrak{K}(\mathfrak{a}, S)$. If w is any element of $\mathfrak{K}(\mathfrak{a}, S)$, then $\nu_\mathfrak{p}(w/u) \geqq 0$ and w/u can be expressed in the form $\sum_{k=1}^d \lambda_k \omega_k + z'$, with $z' \in \mathfrak{p}$, $\lambda_k \in K'$. If $i > 1$, we have $\nu_{\mathfrak{p}_i}(w/u) \geqq 0$, $\nu_{\mathfrak{p}_i}(\omega_k) \geqq 0$ whence $\nu_{\mathfrak{p}_i}(z') \geqq 0$; if $i = 1$, the fact that $z' \in \mathfrak{p}$ implies that $\nu_{\mathfrak{p}_1}(z'u) \geqq e(\mathfrak{p}_1) + 1$. It follows that $z'u$ belongs to $\mathfrak{K}(\mathfrak{b}, S)$ and therefore that every element of $\mathfrak{K}(\mathfrak{a}, S)$ is congruent modulo $\mathfrak{K}(\mathfrak{b}, S)$ to some linear combination of $u\omega_1, \cdots, u\omega_d$ with coefficients in K'. This proves already that $\mathfrak{K}(\mathfrak{a}, S)/\mathfrak{K}(\mathfrak{b}, S)$ is of finite dimension $\leqq d$. To prove that this dimension is d, it is sufficient to show that, $\lambda_1, \cdots, \lambda_d$ being elements not all zero of K', $w = \sum_{k=1}^d \lambda_k \omega_k u$ cannot lie in $\mathfrak{K}(\mathfrak{b}, S)$. Now, the residue class of w/u modulo $\mathfrak{p}$ is $\sum_{k=1}^d \lambda_k \bar{\omega}_k \neq 0$, whence $\nu_\mathfrak{p}(w/u) = 0$, and $\nu_\mathfrak{p}(w) = \nu_\mathfrak{p}(u) = e(\mathfrak{p})$, which proves our assertion since $f(\mathfrak{p}) = e(\mathfrak{p}) + 1$. Lemma 1 is thereby proved.

§8. The divisor of a function

Let K be the field of complex numbers, and let $u = f(x)/g(x)$ be a rational function $\neq 0$ of a complex variable x, f and g being polynomials with coefficients in K with no common factor. Set $R = K\langle x\rangle$; if $a \in K$, denote by $\mathfrak{p}_a$ the place of R at which x takes the value a, and denote by $\mathfrak{p}_\infty$ the unique pole of x ($\mathfrak{p}_\infty$ is therefore the same place which was denoted by $\mathfrak{p}_{1/x}$ in §3). The numbers a for which $\mathfrak{p}_a$ is a zero of u are the roots of the equation $f(x) = 0$. Let $a_1, \cdots, a_h$ be the distinct roots of this equation, and assume that a_i is a root of multiplicity e_i ($1 \leqq i \leqq h$); $\mathfrak{p}_{a_i}$ is then a zero of order e_i of u. On the other hand, $f(x) = \alpha \prod_{i=1}^h (x - a_i)^{e_i}$, $\alpha \in K$, which proves that the sum $\sum_{i=1}^h e_i$ is equal to the degree δ of $f(x)$. Similarly, the numbers a' for which $\mathfrak{p}_{a'}$ is a pole of u are the roots of the equation $g(x) = 0$; if $a_1', \cdots, a_{h'}'$ are these roots, and $e_1', \cdots, e_{h'}'$ their multiplicities, the sum $\sum_{i=1}^{h'} e_i'$ is equal to the degree δ' of $g(x)$. As for the place $\mathfrak{p}_\infty$, we have seen that the order of u at this place is $\delta' - \delta$. Now, we have

$$\delta' - \delta + \sum_{i=1}^{h} e_i - \sum_{i=1}^{h'} e_i' = 0$$

which can be expressed in the form

$$\textstyle\sum_\mathfrak{p} \nu_\mathfrak{p}(u) = 0$$

the sum being extended over all places $\mathfrak{p}$ of the field $K\langle x\rangle$. Thus the sum of the orders of a rational function $\neq 0$ with complex coefficients at all places of $K\langle x\rangle$ is 0.

We shall try to generalize this result. Let us first consider the case where K

is the field of real numbers and $R = K\langle x\rangle$, x transcendental over K. Consider the polynomial $u = x^2 + px + q$, where p and q are real numbers. The only pole of u is $\mathfrak{p}_{1/x}$, and it is a pole of order 2. As for the zeros of u, we must distinguish several cases.

a) If $p^2 - 4q > 0$, then $u = (x - x_1)(x - x_2)$, where x_1 and x_2 are distinct real numbers; this means that u has two zeros both of order 1, viz. the places $\mathfrak{p}_1$ and $\mathfrak{p}_2$ which correspond to the irreducible polynomials $x - x_1$ and $x - x_2$ (cf. §3).

b) If $p^2 - 4q = 0$, then $u = (x - x_3)^2$, with $x_3 = -p/2$. This means that u admits a single zero of order 2, which is the place $\mathfrak{p}_3$ which corresponds to the irreducible polynomial $x - x_3$.

c) If $p^2 - 4q < 0$, then u is irreducible, and admits therefore a single zero of order 1, which is the place $\mathfrak{p}_4$ which corresponds to the irreducible polynomial u.

Thus the statement that the sum of the orders of u at all places is 0 is still true in cases a) and b), but is false in case c). However, we make the following observation. In the case where K was the field of complex numbers, every place was of course of degree 1. In the present case, the places $\mathfrak{p}_1$ and $\mathfrak{p}_2$ in case a) and the place $\mathfrak{p}_3$ in case b) are of degree 1, while the place $\mathfrak{p}_4$ in case c) is of degree 2 (because u is of degree 2; cf. §3). Thus we see that, in all cases considered thus far, it is true that $\sum_{\mathfrak{p}} \nu_{\mathfrak{p}}(u) d(\mathfrak{p}) = 0$; in other terms if we set $\mathfrak{a} = \prod_{\mathfrak{p}} \mathfrak{p}^{\nu_{\mathfrak{p}}(u)}$, then $\mathfrak{a}$ is a divisor of degree 0. It is this last statement which we shall prove to be true in the most general case.

Let then from now on R be a field of algebraic functions of one variable over an arbitrary field, and let x be a non zero element of R. In order for the sum $\sum_{\mathfrak{p}} d(\mathfrak{p})\nu_{\mathfrak{p}}(x)$ to have a meaning, it is necessary that only a finite number of the numbers $\nu_{\mathfrak{p}}(x)$ be $\neq 0$, which means that x has only a finite number of zeros and poles. We shall prove that this is actually the case.

LEMMA 1. *Let R be a field of algebraic functions of one variable, and let x be a non constant element of R. Then x has only a finite number of zeros; if $\mathfrak{p}_1, \cdots, \mathfrak{p}_h$ are the distinct zeros of x, we have $\sum_{i=1}^{h} d(\mathfrak{p}_i)\nu_{\mathfrak{p}_i}(x) \leqq [R : K\langle x\rangle]$, where K is the field of constants of R.*

Let $\mathfrak{p}_1, \cdots, \mathfrak{p}_k$ be distinct zeros of x (not necessarily all of them). Set $\mathfrak{b} = \prod_{i=1}^{k} \mathfrak{p}_i^{b_i}$, where $b_i = \nu_{\mathfrak{p}_i}(x)$, and denote by $\mathfrak{e}$ the unit divisor. Let S be the set composed of $\mathfrak{p}_1, \cdots, \mathfrak{p}_k$. Then, according to Lemma 1, §7, the space $\mathfrak{R}(\mathfrak{e}, S)/\mathfrak{R}(\mathfrak{b}, S)$ is of finite dimension δ equal to $\sum_{i=1}^{k} d(\mathfrak{p}_i)\nu_{\mathfrak{p}_i}(x)$. Let $z_1, \cdots, z_\delta$ be δ elements of $\mathfrak{R}(\mathfrak{e}, S)$ whose residue classes modulo $\mathfrak{R}(\mathfrak{b}, S)$ are linearly independent over K. Then we shall prove that $z_1, \cdots, z_\delta$ themselves are linearly independent over $K\langle x\rangle$. Assume for a moment that they are not, i.e. that there exists a relation

$$(1) \qquad \sum_{j=1}^{\delta} u_j z_j = 0$$

where $u_1, \cdots, u_\delta$ are elements not all zero in $K\langle x\rangle$. Every u_j may be written as a quotient of polynomials in x with coefficients in K. Multiplying by a common denominator, we see that there exists a relation of the form (1) in which the u_j's are polynomials in x not all zero. Dividing if necessary by some power of x, we may furthermore assume that $u_1, \cdots, u_\delta$ are not all divisible by x. Write $u_j = u_j(0) + xv_j$, with $v_j \in K[x]$; then (1) takes the form $\sum_{j=1}^{\delta} u_j(0)z_j = -x \sum_{j=1}^{\delta} v_j z_j$. If $1 \leqq i \leqq k$, we have $\nu_{\mathfrak{p}_i}(v_j) \geqq 0$ because $\nu_{\mathfrak{p}_i}(x) > 0$ and $v_j \in K[x]$; we have also $\nu_{\mathfrak{p}_i}(z_j) \geqq 0$ because $z_j \in \mathfrak{K}(\mathfrak{c}, S)$ and $\mathfrak{p}_i \in S$. Therefore we have $\nu_{\mathfrak{p}_i}(-x \sum_{j=1}^{\delta} v_j z_j) \geqq \nu_{\mathfrak{p}_i}(x)$ which means that $-x \sum_{j=1}^{\delta} v_j z_j \in \mathfrak{K}(\mathfrak{b}, S)$, whence $\sum_{j=1}^{\delta} u_j(0) z_j \in \mathfrak{K}(\mathfrak{b}, S)$. But this conclusion is inadmissible, because the elements $u_j(0)$ of K are not all zero and the residue classes of $z_1, \cdots, z_\delta$ modulo $\mathfrak{K}(\mathfrak{b}, S)$ are linearly independent over K. Thus we have proved that $z_1, \cdots, z_\delta$ are linearly independent over $K\langle x\rangle$. It follows that

$$(2) \qquad \delta = \sum_{i=1}^{k} d(\mathfrak{p}_i)\nu_{\mathfrak{p}_i}(x) \leqq [R:K\langle x\rangle].$$

Since each $\nu_{\mathfrak{p}_i}(x)$ is $\geqq 1$, we have $k \leqq [R:K\langle x\rangle]$. This proves that x cannot have more than $[R:K\langle x\rangle]$ zeros. Taking $k = h$, (2) becomes the inequality stated in Lemma 1.

Our next step will be to prove that the number $\sum_{i=1}^{h} d(\mathfrak{p}_i)\nu_{\mathfrak{p}_i}(x)$ of Lemma 1 is actually equal to $[R : K\langle x\rangle]$.

In order to do this, we introduce the subring $K[x^{-1}]$ of $K\langle x\rangle$. We remind the reader that an element w of R is said to be integral over $K[x^{-1}]$ if it satisfies an equation of the form $w^m + f_1 w^{m-1} + \cdots + f_m = 0$ where $f_1, \cdots, f_m$ belong to $K[x^{-1}]$. If this is the case, w cannot have any pole outside the set $\{\mathfrak{p}_1, \cdots, \mathfrak{p}_h\}$. For, if $\mathfrak{p}$ is a place which is not a zero of x, then the ring $\mathfrak{o}$ of the place $\mathfrak{p}$ contains $K[x^{-1}]$ and is integrally closed (cf. §2), whence $w \in \mathfrak{o}$.

Let n be the number $[R:K\langle x\rangle]$. We shall see that we can find a base $\{w_1, \cdots, w_n\}$ of R with respect to $K\langle x\rangle$ whose elements are integral over $K[x^{-1}]$. Let $\{w_1', \cdots, w_n'\}$ be any base of R with respect to $K\langle x\rangle$. Then each w_ν' satisfies an equation of the form $w_\nu'^{n} + \sum_{\lambda=1}^{n} \varphi_{\nu,\lambda} w_\nu'^{n-\lambda} = 0$ with $\varphi_{\nu,\lambda} \in K\langle x\rangle$. We can write each $\varphi_{\nu,\lambda}$ in the form $f_{\nu,\lambda}/g_\nu$, with $f_{\nu,\lambda}$ and g_ν in $K[x^{-1}]$. Set $w_\nu = g_\nu w_\nu'$; then we have $w_\nu^n + \sum_{\lambda=1}^{n} g_\nu^\lambda \varphi_{\nu,\lambda} w^{n-\lambda} = 0$, and $g_\nu^\lambda \varphi_{\nu,\lambda} = g_\nu^{\lambda-1} f_{\lambda,\nu} \in K[x^{-1}]$, which proves that w_ν is integral over $K[x^{-1}]$. On the other hand, since w_ν is the product of w_ν' by an element $\neq 0$ of $K\langle x\rangle$, it is clear that $w_1, \cdots, w_n$ form a base of R with respect to $K\langle x\rangle$.

Let σ be an integer $\geqq 0$ such that $\nu_{\mathfrak{p}_i}(w_\nu) \geqq -\sigma$ $(1 \leqq i \leqq h, 1 \leqq \nu \leqq n)$, and let μ be an arbitrarily large integer (in particular, we assume that $\mu > \sigma$). We consider the products $x^{-\rho}w_\nu$, with $0 \leqq \rho \leqq \mu - \sigma$, $1 \leqq \nu \leqq n$. We have

$$\nu_{\mathfrak{p}_i}(x^{-\rho}w_\nu) \geqq -\rho\nu_{\mathfrak{p}_i}(x) - \sigma \geqq -\mu\nu_{\mathfrak{p}_i}(x)$$

since $\nu_{\mathfrak{p}_i}(x) \geqq 1$. On the other hand, if $\mathfrak{p}$ is a place distinct from $\mathfrak{p}_1, \cdots, \mathfrak{p}_h$, we have $\nu_{\mathfrak{p}}(x^{-1}) \geqq 0$, $\nu_{\mathfrak{p}}(w_\nu) \geqq 0$ $(1 \leqq \nu \leqq n)$, whence $\nu_{\mathfrak{p}}(x^{-\rho}w_\nu) \geqq 0$. This means that

$$x^{-\rho}w_\nu \equiv 0 \pmod{\mathfrak{b}^{-\mu}}$$

where $\mathfrak{b}$ is the divisor $\prod_{i=1}^{h} \mathfrak{p}_i^{b_i}$, $b_i = \nu_{\mathfrak{p}_i}(x)$.

We shall denote by $\mathfrak{L}_\mu$ the vector space over K composed of all elements of R which are $\equiv 0 \pmod{\mathfrak{b}^{-\mu}}$ and by $\mathfrak{L}'_\mu$ the subspace of $\mathfrak{L}_\mu$ spanned by the elements $x^{-\rho}w_\nu$ $(0 \leqq \rho \leqq \mu - \sigma, 1 \leqq \nu \leqq n)$. The elements w_ν being linearly independent over $K\langle x\rangle$ and the elements $x^{-\rho}$ of $K\langle x\rangle$ being linearly independent over K, it follows that the $n(\mu - \sigma + 1)$ elements $x^{-\rho}w_\nu$ are linearly independent over K and that $\mathfrak{L}'_\mu$ is of dimension $n(\mu - \sigma + 1)$. On the other hand, we have $\mathfrak{L}_\mu \subset \mathfrak{K}(\mathfrak{b}^{-\mu}, S)$, where $S = \{\mathfrak{p}_1, \cdots, \mathfrak{p}_h\}$. We know that, if $\mathfrak{e}$ is the unit divisor, the space $\mathfrak{K}(\mathfrak{b}^{-\mu}, S)/\mathfrak{K}(\mathfrak{e}, S)$ is of dimension $\mu d(\mathfrak{b})$ (Lemma 1, §7), from which it follows that $\mathfrak{L}_\mu/(\mathfrak{L}_\mu \cap \mathfrak{K}(\mathfrak{e}, S))$ is of dimension $\leqq \mu d(\mathfrak{b})$. Now, an element of $\mathfrak{L}_\mu \cap \mathfrak{K}(\mathfrak{e}, S)$ cannot have any pole outside S (because it is in $\mathfrak{L}_\mu$) or in S (because it is in $\mathfrak{K}(\mathfrak{e}, S)$); such an element is therefore a constant (Corollary 3 to Theorem 1, §4), which shows that $\mathfrak{L}_\mu \cap \mathfrak{K}(\mathfrak{e}, S)$ is of dimension $\leqq 1$ and therefore that $\mathfrak{L}_\mu$ is of dimension $\leqq \mu d(\mathfrak{b}) + 1$. Since $\mathfrak{L}'_\mu \subset \mathfrak{L}_\mu$, we have $n(\mu - \sigma + 1) \leqq \mu d(\mathfrak{b}) + 1$, whence

$$d(\mathfrak{b}) \geqq n - \frac{(\sigma - 1)n + 1}{\mu}.$$

This being true for every sufficiently large μ, we have $d(\mathfrak{b}) \geqq n$. We had already proved in Lemma 1 that $d(\mathfrak{b}) \leqq n$; thus we see that $d(\mathfrak{b}) = n$. Moreover, we see that the dimension of $\mathfrak{L}_\mu$, the dimension of $\mathfrak{L}'_\mu$, and the number $\mu d(\mathfrak{b}) = n\mu$ are integers whose mutual differences remain bounded as μ increases indefinitely.

If $\mathfrak{p}_1, \cdots, \mathfrak{p}_h$ are all the distinct zeros of an element $x \neq 0$ of R, and $b_i = \nu_{\mathfrak{p}_i}(x)$, the divisor $\mathfrak{b} = \prod_{i=1}^h \mathfrak{p}_i^{b_i}$ is called the *divisor of zeros* of x. We have proved

THEOREM 4. *Let R be a field of algebraic functions of one variable and let K be the field of constants of R. Then the degree of the divisor of zeros of an element x of R not in K is $[R:K\langle x\rangle]$.*

If we apply our results to x^{-1} instead of x, we see that x can have only a finite number of poles. If $\mathfrak{q}_1, \cdots, \mathfrak{q}_l$ are the distinct poles of x, and $c_i = -\nu_{\mathfrak{q}_i}(x)$, the integral divisor $\mathfrak{c} = \prod_{i=1}^l \mathfrak{q}_i^{c_i}$ is called the *divisor of poles* of x. Since $K\langle x\rangle = K\langle x^{-1}\rangle$, we have the following result:

COROLLARY. *The notation being as in Theorem 4, the degree of the divisor of poles of x is $[R:K\langle x\rangle]$.*

The divisor $\mathfrak{b}\mathfrak{c}^{-1}$ can also be written in the form $\mathfrak{b}\mathfrak{c}^{-1} = \prod_{\mathfrak{p}} \mathfrak{p}^{\nu_{\mathfrak{p}}(x)}$. This divisor is called the *divisor of x*, and is denoted by $\mathfrak{d}(x)$. It is of course defined only if $x \neq 0$; if x is a constant, $\mathfrak{d}(x)$ is the unit divisor. It is clear that, if x and y are elements $\neq 0$ of R, then

$$\mathfrak{d}(xy) = \mathfrak{d}(x)\mathfrak{d}(y), \qquad \mathfrak{d}(x^{-1}) = (\mathfrak{d}(x))^{-1}.$$

Our results establish the truth of

THEOREM 5. *The divisor of an element $\neq 0$ of a field of algebraic functions of one variable is of degree 0.*

Let $\mathfrak{D}$ be the group of all divisors of a field R of algebraic functions of one variable, and let $\mathfrak{D}_0$ be the group of divisors of elements of R. The cosets of $\mathfrak{D}$ modulo $\mathfrak{D}_0$ are called *classes of divisors*; two divisors which are in the same class are said to be *equivalent* to each other. It follows immediately from Theorem 5 that all divisors in a class are of the same degree; this degree is called the *degree of the class.*

To conclude this section, we state in the form of lemmas two results which have been established in the course of the proof of Theorem 4 and which will be useful later.

LEMMA 2. *Let the notation be as in Theorem 4, and let $\mathfrak{b}$ be the divisor of zeros of x. Then the dimension over K of the space of elements of R which are $\equiv 0$ (mod $\mathfrak{b}^{-\mu}$) (where μ is an integer >0) is $\geqq \mu d(\mathfrak{b}) - \tau$, where τ is an integer which does not depend on μ.*

LEMMA 3. *Let the notation be as in Theorem 4, and let $\mathfrak{R}$ be the ring of elements of R which do not have any pole outside the set of zeros of x. Then there exists a finite subset $\{w_1, \cdots, w_N\}$ of $\mathfrak{R}$, which contains a base of R with respect to $K\langle x\rangle$, and which is such that every element of $\mathfrak{R}$ is a linear combination of $w_1, \cdots, w_N$ with coefficients in $K[x^{-1}]$.*

For, the notation being as in the proof of Theorem 4, it is clear that $\mathfrak{R}$ is the union of the spaces $\mathfrak{L}_\mu$ $(1 \leqq \mu < \infty)$. We have seen that the difference between the dimensions of $\mathfrak{L}_\mu$ and $\mathfrak{L}'_\mu$ remains bounded; let m be the largest value of this difference. Let $z_1, \cdots, z_r$ be elements of $\mathfrak{R}$ whose residue classes modulo the space $\sum_{\nu=1}^{n} K[x^{-1}]w_\nu$ are linearly independent over K. Then there exists a $\mu > 0$ such that $z_1, \cdots, z_r$ all belong to $\mathfrak{L}_\mu$; since no linear combination of $z_1, \cdots, z_r$ with coefficients in K can be in $\mathfrak{L}'_\mu$, we have $r \leqq m$. This shows that we can find a finite number of elements $w_{n+1}, \cdots, w_N$ of $\mathfrak{R}$ such that every element of $\mathfrak{R}$ is the sum of an element of the space $\sum_{\nu=1}^{n} K[x^{-1}]w_\nu$ and of a linear combination of $w_{n+1}, \cdots, w_N$ with coefficients in K. Lemma 3 is thereby proved.

CHAPTER II

THE THEOREM OF RIEMANN-ROCH

§1. The genus

Let R be a field of algebraic functions of one variable, and let K be the field of constants of R. We have seen that, being given a finite number of places of R, it is impossible to establish any necessary relation between the modes of behavior of an element of R at these places (cf. Theorem 3, I, §6). The situation is quite different if we consider infinitely many places instead of a finite number of them. In this case, relations can be found between the behaviors of an element of R at these places; for instance, we know that an element of R admits as zeros either all places (and this only in the case where the element is 0) or only a finite number of them. The investigation of these relations will be the main object of this chapter. The problem can also be formulated as follows: given for each place $\mathfrak{p}$ an element $x_\mathfrak{p}$ of R and an integer $m_\mathfrak{p}$, does there exist an $x \in R$ which satisfies all the conditions

$$\nu_\mathfrak{p}(x - x_\mathfrak{p}) \geqq m_\mathfrak{p}, \tag{1}$$

and, if so, what is the structure of the set of elements x which satisfy these conditions?

If all $x_\mathfrak{p}$'s are 0, the elements which satisfy the conditions (1) clearly form a vector space over K. We shall refer to this case as to the "homogeneous case". In the general case, if x is a solution of (1), the other solutions of the same problem are obtained by adding to x the solutions y of the homogeneous problem defined by the conditions $\nu_\mathfrak{p}(y) \geqq m_\mathfrak{p}$. Thus, as usual in linear problems, the study of the system (1) can be decomposed into two parts: 1) does there exist at least one solution?; 2) what are the solutions of the associated homogeneous problem?

Let us first consider the homogeneous problem defined by the conditions

$$\nu_\mathfrak{p}(x) \geqq m_\mathfrak{p}. \tag{2}$$

If infinitely many of the integers $m_\mathfrak{p}$ are >0, the only solution of this problem is the constant 0 (by Lemma 1, I, §8). We shall be primarily interested in the case where all integers $m_\mathfrak{p}$ except a finite number are equal to 0; in this case, the conditions (2) are equivalent to the unique condition $x \equiv 0 \pmod{\mathfrak{a}}$, where $\mathfrak{a}$ is the divisor $\prod_\mathfrak{p} \mathfrak{p}^{m_\mathfrak{p}}$. The main interest of this case is that, as we shall see, the solutions then form a vector space of finite dimension over K.

If $\mathfrak{a}$ is any divisor, we shall denote by $\mathfrak{L}(\mathfrak{a})$ the space of elements $x \in R$ which are $\equiv 0 \pmod{\mathfrak{a}}$.

Let $\mathfrak{b}$ be a multiple of the divisor $\mathfrak{a}$; then $\mathfrak{L}(\mathfrak{b})$ is clearly contained in $\mathfrak{L}(\mathfrak{a})$. We shall first prove that $\mathfrak{L}(\mathfrak{a})/\mathfrak{L}(\mathfrak{b})$ is a vector space of finite dimension, and we shall find an upper bound for the dimension of this space.

Let S be the set of places which occur with exponents $\neq 0$ in either $\mathfrak{a}$ or $\mathfrak{b}$; S is therefore a finite set of places. We define the spaces $\mathfrak{R}(\mathfrak{a}, S)$ and $\mathfrak{R}(\mathfrak{b}, S)$ as in I, §7. It is clear that $\mathfrak{L}(\mathfrak{a}) \subset \mathfrak{R}(\mathfrak{a}, S)$ and $\mathfrak{L}(\mathfrak{b}) \subset \mathfrak{R}(\mathfrak{b}, S)$. Moreover, we have

$$\mathfrak{L}(\mathfrak{b}) = \mathfrak{L}(\mathfrak{a}) \cap \mathfrak{R}(\mathfrak{b}, S). \tag{3}$$

In fact, the inclusion $\mathfrak{L}(\mathfrak{b}) \subset \mathfrak{L}(\mathfrak{a}) \cap \mathfrak{R}(\mathfrak{b}, S)$ is obvious. Conversely, if $x \in \mathfrak{L}(\mathfrak{a}) \cap \mathfrak{R}(\mathfrak{b}, S)$, we have $\nu_{\mathfrak{p}}(x) \geqq 0$ for all places $\mathfrak{p}$ not in S because these places occur in $\mathfrak{a}$ with the exponent 0; since $x \in \mathfrak{R}(\mathfrak{b}, S)$, it follows that $x \in \mathfrak{L}(\mathfrak{b})$, which completes the proof of (3).

The space $\mathfrak{L}(\mathfrak{a})/\mathfrak{L}(\mathfrak{b})$ is therefore equal to $\mathfrak{L}(\mathfrak{a})/(\mathfrak{L}(\mathfrak{a}) \cap \mathfrak{R}(\mathfrak{b}, S))$. According to the first homomorphism theorem, this is isomorphic to $(\mathfrak{L}(\mathfrak{a}) + \mathfrak{R}(\mathfrak{b}, S))/\mathfrak{R}(\mathfrak{b}, S)$, which is a subspace of $\mathfrak{R}(\mathfrak{a}, S)/\mathfrak{R}(\mathfrak{b}, S)$. The latter space being of finite dimension (Lemma 1, I, §7), it follows that $\mathfrak{L}(\mathfrak{a})/\mathfrak{L}(\mathfrak{b})$ is of finite dimension. Since S contains all places which occur with exponents $\neq 0$ in either $\mathfrak{a}$ or $\mathfrak{b}$, it follows immediately from Lemma 1, I, §7 that the dimension of $\mathfrak{R}(\mathfrak{a}, S)/\mathfrak{R}(\mathfrak{b}, S)$ is $d(\mathfrak{b}) - d(\mathfrak{a})$. Thus we see that the dimension of $\mathfrak{L}(\mathfrak{a})/\mathfrak{L}(\mathfrak{b})$ is at most $d(\mathfrak{b}) - d(\mathfrak{a})$.

Now we shall prove that $\mathfrak{L}(\mathfrak{a})$ is of finite dimension. In view of what we just proved, it is sufficient to prove that $\mathfrak{L}(\mathfrak{b})$ is of finite dimension for some multiple $\mathfrak{b}$ of $\mathfrak{a}$. We shall take for $\mathfrak{b}$ an integral divisor which is multiple of $\mathfrak{a}$ and which is different from the unit divisor (it is easily seen that there exists such a divisor). If $x \in \mathfrak{L}(\mathfrak{b})$, then x has at least one zero (because $\mathfrak{b}$ is not the unit divisor) but has no pole (because $\mathfrak{b}$ is integral). Were $x \neq 0$, then $1/x$ would have no zero and would therefore be a constant $\neq 0$ (by Corollary 3 to Theorem 1, I §4) which is impossible since x admits a zero. Thus we see that $\mathfrak{L}(\mathfrak{b})$ contains only 0.

If $\mathfrak{a}$ is any divisor, we shall denote by $l(\mathfrak{a})$ the dimension of the space $\mathfrak{L}(\mathfrak{a})$. If $\mathfrak{b}$ is a multiple of $\mathfrak{a}$, then the dimension of $\mathfrak{L}(\mathfrak{a})/\mathfrak{L}(\mathfrak{b})$ is $l(\mathfrak{a}) - l(\mathfrak{b})$; therefore, we obtain the inequality $l(\mathfrak{a}) - l(\mathfrak{b}) \leqq d(\mathfrak{b}) - d(\mathfrak{a})$, or

$$l(\mathfrak{a}) + d(\mathfrak{a}) \leqq l(\mathfrak{b}) + d(\mathfrak{b}). \tag{4}$$

On the other hand, Lemma 2, I, §8 gives us some information about the number $l(\mathfrak{a})$ for certain divisors $\mathfrak{a}$. Let x be any non constant element of R, and let $\mathfrak{x}$ be the divisor of zeros of x. Then we have $l(\mathfrak{x}^{-\mu}) \geqq \mu d(\mathfrak{x}) - \tau$ for all sufficiently large integers μ, where τ is an integer which depends only on x. This inequality may be written in the form $l(\mathfrak{x}^{-\mu}) + d(\mathfrak{x}^{-\mu}) \geqq -\tau$. Taking (4) into account, we see that

$$l(\mathfrak{b}) + d(\mathfrak{b}) \geqq -\tau \tag{5}$$

whenever $\mathfrak{b}$ is a multiple of some $\mathfrak{x}^{-\mu}$. From this we shall be able to deduce that the numbers $l(\mathfrak{a}) + d(\mathfrak{a})$ for all divisors a have a finite lower bound.

Lemma 1. *If $\mathfrak{b}$ and $\mathfrak{b}'$ are equivalent divisors, we have* $d(\mathfrak{b}) = d(\mathfrak{b}')$ *and* $l(\mathfrak{b}) = l(\mathfrak{b}')$.

The first assertion follows immediately from Theorem 5, I, §8. In order to prove the second, denote by z an element of R such that $\mathfrak{b}' = \mathfrak{d}(z)\mathfrak{b}$. Then the conditions $u \equiv 0 \pmod{\mathfrak{b}}$ and $zu \equiv 0 \pmod{\mathfrak{b}'}$ are clearly equivalent to each other. Thus, if $u \in \mathfrak{L}(\mathfrak{b})$, we have $zu \in \mathfrak{L}(\mathfrak{b}')$; conversely, if $u' \in \mathfrak{L}(\mathfrak{b}')$, then $z^{-1}u' \in \mathfrak{L}(\mathfrak{b})$. The mapping $u \to zu$ induces a one-to-one linear mapping of $\mathfrak{L}(\mathfrak{b})$ onto $\mathfrak{L}(\mathfrak{b}')$, from which we conclude that $l(\mathfrak{b}) = l(\mathfrak{b}')$.

It follows that (5) is still valid for any divisor $\mathfrak{b}$ which is equivalent to some multiple of some $\mathfrak{x}^{-\mu}$. But we shall prove that this is the case for every divisor $\mathfrak{b}$ of R.

Let $\mathfrak{p}$ be any place which is not a zero of x; then $\mathfrak{p}$ is not a pole of x^{-1}. The value ξ taken by x^{-1} at $\mathfrak{p}$ is an element of the residue field of $\mathfrak{p}$, which is algebraic over K (Theorem 2, I, §5). It follows that there exists a polynomial $f_\mathfrak{p}$ with coefficients in K such that $f_\mathfrak{p}(\xi) = 0$, whence $\nu_\mathfrak{p}(f_\mathfrak{p}(x^{-1})) \geqq 1$. Let $\mathfrak{b} = \prod_\mathfrak{p} \mathfrak{p}^{b(\mathfrak{p})}$ be any divisor. Then only a finite number of the exponents $b(\mathfrak{p})$ are <0. We set

$$z = \prod{}' f_\mathfrak{p}^{-b(\mathfrak{p})}(x^{-1})$$

where the product $\prod'$ is extended to those places $\mathfrak{p}$ which are not zeros of x and for which $b(\mathfrak{p}) < 0$ (if there are no such places, set $z = 1$). For these places, we clearly have $\nu_\mathfrak{p}(z) \geqq -b(\mathfrak{p})$, and any such place occurs in $\mathfrak{b}' = \mathfrak{d}(z)\mathfrak{b}$ with an exponent $\geqq 0$. On the other hand, the only poles of $f_\mathfrak{p}(x^{-1})$ are among the zeros of x. It follows that z has no pole outside the set of zeros of x, and therefore that any place which is not a zero of x occurs in $\mathfrak{b}'$ with an exponent $\geqq 0$. Now, let $\mathfrak{q}_1, \cdots, \mathfrak{q}_h$ be the distinct zeros of x, and let μ be an integer such that $-\mu$ is smaller than the exponents with which $\mathfrak{q}_1, \cdots, \mathfrak{q}_h$ enter in $\mathfrak{b}'$. Then, since each $\mathfrak{q}_i$ occurs in $\mathfrak{x}$ with an exponent $\geqq 1$, it is clear that $\mathfrak{b}'$ is a multiple of $\mathfrak{x}^{-\mu}$. Thus our statement that every divisor is equivalent to a multiple of some $\mathfrak{x}^{-\mu}$ is proved. It follows that $l(\mathfrak{b}) + d(\mathfrak{b}) \geqq -\tau$ for every divisor of R.

We can define an integer g by the formula

$$-g + 1 = \min \{l(\mathfrak{b}) + d(\mathfrak{b})\},$$

$\mathfrak{b}$ running over all divisors of the field. This number g, which plays a prominent role in the study of the structure of R, is called the *genus* of R.

The genus of R is always $\geqq 0$. For, if $\mathfrak{e}$ is the unit divisor, we have $d(\mathfrak{e}) = 0$ and the only elements of $\mathfrak{L}(\mathfrak{e})$ are the constants (as follows from Corollary 3 to Theorem 1, I, §4), whence $l(\mathfrak{e}) = 1$ and $l(\mathfrak{e}) + d(\mathfrak{e}) = 1$.

Observing that $d(\mathfrak{a}^{-1}) = -d(\mathfrak{a})$, we obtain

THEOREM 1 (RIEMANN'S THEOREM). *Let R be a field of algebraic functions of one variable, and let g be the genus of R. Let $\mathfrak{a}$ be any divisor of R; denote by $l(\mathfrak{a})$ the dimension of the space of elements x of R which are $\equiv 0 \pmod{\mathfrak{a}}$. Then we have $l(\mathfrak{a}) \geqq d(\mathfrak{a}^{-1}) - g + 1$; the equality $l(\mathfrak{a}) = d(\mathfrak{a}^{-1}) - g + 1$ holds for at least one divisor $\mathfrak{a}_0$, and, if it holds for $\mathfrak{a}_0$, then it holds also for any divisor $\mathfrak{a}$ of which $\mathfrak{a}_0$ is a multiple.*

The last assertion follows immediately from (4).

§2. Fields of genus zero

Let K be any field, and let $R = K\langle x\rangle$, where x is transcendental over K. Then R is a field of algebraic functions of one variable with K as its field of constants (I, §3). We shall see that the genus of R is zero. The element x has a single zero $\mathfrak{x}$ which is of degree 1, and the divisor of zeros of x is $\mathfrak{x}$. We have seen in the course of the proof of Theorem 1, §1 that any divisor $\mathfrak{b}$ of R is equivalent to a multiple $\mathfrak{b}'$ of some $\mathfrak{x}^{-\mu}$, whence $l(\mathfrak{b}) + d(\mathfrak{b}) = l(\mathfrak{b}') + d(\mathfrak{b}') \geqq l(\mathfrak{x}^{-\mu}) + d(\mathfrak{x}^{-\mu})$. Since $d(\mathfrak{x}^{-\mu}) = -\mu$, we see that the genus g of R can be defined by the formula

$$-g + 1 = \min_{0 \leqq \mu < \infty} \{l(\mathfrak{x}^{-\mu}) - \mu\}.$$

We shall compute $l(\mathfrak{x}^{-\mu})$. An element u of R may be written $f(x^{-1})/h(x^{-1})$, where f and h are polynomials with coefficients in K with no common factor. Let us find at which condition u will be $\equiv 0 \pmod{\mathfrak{x}^{-\mu}}$. If φ is any irreducible polynomial with coefficients in K, $\varphi(x^{-1})$ is not a constant and therefore there exists a place $\mathfrak{p}_\varphi$ of R which contains $\varphi(x^{-1})$ (Theorem 1, I, §4); moreover, we have seen (I, §3) that $\mathfrak{p}_\varphi \cap K[x^{-1}]$ consists of the elements of $K[x^{-1}]$ which are divisible by $\varphi(x^{-1})$ in this ring. Since $\mathfrak{p}_\varphi$ is not to be a pole of u, $h(x^{-1})$ cannot be divisible by $\varphi(x^{-1})$. This being true for every irreducible φ, h must be a constant, and we may assume that $h = 1$. If f is of degree m, then $\mathfrak{x}$ is a pole of degree m of $f(x^{-1})$ (I, §3), and we must have $m \leqq \mu$. Thus the elements of R which are $\equiv 0 \pmod{\mathfrak{x}^{-\mu}}$ are the linear combinations with coefficients in K of $1, x^{-1}, \cdots, x^{-\mu}$, whence $l(\mathfrak{x}^{-\mu}) = \mu + 1$, and therefore $g = 0$. Thus we see that *any field of algebraic functions of one variable which is a purely transcendental extension of its field of constants is of genus* 0.

Conversely, let R be a field of algebraic functions of one variable which is of genus 0. We shall see that, if R has at least one place $\mathfrak{p}$ of degree 1, then R is a purely transcendental extension of its field of constants. We have, by Riemann's theorem, $l(\mathfrak{p}^{-1}) \geqq d(\mathfrak{p}) + 1 = 2$, from which it follows that there exists at least one element x of R not contained in the field of constants K which is $\equiv 0 \pmod{\mathfrak{p}^{-1}}$. The divisor of zeros of x^{-1} is then $\mathfrak{p}$, which is of degree 1, and $[R:K\langle x^{-1}\rangle] = 1$ by Theorem 4, I, §8, whence $R = K\langle x^{-1}\rangle$.

The condition that R should have a place of degree 1 is certainly satisfied if K is algebraically closed, because then every place is of degree 1. However, this condition is not always satisfied if K is not algebraically closed. Let for instance K be the field of real numbers; set $R = K\langle x, y\rangle$, where x is transcendental over K and $x^2 + y^2 + 1 = 0$. Then the following facts can be established without difficulty: K is the field of constants of R; every place of R is of degree 2; R is of genus 0. Since R has no place of degree 1, it is not purely transcendental over K.

We shall prove a little later that a field of genus 0 has always a place of degree either 1 or 2 (cf. end of §6 of this chapter).

§3. Fields of genus one

Let R be a field of algebraic functions of one variable of genus one. We shall denote by K the field of constants of R, and we shall assume that R admits at least one place $\mathfrak{p}$ of degree 1 (this condition is certainly satisfied if K is algebraically closed). We have, by Riemann's theorem, $l(\mathfrak{p}^{-2}) \geqq d(\mathfrak{p}^2) = 2$, from which it follows that there exists an element $x \in R$ not contained in K which is $\equiv 0 \pmod{\mathfrak{p}^{-2}}$. The divisor of zeros of x^{-1} is either $\mathfrak{p}$ or $\mathfrak{p}^2$, i.e., it is of degree 1 or 2. If it were of degree 1, we would have $R = K\langle x^{-1}\rangle$ by Theorem 4, I, §8, and R would be of genus zero and not one. We conclude that the divisor of zeros of x^{-1} is $\mathfrak{p}^2$, and therefore that R is of degree 2 with respect to $K\langle x\rangle = K\langle x^{-1}\rangle$.

We have $l(\mathfrak{p}^{-3}) \geqq d(\mathfrak{p}^3) = 3$. It follows that there exists an element $y \in R$ which is $\equiv 0 \pmod{\mathfrak{p}^{-3}}$ and which is such that $1, x, y$ are linearly independent over K. We shall see that the divisor of poles of y is actually $\mathfrak{p}^3$. Let t be a uniformizing variable at $\mathfrak{p}$. Then $\nu_\mathfrak{p}(xt^2) \geqq 0$ and xt^2 takes a value $\alpha \neq 0$ at $\mathfrak{p}$; since $\mathfrak{p}$ is of degree 1, we have $\alpha \in K$. Were the divisor of poles of y either $\mathfrak{p}$ or $\mathfrak{p}^2$, we would have $\nu_\mathfrak{p}(yt^2) \geqq 0$ and yt^2 would take some finite value β at $\mathfrak{p}$. Then $(y - \alpha^{-1}\beta x)t^2$ would take the value 0 at $\mathfrak{p}$, whence $\nu_\mathfrak{p}(y - \alpha^{-1}\beta x) \geqq -1$. The element $z = y - \alpha^{-1}\beta x$ would be $\equiv 0 \pmod{\mathfrak{p}^{-1}}$ but would not belong to K because $1, x, y$ are linearly independent over K; the divisor of poles of z would therefore be $\mathfrak{p}^{-1}$ and we would have $[R:K\langle z^{-1}\rangle] = 1$, which is impossible.

Since the divisor of poles of y is $\mathfrak{p}^3$, of degree 3, we have $[R:K\langle y\rangle] = 3$. It follows that y does not belong to $K\langle x\rangle$, since otherwise $[R:K\langle y\rangle]$ would be divisible by $[R:K\langle x\rangle]$. Since R is quadratic over $K\langle x\rangle$, we conclude that $R = K\langle x, y\rangle$. The element y satisfies an equation of the form

$$C(x)y^2 + A(x)y + B(x) = 0$$

where A, B, C are polynomials with coefficients in K which we may assume to have no common factor. Since y is not in $K\langle x\rangle$, B and C are both $\neq 0$; furthermore, the polynomial $C(x)Y^2 + A(x)Y + B(x)$ in a letter Y with coefficients in $K\langle x\rangle$ is irreducible. Since A, B, C have no common factor, the polynomial $C(X)Y^2 + A(X)Y + B(X)$ in the letters X and Y with coefficients in K is irreducible in $K[X, Y]$. It follows that $C(X)y^2 + A(X)y + B(X)$, considered as a polynomial in X with coefficients in $K\langle y\rangle$, is irreducible. But $K\langle x, y\rangle$ is of degree 3 over $K\langle y\rangle$; it follows that the degrees of A, B, C are at most 3, one of them being 3.

Now, if $M(X) = a_0X^m + a_1X^{m-1} + \cdots + a_m$ is a polynomial of degree m with coefficients in K, we may write $M(x) = x^m(a_0 + a_1x^{-1} + \cdots + a_mx^{-m})$; since $\nu_\mathfrak{p}(x^{-1}) = 2$, it follows that $\nu_\mathfrak{p}(M(x)) = -2m$. In particular, we see that $\nu_\mathfrak{p}(B(x))$ is even; if $A \neq 0$, then $\nu_\mathfrak{p}(A(x)y) = -3 + \nu_\mathfrak{p}(A(x))$ is odd, and is therefore different from $\nu_\mathfrak{p}(B(x))$. It follows that

$$\nu_\mathfrak{p}(-A(x)y - B(x)) = \min\{\nu_\mathfrak{p}(A(x)y), \nu_\mathfrak{p}(B(x))\}$$

(cf. I, §3). Since $\nu_\mathfrak{p}(C(x)y^2)$ is even, we have $\nu_\mathfrak{p}(C(x)y^2) = \nu_\mathfrak{p}(B(x)) < \nu_\mathfrak{p}(A(x)y)$. Let β and γ be the degrees of B and C respectively; then $-2\gamma - 6 = -2\beta$, whence $\beta = \gamma + 3$. Since $\beta \leqq 3$, $\gamma \leqq 3$, we have $\gamma = 0$, $\beta = 3$. Since $\gamma = 0$, we may assume that $C = 1$. If $A \neq 0$, let α be its degree; then $-6 < -2\alpha - 3$, whence $\alpha \leqq 1$.

Now, assume that the characteristic of R is not 2. Set then $y_1 = y + A/2$; we have $y_1^2 = B_1(x)$, where $B_1 = A^2/4 - B$ is of degree 3. We may add that the equation $B_1(X) = 0$ cannot have a double root. In fact, were this the case, we could write $B_1 = B_2^2 B_3$, with B_2 and B_3 of degree 1 with coefficients in K, whence $(y_1/B_2(x))^2 = B_3(x)$. Since B_3 is of degree 1, x would be contained in $K\langle y_1/B_2(x)\rangle$; the same would be true of $y_1 = B_2(x)\cdot y_1/B_2(x)$ and of $y = y_1 - (1/2)A(x)$, and we would have $R = K\langle y_1/B_2(x)\rangle$, which is impossible.

Conversely, it can be proved that, if K is a field of characteristic $\neq 2$ and $R = K\langle x, y\rangle$, where x is transcendental over K and $y^2 = B(x)$, B being a polynomial of degree 3 with no multiple factor, then R is of genus one. We shall not give the proof here.

§4. Repartitions

Let R be a field of algebraic functions of one variable, and let K be the field of constants of R. The problem stated in §1 was the following: given for each place $\mathfrak{p}$ of R an element $x_\mathfrak{p}$ of R and an integer $m_\mathfrak{p}$, find the conditions which must be satisfied in order that there should exist an $x \in R$ such that

$$\nu_\mathfrak{p}(x - x_\mathfrak{p}) \geqq m_\mathfrak{p} \tag{1}$$

for all places $\mathfrak{p}$.

In order to treat this problem, it will be convenient to introduce the following notion: a mapping $\mathfrak{x}$ which assigns to every place $\mathfrak{p}$ of R an element $\mathfrak{x}(\mathfrak{p})$ of R will be called a *repartition* if the following condition is satisfied: there are only a finite number of places $\mathfrak{p}$ for which $\nu_\mathfrak{p}(\mathfrak{x}(\mathfrak{p})) < 0$.

The set $\mathfrak{X}$ of all repartitions is a subset of the set $\mathfrak{X}_0$ of all mappings $\mathfrak{p} \to \mathfrak{x}_0(\mathfrak{p})$ of the set S of all places into R. Since R is a ring, $\mathfrak{X}_0$ is itself a ring, in which the laws of composition are defined by

$$(\mathfrak{x}_0 + \mathfrak{x}_0')(\mathfrak{p}) = \mathfrak{x}_0(\mathfrak{p}) + \mathfrak{x}_0'(\mathfrak{p}),$$

$$(\mathfrak{x}_0\mathfrak{x}_0')(\mathfrak{p}) = \mathfrak{x}_0(\mathfrak{p})\mathfrak{x}_0'(\mathfrak{p})$$

for all places $\mathfrak{p}$. If $\mathfrak{x}$ and $\mathfrak{x}'$ are repartitions, there are only a finite number of places $\mathfrak{p}$ for which $\nu_\mathfrak{p}(\mathfrak{x}(\mathfrak{p}))$ and $\nu_\mathfrak{p}(\mathfrak{x}'(\mathfrak{p}))$ are not both $\geqq 0$; it follows immediately that $\mathfrak{x} - \mathfrak{x}'$ and $\mathfrak{x}\mathfrak{x}'$ are also repartitions. This means that the set $\mathfrak{X}$ of repartitions is a subring of $\mathfrak{X}_0$.

If $x \in R$ we may associate with x the repartition $\mathfrak{x}_x$ which maps every place $\mathfrak{p}$ upon the element x. It is clear that the mapping $x \to \mathfrak{x}_x$ is an isomorphism of the field R with a subring of the ring of repartitions. We shall allow ourselves, whenever convenient, to denote by x the repartition $\mathfrak{x}_x$ which corresponds to the element x of R. However, one should not confuse the value taken by the element x at $\mathfrak{p}$ (which is an element of the residue field of $\mathfrak{p}$) with the value taken at $\mathfrak{p}$

by the repartition $\mathfrak{x}_x$ (which is x itself). In order to avoid this confusion, we shall call the value assigned to a place $\mathfrak{p}$ by a repartition $\mathfrak{x}$ the $\mathfrak{p}$-*component* of $\mathfrak{x}$.

The unit element of R is also the unit element of $\mathfrak{X}$. Since we have mapped R isomorphically into $\mathfrak{X}$, it follows that we may consider $\mathfrak{X}$ as a vector space over R, and, *a fortiori*, as a vector space over K, which is a subfield of R. *Whenever we shall speak of $\mathfrak{X}$ as of a vector space, it is its structure of vector space over K (and not over R!) which we shall have in mind.*

Let $\mathfrak{x}$ be a repartition, and let $\mathfrak{p}$ be a place of R; then we set

$$\nu_\mathfrak{p}(\mathfrak{x}) = \nu_\mathfrak{p}(\mathfrak{x}(\mathfrak{p}))$$

(it should be observed that, if $\mathfrak{x} = x \in R$, then this definition gives to the symbol $\nu_\mathfrak{p}(x)$ the same meaning it had already); $\nu_\mathfrak{p}(\mathfrak{x})$ is called the *order of the repartition* $\mathfrak{x}$ *at* $\mathfrak{p}$. It is clear that, if $\mathfrak{x}$ and $\mathfrak{y}$ are repartitions, then

$$\nu_\mathfrak{p}(\mathfrak{x} \pm \mathfrak{y}) \geqq \min \{\nu_\mathfrak{p}(\mathfrak{x}), \nu_\mathfrak{p}(\mathfrak{y})\}; \qquad \nu_\mathfrak{p}(\mathfrak{x}\mathfrak{y}) = \nu_\mathfrak{p}(\mathfrak{x}) + \nu_\mathfrak{p}(\mathfrak{y}).$$

Moreover, if $\nu_\mathfrak{p}(\mathfrak{x}) < \nu_\mathfrak{p}(\mathfrak{y})$, we have $\nu_\mathfrak{p}(\mathfrak{x} \pm \mathfrak{y}) = \nu_\mathfrak{p}(\mathfrak{x})$.

Let $\mathfrak{a} = \prod_\mathfrak{p} \mathfrak{p}^{m_\mathfrak{p}}$ be a divisor. We shall say that the repartitions $\mathfrak{x}$ and $\mathfrak{y}$ are congruent to each other modulo $\mathfrak{a}$, and we shall write $\mathfrak{x} \equiv \mathfrak{y} \pmod{\mathfrak{a}}$, if the conditions $\nu_\mathfrak{p}(\mathfrak{x} - \mathfrak{y}) \geqq m_\mathfrak{p}$ are satisfied for all places $\mathfrak{p}$ (this definition agrees with the definition of congruence for elements of R if $\mathfrak{x}$ and $\mathfrak{y}$ are in R).

Now, let us return to the problem defined by the conditions (1) above. We shall make two further assumptions: 1) we shall assume that $\nu_\mathfrak{p}(x_\mathfrak{p}) \geqq 0$ for almost all places $\mathfrak{p}$, i.e., that the mapping $\mathfrak{p} \to x_\mathfrak{p}$ is a repartition $\mathfrak{x}$; 2) we shall assume that the integers $m_\mathfrak{p}$ are almost all zero, i.e., that the symbol $\prod_\mathfrak{p} \mathfrak{p}^{m_\mathfrak{p}}$ represents a divisor $\mathfrak{a}$. Under these conditions, our problem can be formulated as follows: does there exist an $x \in R$ which is $\equiv \mathfrak{x} \pmod{\mathfrak{a}}$?

Denote by $\mathfrak{X}(\mathfrak{a})$ the set of all repartitions which are $\equiv 0 \pmod{\mathfrak{a}}$; it is clear that $\mathfrak{X}(\mathfrak{a})$ is a vector subspace of $\mathfrak{X}$. Then our problem amounts to the following: does there exist an $x \in R$ which can be written in the form $\mathfrak{x} + \mathfrak{y}$ with $\mathfrak{y} \in \mathfrak{X}(\mathfrak{a})$? Now, if $x = \mathfrak{x} + \mathfrak{y}$, then $\mathfrak{x} = x - \mathfrak{y}$ belongs to the space $\mathfrak{X}(\mathfrak{a}) + R$ composed of the sums of elements of $\mathfrak{X}(\mathfrak{a})$ and R; and conversely, if $\mathfrak{x}$ belongs to $\mathfrak{X}(\mathfrak{a}) + R$, then our problem is possible. Now, we shall prove that *the vector space $\mathfrak{X}/(\mathfrak{X}(\mathfrak{a}) + R)$ is of finite dimension.*

Let $\mathfrak{a}'$ be a divisor which divides $\mathfrak{a}$. Then $\mathfrak{X}(\mathfrak{a}')$ clearly contains $\mathfrak{X}(\mathfrak{a})$. Moreover, given any $\mathfrak{x} \in \mathfrak{X}$, we can always find a divisor $\mathfrak{a}'$ which divides $\mathfrak{a}$ and which is such that $\mathfrak{x} \in \mathfrak{X}(\mathfrak{a}')$. For, set $\mu(\mathfrak{p}) = \min \{0, \nu_\mathfrak{p}(\mathfrak{x})\}$; since $\nu_\mathfrak{p}(\mathfrak{x}) \geqq 0$ for almost all $\mathfrak{p}$'s, we have $\mu(\mathfrak{p}) = 0$ for almost all $\mathfrak{p}$'s; moreover, if $\mathfrak{a}_1' = \prod_\mathfrak{p} \mathfrak{p}^{\mu(\mathfrak{p})}$, we have $\mathfrak{x} \equiv 0 \pmod{\mathfrak{a}_1'}$. If $\mathfrak{a}'$ is the highest common divisor of $\mathfrak{a}$ and $\mathfrak{a}_1'$, then $\mathfrak{a}'$ divides $\mathfrak{a}$ and $\mathfrak{x}$ belongs to $\mathfrak{X}(\mathfrak{a}')$. Thus, every element of $\mathfrak{X}/(\mathfrak{X}(\mathfrak{a}) + R)$ belongs to some $(\mathfrak{X}(\mathfrak{a}') + R)/(\mathfrak{X}(\mathfrak{a}) + R)$; we shall prove that $(\mathfrak{X}(\mathfrak{a}') + R)/(\mathfrak{X}(\mathfrak{a}) + R)$ is of finite dimension, and that its dimension does not exceed a certain number which depends on $\mathfrak{a}$ only.

LEMMA 1. *If $\mathfrak{a}'$ divides $\mathfrak{a}$, the space $\mathfrak{X}(\mathfrak{a}')/\mathfrak{X}(\mathfrak{a})$ is of finite dimension equal to $d(\mathfrak{a}) - d(\mathfrak{a}')$.*

Let S_0 be a finite set of places which contains all the places which occur with exponents $\neq 0$ in either $\mathfrak{a}$ or $\mathfrak{a}'$. Let u be any element of the space $\mathfrak{K}(\mathfrak{a}', S_0)$ (cf. I, §7); we associate to u the repartition $\mathfrak{z}_u$ defined by $\mathfrak{z}_u(\mathfrak{p}) = u$ if $\mathfrak{p} \in S_0$, $\mathfrak{z}_u(\mathfrak{p}) = 0$ if $\mathfrak{p} \notin S_0$. It is clear that $\mathfrak{z}_u$ belongs to the space $\mathfrak{X}(\mathfrak{a}')$ and that the mapping $u \to \mathfrak{z}_u$ of $\mathfrak{K}(\mathfrak{a}', S_0)$ into $\mathfrak{X}(\mathfrak{a}')$ is linear. If u belongs to $\mathfrak{K}(\mathfrak{a}, S_0)$, then $\mathfrak{z}_u$ belongs to $\mathfrak{X}(\mathfrak{a})$, and conversely. It follows that the mapping $u \to \mathfrak{z}_u$ defines an isomorphic linear mapping of $\mathfrak{K}(\mathfrak{a}', S_0)/\mathfrak{K}(\mathfrak{a}, S_0)$ into $\mathfrak{X}(\mathfrak{a}')/\mathfrak{X}(\mathfrak{a})$. We shall see that this mapping is onto. Let $\mathfrak{z}$ be any element of $\mathfrak{X}(\mathfrak{a}')$. If $\mathfrak{p} \in S_0$, let $m_\mathfrak{p}$ be the exponent with which $\mathfrak{p}$ enters in $\mathfrak{a}$. Making use of Theorem 3, I, §6, we see that there exists an element $u \in R$ such that $\nu_\mathfrak{p}(u - \mathfrak{z}(\mathfrak{p})) \geqq m_\mathfrak{p}$ for all places $\mathfrak{p} \in S_0$. If $m_\mathfrak{p}'$ is the exponent with which $\mathfrak{p}$ enters in $\mathfrak{a}'$, we have $\nu_\mathfrak{p}(\mathfrak{z}(\mathfrak{p})) \geqq m_\mathfrak{p}'$ (because $\mathfrak{z}$ is in $\mathfrak{X}(\mathfrak{a}')$) and $m_\mathfrak{p} \geqq m_\mathfrak{p}'$ (because $\mathfrak{a}'$ divides $\mathfrak{a}$). Since $u = (u - \mathfrak{z}(\mathfrak{p})) + \mathfrak{z}(\mathfrak{p})$, we have $\nu_\mathfrak{p}(u) \geqq m_\mathfrak{p}'$, which proves that u is in $\mathfrak{K}(\mathfrak{a}', S_0)$. For the places $\mathfrak{p}$ in S_0, we have $\mathfrak{z}_u(\mathfrak{p}) = u$, whence $\nu_\mathfrak{p}(\mathfrak{z}_u(\mathfrak{p}) - \mathfrak{z}(\mathfrak{p})) \geqq m_\mathfrak{p}$. If $\mathfrak{q}$ is a place not in S_0, we have $\mathfrak{z}_u(\mathfrak{q}) = 0$ and therefore $\nu_\mathfrak{q}(\mathfrak{z}_u(\mathfrak{q}) - \mathfrak{z}(\mathfrak{q})) = \nu_\mathfrak{q}(\mathfrak{z}(\mathfrak{q})) \geqq 0$ because $\mathfrak{z} \in \mathfrak{X}(\mathfrak{a}')$. Since the places $\mathfrak{q}$ not in S_0 occur with exponent 0 in $\mathfrak{a}$, we have $\mathfrak{z}_u \equiv \mathfrak{z} \pmod{\mathfrak{a}}$. This proves that our mapping of $\mathfrak{K}(\mathfrak{a}', S_0)/\mathfrak{K}(\mathfrak{a}, S_0)$ into $\mathfrak{X}(\mathfrak{a}')/\mathfrak{X}(\mathfrak{a})$ is onto. Lemma 1 therefore follows from Lemma 1, I, §7.

Now, we have to estimate the dimension of $(\mathfrak{X}(\mathfrak{a}') + R)/(\mathfrak{X}(\mathfrak{a}) + R)$. According to the first Noether homomorphism theorem, our vector space is isomorphic to the space $\mathfrak{X}(\mathfrak{a}')/(\mathfrak{X}(\mathfrak{a}') \cap (\mathfrak{X}(\mathfrak{a}) + R))$. The elements u of R such that $u + \mathfrak{X}(\mathfrak{a}) \subset \mathfrak{X}(\mathfrak{a}')$ are obviously those of $\mathfrak{X}(\mathfrak{a}')$, i.e., those which are $\equiv 0 \pmod{\mathfrak{a}'}$. The set of these elements has been denoted in §1 by $\mathfrak{L}(\mathfrak{a}')$; thus we see that $\mathfrak{X}(\mathfrak{a}') \cap (\mathfrak{X}(\mathfrak{a}) + R) = \mathfrak{X}(\mathfrak{a}) + \mathfrak{L}(\mathfrak{a}')$. Making use of the second Noether homomorphism theorem, we see that $(\mathfrak{X}(\mathfrak{a}) + \mathfrak{L}(\mathfrak{a}'))/\mathfrak{X}(\mathfrak{a})$ is isomorphic with $\mathfrak{L}(\mathfrak{a}')/(\mathfrak{L}(\mathfrak{a}') \cap \mathfrak{X}(\mathfrak{a}))$. Now, $\mathfrak{L}(\mathfrak{a}') \cap \mathfrak{X}(\mathfrak{a})$ is obviously identical with $\mathfrak{L}(\mathfrak{a})$. We have seen (Theorem 1, §1) that $\mathfrak{L}(\mathfrak{a})$ is of finite dimension $l(\mathfrak{a})$; $\mathfrak{L}(\mathfrak{a}')/\mathfrak{L}(\mathfrak{a})$ is therefore of dimension $l(\mathfrak{a}') - l(\mathfrak{a})$. Making use of Lemma 1, we see that the space $(\mathfrak{X}(\mathfrak{a}') + R)/(\mathfrak{X}(\mathfrak{a}) + R)$ is of dimension

$$(d(\mathfrak{a}) - d(\mathfrak{a}')) - (l(\mathfrak{a}') - l(\mathfrak{a})) = (d(\mathfrak{a}) + l(\mathfrak{a})) - (d(\mathfrak{a}') + l(\mathfrak{a}')).$$

The sum $d(\mathfrak{a}) + l(\mathfrak{a})$ depends only on $\mathfrak{a}$; the sum $d(\mathfrak{a}') + l(\mathfrak{a}')$ is always $\geqq -g + 1$, where g is the genus of R (Theorem 1, §1). It follows that $(\mathfrak{X}(\mathfrak{a}') + R)/(\mathfrak{X}(\mathfrak{a}) + R)$ is of dimension $\leqq d(\mathfrak{a}) + l(\mathfrak{a}) + g - 1$, and this bound depends only on $\mathfrak{a}$. Set

$$\lambda(\mathfrak{a}) = d(\mathfrak{a}) + l(\mathfrak{a}) + g - 1.$$

We can now prove that $\mathfrak{X}/(\mathfrak{X}(\mathfrak{a}) + R)$ is of finite dimension $\leqq \lambda(\mathfrak{a})$. Assume for a moment that there would exist $k > \lambda(\mathfrak{a})$ elements $\mathfrak{x}_1, \cdots, \mathfrak{x}_k$ of $\mathfrak{X}$ which would be linearly independent modulo $\mathfrak{X}(\mathfrak{a}) + R$. For each i $(1 \leqq i \leqq k)$ we could find a divisor $\mathfrak{a}_i'$ which divides $\mathfrak{a}$ such that $\mathfrak{x}_i \in \mathfrak{X}(\mathfrak{a}_i')$. There would exist a divisor $\mathfrak{a}'$ which would divide $\mathfrak{a}_1', \cdots, \mathfrak{a}_k'$ and the elements $\mathfrak{x}_i$ would all be in $\mathfrak{X}(\mathfrak{a}')$, which is impossible.

We may even assert that the dimension of $\mathfrak{X}/(\mathfrak{X}(\mathfrak{a}) + R)$ is exactly $\lambda(\mathfrak{a})$. In fact, we know that there exists a divisor $\mathfrak{a}_0$ such that $l(\mathfrak{a}_0) + d(\mathfrak{a}_0) = -g + 1$, and that the equality $l(\mathfrak{a}') + d(\mathfrak{a}') = -g + 1$ holds for any divisor $\mathfrak{a}'$ which divides $\mathfrak{a}_0$ (Theorem 1, §1). Thus we see that, if $\mathfrak{a}'$ is the highest common divisor of $\mathfrak{a}_0$ and $\mathfrak{a}$, then $(\mathfrak{X}(\mathfrak{a}') + R)/(\mathfrak{X}(\mathfrak{a}) + R)$ is of dimension $\lambda(\mathfrak{a})$, which proves our assertion. Thus we have proved

LEMMA 2. *Let $\mathfrak{a}$ be a divisor, and let $\mathfrak{X}(\mathfrak{a})$ be the space of repartitions which are $\equiv 0$ (mod $\mathfrak{a}$). If $\mathfrak{X}$ is the space of all repartitions, the space $\mathfrak{X}/(\mathfrak{X}(\mathfrak{a}) + R)$ is of dimension $\lambda(\mathfrak{a}) = l(\mathfrak{a}) + d(\mathfrak{a}) + g - 1$, where $l(\mathfrak{a})$ is the dimension of the space of elements of R which are $\equiv 0$ (mod $\mathfrak{a}$), while g is the genus.*

This being said, we have seen that the problem proposed at the beginning of this section is equivalent to the following: determine whether a given repartition $\mathfrak{x}$ belongs to the space $\mathfrak{X}(\mathfrak{a}) + R$. Let $\bar{\mathfrak{x}}$ be the residue class of $\mathfrak{x}$ modulo $\mathfrak{X}(\mathfrak{a}) + R$. Since $\mathfrak{X}/(\mathfrak{X}(\mathfrak{a}) + R)$ is a vector space, in order for $\bar{\mathfrak{x}}$ to be 0, it is necessary and sufficient that $\bar{\omega}(\bar{\mathfrak{x}}) = 0$ for every linear function $\bar{\omega}$ on $\mathfrak{X}/(\mathfrak{X}(\mathfrak{a}) + R)$. This can also be expressed in the form that we must have $\omega(\mathfrak{x}) = 0$ for every linear function ω on $\mathfrak{X}$ which vanishes everywhere on $\mathfrak{X}(\mathfrak{a}) + R$. Thus we see that the condition of possibility of our problem can be expressed by a system of $\lambda(\mathfrak{a})$ linearly independent linear conditions on $\mathfrak{x}$.

In order for this result to be useful, we have to learn more about the linear functions ω which were introduced above. In the next section, we shall determine the nature of these functions in the case where $R = K\langle x\rangle$, K being the field of complex numbers.

§5. DIFFERENTIALS

Let K be the field of complex numbers, and let $R = K\langle x\rangle$ be the field of rational functions of x with complex coefficients.

If u is any element of R, we may consider the symbol udx as an element of integral: if γ is any path which does not go through any pole of u, the symbol $\int_\gamma udx$ has a meaning. We shall say that udx is a differential of the field R.

Let a be any complex number, and let $\mathfrak{p}(a)$ be the place at which x takes the value a. We shall call the number $\nu_{\mathfrak{p}(a)}(u)$ the order of udx at $\mathfrak{p}(a)$. Denote by $\mathfrak{p}(\infty)$ the unique pole of x. If we set $x' = x^{-1}$ and consider u as a rational function of x', then the element of integral udx is the same as $-ux'^{-2}dx'$; it is therefore natural to call the number $\nu_{\mathfrak{p}(\infty)}(u) - 2$ the order of udx at $\mathfrak{p}(\infty)$.

If a is any complex number, and if γ is a simple closed curve including a, but no other pole of u than perhaps a itself, in its inside, then the number

$$\frac{1}{2\pi(-1)^{1/2}}\int_\gamma u\,dx$$

is called the residue of udx at $\mathfrak{p}(a)$; it is clear that this residue is 0 if $\mathfrak{p}(a)$ is not a pole of u. The residue of udx at $\mathfrak{p}(\infty)$ is defined to be the number

$$\frac{1}{2\pi(-1)^{1/2}}\int_\Gamma u\,dx$$

where Γ is a simple closed curve including in its inside all the poles of u at a finite distance. Let $a_1, \cdots, a_h$ be these poles; for each i, let γ_i be a simple closed curve contained in the interior of Γ and including a_i (but not any a_j with $j \neq i$) in its inside. Then Cauchy's integral theorem gives

$$\int_\Gamma u\,dx + \sum_{i=1}^h \int_{\gamma_i} u\,dx = 0,$$

which shows that *the sum of all residues of a differential is* 0.

Let $\mathfrak{x}$ be any repartition of the field R. If $\mathfrak{p}$ is any place of R, set

$$\omega^{\mathfrak{p}}(\mathfrak{x}) = \text{residue at } \mathfrak{p} \text{ of } \mathfrak{x}(\mathfrak{p})u\,dx. \tag{1}$$

Then, clearly, $\omega^{\mathfrak{p}}(\mathfrak{x}) = 0$ for almost all $\mathfrak{p}$'s. Set

$$\omega(\mathfrak{x}) = \sum \omega^{\mathfrak{p}}(\mathfrak{x}),$$

the sum being extended over all places. The function ω defined by this formula is a linear function on the space $\mathfrak{X}$ of all repartitions. If $\mathfrak{x} = x_1 \in R$, then $\omega(\mathfrak{x})$ is the sum of the residues of the differential x_1udx, and is therefore 0. Thus, ω vanishes on the subspace R of $\mathfrak{X}$.

Let $\mathfrak{a} = \prod_{\mathfrak{p}} \mathfrak{p}^{m(\mathfrak{p})}$ be a divisor. We shall say that the differential udx is $\equiv 0$ (mod $\mathfrak{a}$) if, for every place $\mathfrak{p}$, the order of udx at $\mathfrak{p}$ is $\geqq m(\mathfrak{p})$. Assume that this is the case, and let $\mathfrak{X}(\mathfrak{a}^{-1})$ be the space of repartitions which are $\equiv 0$ (mod $\mathfrak{a}^{-1}$). If $\mathfrak{x} \in \mathfrak{X}(\mathfrak{a}^{-1})$, and $\mathfrak{p}$ is any place, then we have $\nu_{\mathfrak{p}}(\mathfrak{x}(\mathfrak{p})) \geqq -m(\mathfrak{p})$. The order of udx at $\mathfrak{p}$ being $\geqq m(\mathfrak{p})$, we see that the order of the differential $\mathfrak{x}(\mathfrak{p})udx$ at $\mathfrak{p}$ is $\geqq 0$, whence $\omega^{\mathfrak{p}}(\mathfrak{x}) = 0$. This means that ω vanishes on the set $\mathfrak{X}(\mathfrak{a}^{-1}) + R$.

Replacing $\mathfrak{a}$ by $\mathfrak{a}^{-1}$, we see that any differential udx which is $\equiv 0$ (mod $\mathfrak{a}^{-1}$) gives rise to a linear function ω on the space of repartitions which vanishes on $\mathfrak{X}(\mathfrak{a}) + R$: this is precisely the type of linear functions to the consideration of which we were led at the end of the last section.

Now let us count how many linearly independent differentials udx we have which are $\equiv 0$ (mod $\mathfrak{a}^{-1}$). Write $\mathfrak{a} = \prod_{\mathfrak{p}} \mathfrak{p}^{m(\mathfrak{p})}$. If udx is to be $\equiv 0$ (mod $\mathfrak{a}^{-1}$), we must first have, for any complex number a, $\nu_{\mathfrak{p}(a)}(u) \geqq -m(\mathfrak{p}(a))$. Set $h = \prod_a (x-a)^{m(\mathfrak{p}(a))}$; then we must have $\nu_{\mathfrak{p}(a)}(hu) \geqq 0$ for every $a \in K$, which means that $hu = f$ must be a polynomial in x. On the other hand, we must have $\nu_{\mathfrak{p}(\infty)}(u) - 2 \geqq -m(\mathfrak{p}(\infty))$, i.e., $\nu_{\mathfrak{p}(\infty)}(f) \geqq 2 - m(\mathfrak{p}(\infty)) + \nu_{\mathfrak{p}(\infty)}(h)$. The number $\nu_{\mathfrak{p}(\infty)}(h)$ is equal to $-\sum_a m(\mathfrak{p}(a))$, and $-\sum_a m(\mathfrak{p}(a)) - m(\mathfrak{p}(\infty))$ is $-d(\mathfrak{a})$. Thus we find the condition $\nu_{\mathfrak{p}(\infty)}(f) \geqq 2 - d(\mathfrak{a})$. If $d(\mathfrak{a}) \geqq 2$, then f may be an arbitrary polynomial of degree $\leqq d(\mathfrak{a}) - 2$; if $d(\mathfrak{a}) < 2$, then f must be 0. In the first case, we see that there are exactly $d(\mathfrak{a}) - 1$ linearly independent differentials which are $\equiv 0$ (mod $\mathfrak{a}^{-1}$).

Now, we observe that the function ω which corresponds to a differential udx is $\neq 0$ if $u \neq 0$. For, let a be a number which is neither a zero nor a pole of u, and let $\mathfrak{x}$ be the repartition defined by $\mathfrak{x}(\mathfrak{p}(a)) = (x-a)^{-1}$, $\mathfrak{x}(\mathfrak{p}) = 0$ if $\mathfrak{p} \neq \mathfrak{p}(a)$. Then $\omega^{\mathfrak{p}(a)}(\mathfrak{x}) = u(a) \neq 0$, while $\omega^{\mathfrak{p}}(\mathfrak{x}) = 0$ if $\mathfrak{p} \neq \mathfrak{p}(a)$, and therefore $\omega(\mathfrak{x}) = u(a) \neq 0$. Thus we see that the differentials udx which are $\equiv 0$ (mod $\mathfrak{a}^{-1}$) yield exactly max $\{0, d(\mathfrak{a}) - 1\}$ linearly independent linear functions on the space $\mathfrak{X}$

of representations which vanish on $\mathfrak{X}(\mathfrak{a}) + R$. On the other hand, the genus of R being 0, the maximum number of such linear functions is $\lambda(\mathfrak{a}) = d(\mathfrak{a}) + l(\mathfrak{a}) - 1$ (cf. §4). Let us compute $l(\mathfrak{a})$. In the computation which was made above, the functions u were submitted to the condition $u \equiv 0 \pmod{\mathfrak{a}^{-1}(\mathfrak{p}(\infty))^2}$, and we therefore have $l(\mathfrak{a}^{-1}(\mathfrak{p}(\infty))^2) = \max\{0, d(\mathfrak{a}) - 1\}$. This being true for every divisor $\mathfrak{a}$, we have $l(\mathfrak{a}) = \max\{0, d(\mathfrak{a}^{-1}) + 1\}$, whence $d(\mathfrak{a}) + l(\mathfrak{a}) - 1 = \max\{d(\mathfrak{a}) - 1, 0\}$. We conclude that *every linear function on the space of repartitions which vanishes on* $\mathfrak{X}(\mathfrak{a}) + R$ *is generated by some differential* $u\,dx$ *which is* $\equiv 0 \pmod{\mathfrak{a}^{-1}}$.

In view of this result, it is natural to set up the following definition. Let R be any field of algebraic functions of one variable, let K be the field of constants of R and let $\mathfrak{X}$ be the space of repartitions. By a *differential* of the field R is meant a linear function ω on $\mathfrak{X}$ which has the following property: there exists a divisor $\mathfrak{a}$ such that ω vanishes identically on $\mathfrak{X}(\mathfrak{a}^{-1}) + R$ (where $\mathfrak{X}(\mathfrak{a}^{-1})$ is the space of repartitions which are $\equiv 0 \pmod{\mathfrak{a}^{-1}}$). If $\mathfrak{a}$ is any divisor which has the stated property, then we say that the differential ω is *multiple* of $\mathfrak{a}$, and we write $\omega \equiv 0 \pmod{\mathfrak{a}}$.

It is clear that, if a divisor $\mathfrak{a}'$ divides $\mathfrak{a}$, a differential which is multiple of $\mathfrak{a}$ is also multiple of $\mathfrak{a}'$. If ω and ω' are differentials and λ, λ' elements of K, then $\lambda\omega + \lambda'\omega'$ is a differential; if ω is multiple of $\mathfrak{a}$ and ω' of $\mathfrak{a}'$, then $\lambda\omega + \lambda'\omega'$ is multiple of the highest common divisor of $\mathfrak{a}$ and $\mathfrak{a}'$. The differentials of R clearly form a vector space over K; those differentials which are multiples of a given divisor $\mathfrak{a}$ form a subspace of the space of all differentials.

We can now formulate as follows the result which was obtained in §4:

THEOREM 2. *Let R be a field of algebraic functions of one variable. Let $\mathfrak{a}$ be a divisor of R and let $\mathfrak{x}$ be a repartition in R. In order that there should exist an element x of R such that* $x \equiv \mathfrak{x} \pmod{\mathfrak{a}}$, *it is necessary and sufficient that* $\omega(\mathfrak{x}) = 0$ *for every differential ω which is multiple of* $\mathfrak{a}^{-1}$.

From now on, we shall denote by $\delta(\mathfrak{a})$ the dimension of the space of differentials which are $\equiv 0 \pmod{\mathfrak{a}}$. From what we have seen in §4, it follows that $\delta(\mathfrak{a}) = l(\mathfrak{a}^{-1}) + d(\mathfrak{a}^{-1}) + g - 1$, where g is the genus of the field. Thus we have proved

THEOREM 3 (THEOREM OF RIEMANN-ROCH). *Let R be a field of algebraic functions of one variable of genus g, and let $\mathfrak{a}$ be a divisor of R. Denote by $l(\mathfrak{a})$ the dimension of the space of elements of R which are* $\equiv 0 \pmod{\mathfrak{a}}$ *and by $\delta(\mathfrak{a}^{-1})$ the dimension of the space of differentials which are* $\equiv 0 \pmod{\mathfrak{a}^{-1}}$. *Then we have*

$$l(\mathfrak{a}) = d(\mathfrak{a}^{-1}) - g + 1 + \delta(\mathfrak{a}^{-1}).$$

A differential ω is said to be of *the first kind* when it is a multiple of the unit divisor $\mathfrak{e}$. We have $l(\mathfrak{e}) = 1$, $d(\mathfrak{e}) = 0$, which proves

THEOREM 4. *The space of differentials of the first kind of a field of algebraic functions of one variable is of dimension equal to the genus of the field.*

§6. The canonical class

Let R be a field of algebraic functions of one variable and let K be the field of constants of R. Denote by ω a differential of R. Let also $\mathfrak{X}$ be the space of repartitions of R, and x an element $\neq 0$ of R. Then the mapping $\mathfrak{x} \to \omega(x\mathfrak{x})$ of $\mathfrak{X}$ into K is linear and assigns 0 to every element of R. If $\mathfrak{a}$ is a divisor such that $\omega \equiv 0 \pmod{\mathfrak{a}}$, we have $\omega(x\mathfrak{x}) = 0$ whenever $x\mathfrak{x} \equiv 0 \pmod{\mathfrak{a}^{-1}}$, i.e., whenever $\mathfrak{x} \equiv 0$ (mod $\mathfrak{a}^{-1}\mathfrak{d}^{-1}(x)$). It follows that our mapping is a differential and is a multiple of $\mathfrak{a}\mathfrak{d}(x)$; we shall denote this differential by $x\omega$; if $x = 0$ we denote of course by $0 \cdot \omega$ the zero differential. It is clear that

$$(x + x')\omega = x\omega + x'\omega; \qquad x(\omega + \omega') = x\omega + x\omega';$$

$$(xx')\omega = x(x'\omega); \qquad 1 \cdot \omega = \omega$$

(where x, x' are in R and ω, ω' differentials of R).

Theorem 5. *Let ω be any differential $\neq 0$ of a field R of algebraic functions of one variable. Then every differential of R is of the form $x\omega$, with $x \in R$.*

Let ω' be any differential of R, and let $\mathfrak{a}$ and $\mathfrak{a}'$ be divisors such that $\omega \equiv 0 \pmod{\mathfrak{a}}$ and $\omega' \equiv 0 \pmod{\mathfrak{a}'}$. Denote by g the genus of R; let $\mathfrak{b}$ be an integral divisor whose degree satisfies the following conditions: $d(\mathfrak{b}) > g - 1 - d(\mathfrak{a})$, $d(\mathfrak{b}) > g - 1 - d(\mathfrak{a}')$, $d(\mathfrak{b}) > 3g - 2 - d(\mathfrak{a}) - d(\mathfrak{a}')$ (we may take $\mathfrak{b} = \mathfrak{p}^m$, where $\mathfrak{p}$ is a place and m is sufficiently large). We set $a = l(\mathfrak{a}^{-1}\mathfrak{b}^{-1})$, $a' = l(\mathfrak{a}'^{-1}\mathfrak{b}^{-1})$; by Riemann's theorem, we have $a \geqq d(\mathfrak{a}\mathfrak{b}) - g + 1 > 0$ and similarly $a' > 0$. Let $x_1, \cdots, x_a$ be a linearly independent elements of R which are $\equiv 0$ (mod $\mathfrak{a}^{-1}\mathfrak{b}^{-1}$); let $x_1', \cdots, x_{a'}'$ be a' linearly independent elements of R which are $\equiv 0$ (mod $\mathfrak{a}'^{-1}\mathfrak{b}^{-1}$). Then the differentials $x_i\omega$, $x_j'\omega'$ $(1 \leqq i \leqq a, 1 \leqq j \leqq a')$ are all $\equiv 0$ (mod $\mathfrak{b}^{-1}$). Now we have, by Riemann-Roch's theorem, $\delta(\mathfrak{b}^{-1}) = l(\mathfrak{b}) + d(\mathfrak{b}) + g - 1$, and $l(\mathfrak{b}) \leqq 1$, because $\mathfrak{b}$ is integral, whence $\delta(\mathfrak{b}^{-1}) \leqq d(\mathfrak{b}) + g$. On the other hand, we have $a + a' \geqq d(\mathfrak{a}) + d(\mathfrak{a}') + 2d(\mathfrak{b}) - 2g + 2$. Since $d(\mathfrak{b}) > 3g - 2 - d(\mathfrak{a}) - d(\mathfrak{a}')$, we have $a + a' > \delta(\mathfrak{b}^{-1})$. It follows that the differentials $x_i\omega$, $x_j\omega'$ are linearly dependent. Therefore, we have a relation of the form

$$\left(\sum_{i=1}^{a} \lambda_i x_i\right)\omega + \left(\sum_{j=1}^{a'} \lambda_j' x_j'\right)\omega' = 0$$

where $\lambda_1, \cdots, \lambda_a, \lambda_1', \cdots, \lambda_{a'}'$ are elements not all zero of the field of constants of R. Set $y = \sum_{i=1}^{a} \lambda_i x_i$, $y' = \sum_{j=1}^{a'} \lambda_j' x_j'$; then y and y' are not both zero and $y\omega + y'\omega' = 0$. If y' were 0, y would be $\neq 0$ but $y\omega$ would be 0, which is impossible since $\omega = y^{-1}y\omega \neq 0$. Thus $y' \neq 0$ and $\omega' = (yy'^{-1})\omega$, which proves Theorem 5.

Let ω be any differential $\neq 0$ of R. We shall prove that the degree of a divisor $\mathfrak{a}$ such that $\omega \equiv 0 \pmod{\mathfrak{a}}$ cannot exceed $2g - 2$, where g is the genus of R. If $x_1, \cdots, x_h$ are linearly independent elements of R which are $\equiv 0$ (mod $\mathfrak{a}^{-1}$), $x_1\omega, \cdots, x_h\omega$ are linearly independent differentials of the first kind, whence

$h \leqq g$ by Theorem 4. Thus we have $l(\mathfrak{a}^{-1}) \leqq g$. On the other hand, we have $l(\mathfrak{a}^{-1}) = d(\mathfrak{a}) - g + 1 + \delta(\mathfrak{a})$, and $\delta(\mathfrak{a}) \geqq 1$ since $\omega \equiv 0 \pmod{\mathfrak{a}}$. It follows that $d(\mathfrak{a}) \leqq l(\mathfrak{a}^{-1}) + g - 2 = 2g - 2$, which proves our assertion.

Next we observe that, if a differential ω is multiple of both the divisors $\mathfrak{a}$ and $\mathfrak{a}'$, then it is multiple of the least common multiple $\mathfrak{b}$ of $\mathfrak{a}$ and $\mathfrak{a}'$. To prove this, it will be sufficient to show that $\mathfrak{X}(\mathfrak{b}^{-1}) \subset \mathfrak{X}(\mathfrak{a}^{-1}) + \mathfrak{X}(\mathfrak{a}'^{-1})$. Let $m_\mathfrak{p}$ and $m'_\mathfrak{p}$ be the exponents with which an arbitrary place $\mathfrak{p}$ enters in $\mathfrak{a}$ and $\mathfrak{a}'$ respectively; then $\mathfrak{p}$ enters in $\mathfrak{b}$ with the exponent $n_\mathfrak{p} = \max\{m_\mathfrak{p}, m'_\mathfrak{p}\}$. If $\mathfrak{x}$ is any repartition in $\mathfrak{X}(\mathfrak{b}^{-1})$, we set $\mathfrak{y}(\mathfrak{p}) = \mathfrak{x}(\mathfrak{p})$ if $n_\mathfrak{p} = m_\mathfrak{p}$, $\mathfrak{y}(\mathfrak{p}) = 0$ if $n_\mathfrak{p} = m'_\mathfrak{p} > m_\mathfrak{p}$; we set also $\mathfrak{y}'(\mathfrak{p}) = \mathfrak{x}(\mathfrak{p}) - \mathfrak{y}(\mathfrak{p})$. It is clear that the mappings $\mathfrak{p} \to \mathfrak{y}(\mathfrak{p})$ and $\mathfrak{p} \to \mathfrak{y}'(\mathfrak{p})$ are repartitions $\mathfrak{y}$ and $\mathfrak{y}'$, that $\mathfrak{x} = \mathfrak{y} + \mathfrak{y}'$, that $\mathfrak{y}$ is in $\mathfrak{X}(\mathfrak{a}^{-1})$ and $\mathfrak{y}'$ in $\mathfrak{X}(\mathfrak{a}'^{-1})$.

In view of what we proved before, among all divisors $\mathfrak{a}$ such that $\omega \equiv 0 \pmod{\mathfrak{a}}$, we can select one, say $\mathfrak{a}_0$, of largest possible degree. If $\mathfrak{a}$ is any divisor such that $\omega \equiv 0 \pmod{\mathfrak{a}}$, then ω is also multiple of the least common multiple of $\mathfrak{a}$ and $\mathfrak{a}_0$. This least common multiple cannot be of higher degree than $\mathfrak{a}_0$ and must therefore coincide with $\mathfrak{a}_0$. Thus we see that to every differential $\omega \neq 0$ we can associate a divisor $\mathfrak{d}(\omega)$ of which ω is a multiple and which is a multiple of every divisor of which ω is a multiple. The divisor $\mathfrak{d}(\omega)$ is obviously uniquely determined by these properties. It is called the *divisor of the differential* ω. If $\mathfrak{p}$ is any place, the exponent with which $\mathfrak{p}$ enters in $\mathfrak{d}(\omega)$ is called the *order of* ω *at* $\mathfrak{p}$ and is denoted by $\nu_\mathfrak{p}(\omega)$. If $\nu_\mathfrak{p}(\omega) > 0$, then $\mathfrak{p}$ is said to be a *zero* (of order $\nu_\mathfrak{p}(\omega)$) of ω; if $\nu_\mathfrak{p}(\omega) < 0$ then $\mathfrak{p}$ is said to be a *pole* (of order $-\nu_\mathfrak{p}(\omega)$) of ω. If x is an element $\neq 0$ of R, it is clear that the divisor of the differential $x\omega$ is $\mathfrak{d}(x)\mathfrak{d}(\omega)$.

It follows from Theorem 5 that the divisors of all differentials $\neq 0$ of R are in the same class; this class is called the *canonical class* of the field R. (It should be observed that there always exists at least one differential $\neq 0$ of the field R. In fact, let $\mathfrak{a}$ be any integral divisor of degree >1; then, by the Riemann-Roch theorem, we have $\delta(\mathfrak{a}^{-1}) = l(\mathfrak{a}) + d(\mathfrak{a}) + g - 1 > 0$).

THEOREM 6. *The degree of the canonical class of a field R of algebraic functions of one variable is $2g - 2$, where g is the genus of R.*

Let ω be any differential $\neq 0$ of R; then, of course, $\delta(\mathfrak{d}(\omega)) \geqq 1$. We shall prove that this number is equal to 1. If ω' is any differential $\neq 0$ which is $\equiv 0 \pmod{\mathfrak{d}(\omega)}$, then $\mathfrak{d}(\omega')$ is a multiple of $\mathfrak{d}(\omega)$. Since $\mathfrak{d}(\omega')$ has the same degree as $\mathfrak{d}(\omega)$, we have $\mathfrak{d}(\omega') = \mathfrak{d}(\omega)$. If we write $\omega' = x\omega$, $x \in R$, then $\mathfrak{d}(x)$ must be the unit divisor, whence $x \in K$, and this proves that $\delta(\mathfrak{d}(\omega)) = 1$. On the other hand, the differentials of the first kind are the differentials $x\omega$ for which x is $\equiv 0 \pmod{(\mathfrak{d}(\omega))^{-1}}$, whence $l((\mathfrak{d}(\omega))^{-1}) = g$, by Theorem 4, §5. By the Riemann-Roch theorem, we have $l((\mathfrak{d}(\omega))^{-1}) = d(\mathfrak{d}(\omega)) - g + 1 + \delta(\mathfrak{d}(\omega))$, whence $d(\mathfrak{d}(\omega)) = 2g - 2$.

COROLLARY. *Let R be a field of algebraic functions of one variable of genus g. If $\mathfrak{a}$ is a divisor of R of degree $>2g - 2$, then we have $l(\mathfrak{a}^{-1}) = d(\mathfrak{a}) - g + 1$.*

In fact, no divisor of the canonical class can be a multiple of $\mathfrak{a}$, from which it follows that no differential $\neq 0$ can be multiple of $\mathfrak{a}$. The corollary follows therefore immediately from the theorem of Riemann-Roch.

We may now give an equivalent formulation of the theorem of Riemann-Roch (this new formulation was actually the original one):

THEOREM 7. *Let $\mathfrak{a}$ be a divisor of a field R of algebraic functions of one variable, and let $\mathfrak{a}'$ be a divisor such that $\mathfrak{a}\mathfrak{a}'$ belongs to the canonical class. Then we have*

$$l(\mathfrak{a}^{-1}) = d(\mathfrak{a}) - g + 1 + l(\mathfrak{a}'^{-1}).$$

For, let ω be any differential $\neq 0$. The differentials which are $\equiv 0 \pmod{\mathfrak{a}}$ are the differentials $x\omega$ for which $x \equiv 0 \pmod{\mathfrak{a}(\mathfrak{d}(\omega))^{-1}}$. Thus we have $\delta(\mathfrak{a}) = l(\mathfrak{a}(\mathfrak{d}(\omega))^{-1})$. But $(\mathfrak{a}(\mathfrak{d}(\omega))^{-1})^{-1}\mathfrak{a} = \mathfrak{d}(\omega)$ belongs to the canonical class, which proves that $\mathfrak{a}(\mathfrak{d}(\omega))^{-1}$ is equivalent to $\mathfrak{a}'^{-1}$ and therefore that $l(\mathfrak{a}(\mathfrak{d}(\omega))^{-1}) = l(\mathfrak{a}'^{-1})$. Theorem 7 therefore follows immediately from the theorem of Riemann-Roch.

As an application of Theorem 6, we shall prove that a field R of algebraic functions of one variable which is of genus zero admits a place which is of degree either 1 or 2. Denote by K the field of constants of R and by d the smallest of the degrees of the places of R. Then we shall see that the degree of any divisor $\mathfrak{a}$ of R is divisible by d. Let $n = d(\mathfrak{a})$, and assume for a moment that $n \not\equiv 0 \pmod{d}$; then we may write $n = kd - n'$, with $0 < n' < d$. Let $\mathfrak{p}$ be a place of degree d. Making use of the theorem of Riemann-Roch, we have $l(\mathfrak{p}^{-k}\mathfrak{a}) \geqq kd - n = n'$. Since $n' > 0$, there exists an element $y \neq 0$ of R which is multiple of $\mathfrak{p}^{-k}\mathfrak{a}$. If we write $\mathfrak{d}(y) = \mathfrak{p}^{-k}\mathfrak{a}\mathfrak{b}$, $\mathfrak{b}$ is an integral divisor of degree n'. Since $n' > 0$, there exists a place $\mathfrak{q}$ which occurs with an exponent $\neq 0$ in $\mathfrak{b}$; but this is impossible, since it would imply $d(\mathfrak{q}) \leqq n' < d$. Thus we see that the degrees of all divisors of R are multiples of d. But, since R is of genus 0, every divisor of the canonical class of R is of degree -2; therefore d is either 1 or 2.

If R has a place of degree 1, then it is a purely transcendental extension of K (cf. §2). If not, let $\mathfrak{p}$ be a place of degree 2; then, by Riemann's theorem, $l(\mathfrak{p}^{-1}) \geqq 3$, and R contains a non constant element x which is $\equiv 0 \pmod{\mathfrak{p}^{-1}}$. The divisor of poles of x is clearly $\mathfrak{p}$. Making use of the corollary to Theorem 4, I, §8, we see that R is quadratic over $K\langle x\rangle$.

§7. THE LOCAL COMPONENTS OF A DIFFERENTIAL

Let us return for a moment to the notation of the beginning of §5, where we had introduced directly the differentials ω of a field of the form $K\langle x\rangle$, K being the field of complex numbers and x transcendental over K. To any $u \in K\langle x\rangle$ we have associated a differential $\omega = u\,dx$ which was defined by the formula

$$\omega(\mathfrak{x}) = \sum_{\mathfrak{p}} \omega^{\mathfrak{p}}(\mathfrak{x}), \tag{1}$$

the summation being extended over all places $\mathfrak{p}$; $\omega^{\mathfrak{p}}(\mathfrak{x})$ was a function which depended only on the $\mathfrak{p}$-component $\mathfrak{x}(\mathfrak{p})$ of $\mathfrak{x}$.

Now, let R be any field of algebraic functions of one variable and let ω be a differential of R. It will sometimes be convenient to have a representation of ω by a formula of the type (1) above. We shall now derive such a representation.

Let $\mathfrak{p}$ be any place of R. We define the $\mathfrak{p}$-component of ω by the following formula:

$$\omega^{\mathfrak{p}}(\mathfrak{x}) = \omega(\mathfrak{x}^{\mathfrak{p}})$$

where, $\mathfrak{x}$ being any repartition, $\mathfrak{x}^{\mathfrak{p}}$ is the repartition which assigns $\mathfrak{x}(\mathfrak{p})$ to $\mathfrak{p}$ and 0 to every other place. It is clear that $\omega^{\mathfrak{p}}$ is a linear function on the space of repartitions and that $\omega^{\mathfrak{p}}(\mathfrak{x})$ depends only on the $\mathfrak{p}$-component of $\mathfrak{x}$; we may therefore consider $\omega^{\mathfrak{p}}$ as a function on R. This function is called the $\mathfrak{p}$-*component of* ω.

LEMMA 1. *Let* ω *be a differential of* R, *and let, for each place* $\mathfrak{p}$, $\omega^{\mathfrak{p}}$ *be the* $\mathfrak{p}$-*component of* ω. *Then, if* $\mathfrak{x}$ *is any repartition, there are only a finite number of places* $\mathfrak{p}$ *for which* $\omega^{\mathfrak{p}}(\mathfrak{x}) \neq 0$, *and we have* $\omega(\mathfrak{x}) = \sum_{\mathfrak{p}} \omega^{\mathfrak{p}}(\mathfrak{x})$.

Let $\mathfrak{a} = \prod_{\mathfrak{p}} \mathfrak{p}^{a(\mathfrak{p})}$ be a divisor of which ω is a multiple. There are only a finite number of places $\mathfrak{q}$ which fail to satisfy at the same time the two conditions $a(\mathfrak{q}) = 0$ and $\nu_{\mathfrak{q}}(\mathfrak{x}) \geqq 0$; let $\mathfrak{q}_1, \cdots, \mathfrak{q}_h$ be these places. If a place $\mathfrak{p}$ does not occur among them, then the repartition which assigns $\mathfrak{x}(\mathfrak{p})$ to $\mathfrak{p}$ and 0 to every other place is a multiple of $\mathfrak{a}^{-1}$, whence $\omega^{\mathfrak{p}}(\mathfrak{x}) = 0$. Let $\mathfrak{x}_i$ be the repartition which assigns $\mathfrak{x}(\mathfrak{q}_i)$ to $\mathfrak{q}_i$ and 0 to every other place; set $\mathfrak{x}' = \sum_{i=1}^{h} \mathfrak{x}_i$. Then $\mathfrak{x} - \mathfrak{x}'$ is a multiple of $\mathfrak{a}^{-1}$, whence $\omega(\mathfrak{x}) = \omega(\mathfrak{x}') = \sum_{i=1}^{h} \omega(\mathfrak{x}_i) = \sum_{\mathfrak{p}} \omega^{\mathfrak{p}}(\mathfrak{x})$. Lemma 1 is thereby proved.

LEMMA 2. *The exponent with which a place* $\mathfrak{p}$ *occurs in the divisor* $\mathfrak{d}(\omega)$ *of a differential* $\omega \neq 0$ *of* R *is the largest integer* m *for which it is true that* $\omega^{\mathfrak{p}}(x) = 0$ *for every* $x \in R$ *such that* $\nu_{\mathfrak{p}}(x) \geqq -m$.

Set $\mathfrak{d}(\omega) = \prod_{\mathfrak{q}} \mathfrak{q}^{a(\mathfrak{q})}$. Let x be an element of R such that $\nu_{\mathfrak{p}}(x) \geqq -a(\mathfrak{p})$, and let $x^{\mathfrak{p}}$ be the repartition which assigns x to $\mathfrak{p}$ and 0 to every other place. Then $x^{\mathfrak{p}}$ is clearly a multiple of $(\mathfrak{d}(\omega))^{-1}$, whence $\omega(x^{\mathfrak{p}}) = \omega^{\mathfrak{p}}(x) = 0$. On the other hand, there exists a repartition $\mathfrak{x}$ which is a multiple of $\mathfrak{p}^{-1}(\mathfrak{d}(\omega))^{-1}$ but for which $\omega(\mathfrak{x}) \neq 0$. If $\mathfrak{q}$ is a place $\neq \mathfrak{p}$, then $\nu_{\mathfrak{q}}(\mathfrak{x}(\mathfrak{q})) \geqq -a(\mathfrak{q})$, whence $\omega^{\mathfrak{q}}(\mathfrak{x}) = \omega^{\mathfrak{q}}(\mathfrak{x}(\mathfrak{q})) = 0$. It follows that $\omega(\mathfrak{x}) = \omega^{\mathfrak{p}}(\mathfrak{x}) = \omega^{\mathfrak{p}}(\mathfrak{x}(\mathfrak{p})) \neq 0$; on the other hand, we have $\nu_{\mathfrak{p}}(\mathfrak{x}(\mathfrak{p})) \geqq -a(\mathfrak{p}) - 1$. This proves that $a(\mathfrak{p}) = m$.

COROLLARY. *If* ω *is a differential* $\neq 0$, *we have* $\omega^{\mathfrak{p}} \neq 0$ *for every place* $\mathfrak{p}$.

§8. FIELDS OF ELLIPTIC FUNCTIONS

Let ζ and ζ' be two complex numbers $\neq 0$ whose ratio ζ'/ζ is not a real number. We shall consider the *elliptic functions* $f(z)$ with the periods ζ, ζ', i.e., the functions $f(z)$ which are meromorphic in the whole z-plane and which admit ζ and ζ' as periods. If f and g are elliptic functions, then the functions $f + g$, $f - g$, fg and (if $g \neq 0$) f/g are likewise elliptic. It follows that the elliptic functions form a field R. Every constant function is elliptic; identifying any constant function with its constant value, we see that R contains the field K of complex numbers.

We shall prove that there exists at least one non constant elliptic function. If m and n are integers, we set $\zeta_{m,n} = m\zeta + n\zeta'$. By the sign $\sum'_{m,n}$ we shall understand a summation extended to all integral values of m and n except $m = n = 0$. We first show that the series

$$\sum_{m,n}' \frac{1}{|\zeta_{m,n}|^3} \tag{1}$$

is convergent. Denote by P the parallelogram of vertices $0, \zeta, \zeta', \zeta + \zeta'$; P is therefore the set of complex numbers of the form $t\zeta + t'\zeta'$, where t, t' are real numbers from the closed interval $[0, 1]$. Let $P_{m,n}$ be the parallelogram deduced from P by the translation of the complex plane which brings 0 at $\zeta_{m,n}$. It is clear that no two of the sets $P_{m,n}$ have an interior point in common. Denote by d the diameter and by A the area of P. Let $\Gamma(R)$ be the closed circular disc of center 0 and radius R, and let $\nu(R)$ be the number of points $\zeta_{m,n}$ which lie in $\Gamma(R)$. It is clear that $\zeta_{m,n} \in \Gamma(R)$ implies $P_{m,n} \subset \Gamma(R + d)$; it follows that $A\nu(R) \leqq \pi(R + d)^2$, whence $\nu(R) \leqq (4\pi/A)R^2$ if $R \geqq d$. Let $(R_1, \cdots, R_k, \cdots)$ be an increasing sequence of numbers, all $\geqq d$, such that the series $\sum_{k=1}^{\infty} R_{k+1}^2/R_k^3$ is convergent (for instance, $R_k = dk^2$). The number of combinations (m, n) such that $R_k \leqq |\zeta_{m,n}| < R_{k+1}$ is $\leqq \nu(R_{k+1})$, whence

$$\sum_{m,n}' \frac{1}{|\zeta_{m,n}|^3} \leqq \alpha + \sum_1^\infty \frac{4\pi}{A} \frac{R_{k+1}^2}{R_k^3} < \infty$$

where α is the sum of the finite number of terms of the series which are $> R_1^{-2}$.

Now we consider the series

$$\sum_{m,n}' \left\{ \frac{1}{(z - \zeta_{m,n})^2} - \frac{1}{\zeta_{m,n}^2} \right\}. \tag{2}$$

Let ε be a number $< d/2$; we surround each $\zeta_{m,n} \neq 0$ by an open circular disc of center $\zeta_{m,n}$ and radius ε, and we denote by D the set which remains after exclusion of all these discs from the plane. We shall see that our series converges uniformly on every bounded subset E of D. We have

$$\frac{1}{(z - \zeta_{m,n})^2} - \frac{1}{\zeta_{m,n}^2} = - \frac{z^2 - 2z\zeta_{m,n}}{\zeta_{m,n}^2(z - \zeta_{m,n})^2}.$$

Let ρ be an upper bound for the absolute values of the points of E. If $|\zeta_{m,n}| \geqq 2\rho$, $z \in E$, we have $|z - \zeta_{m,n}| \geqq (1/2)\,|\zeta_{m,n}|$, whence

$$\left| \frac{1}{(z - \zeta_{m,n})^2} - \frac{1}{\zeta_{m,n}^2} \right| \leqq \frac{4\rho^2}{|\zeta_{m,n}|^4} + \frac{8\rho}{|\zeta_{m,n}|^3}.$$

There are only a finite number of combinations (m, n) for which $|\zeta_{m,n}| < 2\rho$. On the other hand, the convergence of the series (1) implies that of the series $\sum'_{m,n} |\zeta_{m,n}|^{-4}$; this proves the uniform convergence of (2) on E.

It follows that the formula

$$f(z) = \frac{1}{z^2} + \sum_{m,n}' \left\{\frac{1}{(z - \zeta_{m,n})^2} - \frac{1}{\zeta_{m,n}^2}\right\}$$

defines a function $f(z)$ which is meromorphic on the whole plane except perhaps at the points $\zeta_{m,n}$, and which has 0 as a pole of order 2. Furthermore, it follows from general theorems on analytic functions that the derivative $f'(z)$ of $f(z)$ is given by the formula

$$f'(z) = -\frac{2}{z^3} - 2\sum_{m,n}' \frac{1}{(z - \zeta_{m,n})^3}.$$

If we replace z by $z + \zeta$ or $z + \zeta'$, the terms of the series on the right side of this formula are permuted with each other. It follows that $f'(z)$ admits the periods ζ and ζ', and therefore that the functions $f(z + \zeta) - f(z)$ and $f(z + \zeta') - f(z)$ are constants. On the other hand, it is clear from the expression of $f(z)$ by a series that $f(z)$ is an even function (i.e., that $f(-z) = f(z)$). Since $f(z + \zeta) - f(z) = f(\zeta/2) - f(-\zeta/2)$, we have $f(z + \zeta) = f(z)$, and we would see in the same way that $f(z + \zeta') = f(z)$. It follows that $f(z + \zeta_{m,n}) = f(z)$ for each (m, n) and therefore that $\zeta_{m,n}$ is a pole of order 2 of $f(z)$. This proves that $f(z)$ is an elliptic function, which is not constant since it has poles.

Now we wish to prove that R is a field of algebraic functions over K. Since K is algebraically closed, $f(z)$ is transcendental over K; it will therefore be sufficient to show that R is algebraic of finite degree over $K\langle f(z)\rangle$; in fact, we shall prove that $[\mathrm{R}:K\langle f(z)\rangle] = 2$.

LEMMA 1. *An elliptic function $h(z)$ which has no pole is constant.*

For, $h(z)$ is obviously an entire function. It is bounded on the parallelogram P. On the other hand, if z is any complex number, we can find integers m and n such that $z + \zeta_{m,n} \in P$. Since $h(z)$ is elliptic, we have $h(z + \zeta_{m,n}) = h(z)$. It follows that $h(z)$ is bounded in the whole plane and is therefore a constant.

Now, we prove that any even elliptic function $g(z)$ belongs to $K\langle f(z)\rangle$. Since $g(z)$ is meromorphic in the whole plane, it has only a finite number of poles in P. From this finite number, we select a maximal subsystem $\{a_1, \cdots, a_h\}$ with the following properties: no two of them differ from each other by either ζ or ζ'; none of them is either 0, ζ, ζ', or $\zeta + \zeta'$. Let e_i be the order of the pole a_i. The only poles of $f(z)$ in P being 0, ζ, ζ', and $\zeta + \zeta'$, none of the numbers a_i is a pole of $f(z)$. We construct the function $g_1(z) = g(z)\prod_{i=1}^{h}(f(z) - f(a))^{e_i}$ Then none. of the points $a_1, \cdots, a_h$ is a pole of $g_1(z)$. If a' is a pole of $g(z)$ in P which is distinct from 0, ζ, ζ', and $\zeta + \zeta'$, then a' is either of the form $a_i \pm \zeta$ or $a_i \pm \zeta'$; since $g_1(z)$ is elliptic, a' is not a pole of $g_1(z)$. Thus we see that the only possible poles of $g_1(z)$ in P are 0, ζ, ζ', and $\zeta + \zeta'$. On the other hand, $g_1(z)$ is even. It follows that the Laurent expansion of $g_1(z)$ around 0 is of the form

$$g_1(z) = \sum_{k=1}^{r} \frac{c_k}{z^{2k}} + \sum_{k=0}^{\infty} d_k z^{2k}.$$

We shall prove that $g_1(z)$ can be written in the form $\gamma_0 f^r(z) + \gamma_1 f^{r-1}(z) + \cdots + \gamma_r$, with constant $\gamma_0, \gamma_1, \cdots, \gamma_r$. Set $\gamma_0 = c_r$; then, clearly, $g_1(z) - \gamma_0 f^r(z)$ admits 0 as a pole of order $< 2r$. Suppose that, for some $h < r - 1$, constants $\gamma_0, \cdots, \gamma_h$ have already been determined such that $g_1(z) - \sum_{i=0}^{h} \gamma_i f^{r-i}(z)$ admits 0 as a pole of order $<2(r - h)$. Since this function is even, its Laurent expansion around 0 is of the form

$$\sum_{k=0}^{r-h-1} \frac{c_{k,h}}{z^{2k}} + \sum_{k=0}^{\infty} d_{k,h} z^{2k}$$

and, if we set $\gamma_{h+1} = c_{r-h-1,h}$, the function $g_1(z) - \sum_{i=0}^{h+1} \gamma_i f^{r-i}(z)$ admits 0 as a pole of order $<2(r - h - 1)$. By this process, we eventually get a function of the form $g_1(z) - \sum_0^{r-1} \gamma_i f^{r-i}(z)$ which admits 0 as a pole of order <2. This function, being even, has no pole at all at 0. Being elliptic, it does not have any of the points $\zeta_{m,n}$ as a pole. On the other hand, this function has no pole in P outside of 0, ζ, ζ', or $\zeta + \zeta'$. Therefore it has no pole at all, and it is a constant γ_r. This proves that our original function $g(z)$ can be expressed as a rational function of $f(z)$.

Now, if $\varphi(z)$ is any element of R, so is $\varphi(-z)$, and the mapping $\sigma: \varphi(z) \to \varphi(-z)$ is obviously an automorphism of order 2 of R. It follows from what was proved above that the field of elements left fixed by σ is $K\langle f(z)\rangle$; we conclude that R is algebraic of degree 2 over $K\langle f(z)\rangle$.

Let a be a complex number. The set $\mathfrak{o}_a$ of elliptic functions which do not admit a as a pole is clearly a subring of R. This ring is $\neq R$. For, if a is not one of the points $\zeta_{m,n}$, then $(f(z) - f(a))^{-1}$ does not belong to $\mathfrak{o}_a$, while if $a = \zeta_{m,n}$, then $f(z) \notin \mathfrak{o}_a$. On the other hand, it is obvious that the inverse of an elliptic function which does not belong to $\mathfrak{o}_a$ is in $\mathfrak{o}_a$. Thus we see that $\mathfrak{o}_a$ is a V-ring in R (over K); we shall denote by $\mathfrak{p}_a$ the corresponding place of R. The places which correspond in this manner to a and to $a + \zeta_{m,n}$ are obviously identical; on the other hand, if a, a' are complex numbers whose difference is not one of the numbers $\zeta_{m,n}$, the places $\mathfrak{p}_a$ and $\mathfrak{p}_{a'}$ are distinct, because $f(z - a)$ belongs to $\mathfrak{o}_{a'}$ but not to $\mathfrak{o}_a$.

Now we shall prove that every place $\mathfrak{p}$ of R is of the form $\mathfrak{p}_a$ for some a. Let m be an integer greater than the genus of R; since $d(\mathfrak{p}^m) = m$, it follows from Riemann's theorem that $l(\mathfrak{p}^{-m}) \geqq 2$. Thus, there exists a function $u(z) \in R$ which is not constant and whose order at every place $\neq \mathfrak{p}$ is $\geqq 0$. If $\mathfrak{p}$ were distinct from all places $\mathfrak{p}_a$, the elliptic function $u(z)$ would not have any pole, which is impossible in virtue of Lemma 1. If a is any complex number, there exists an elliptic function $p_a(z)$ which has a simple zero at a, viz. $f(z - a)/f'(z - a)$. If $h(z)$ is any elliptic function, and ν the order at a of the meromorphic function $h(z)$, then $h(z) = (p_a(z))^\nu h'(z)$, where $h'(z)$ is an elliptic function which has a neither as a zero nor a pole, and which is therefore a unit in the ring of the place $\mathfrak{p}_a$. It follows immediately that $\nu_{\mathfrak{p}_a}(h(z))$ is equal to the order of the function $h(z)$ at a.

Now, we shall determine the genus g of the field R. If $\mathfrak{p}_0$ is the place which corresponds to the number 0, it is clear that $f(z)$, considered as an element of R, is multiple of the divisor $\mathfrak{p}_0^{-2}$, and that $f'(z)$ is multiple of the divisor $\mathfrak{p}_0^{-3}$. Thus, if a and b are integers $\geqq 0$, $(f(z))^a f'(z)$ is multiple of $\mathfrak{p}_0^{-2a-3}$, while $(f(z))^b$ is multiple of $\mathfrak{p}_0^{-2b}$. Let m be an integer which we shall choose larger than the smallest of the numbers $2, 2g - 2$. Denote by m_1 the largest integer which is $\leqq (m-3)/2$ and by m_2 the largest integer which is $\leqq m/2$. If a, b are such that $0 \leqq a \leqq m_1$, $0 \leqq b \leqq m_2$, then $(f(z))^a f'(z)$ and $(f(z))^b$ are multiples of $\mathfrak{p}_0^{-m}$. The $m_1 + m_2 + 2 = m$ functions obtained in this way are linearly independent over K. For, let P and Q be polynomials with complex coefficients such that $P(f(z))f'(z) + Q(f(z)) = 0$. Replacing z by $-z$ in this identity and observing that $f'(z)$ is an odd function, we get $-P(f(z))f'(z) + Q(f(z)) = 0$; since $f'(z) \neq 0$ and $f(z)$ is transcendental over K, this implies $P = Q = 0$, which proves our assertion. Thus we see that $l(\mathfrak{p}_0^{-m}) \geqq m$. On the other hand, it follows from the corollary to Theorem 6, §6 that $l(\mathfrak{p}_0^{-m}) = d(\mathfrak{p}_0^{m}) - g + 1 = m - g + 1$; therefore we have $g \leqq 1$. We shall prove that g cannot be 0. In order to do this, it will be sufficient to prove that an elliptic function $h(z)$ which is multiple of $\mathfrak{p}_0^{-1}$ is a constant. The Laurent expansion of $h(z)$ around 0 is of the form $h(z) = a/z + \sum_{k=0}^{\infty} a_k z^k$. It follows that 0 is not a pole of the elliptic function $h(z) + h(-z)$, and therefore that this function has no pole whatsoever and is a constant C. If we set $h_1(z) = h(z) - C/2$, then $h_1(z)$ is an odd elliptic function and is multiple of $\mathfrak{p}_0^{-1}$. We have $h_1(-\zeta/2) = -h_1(\zeta/2)$; on the other hand, since $-\zeta/2 = \zeta/2 - \zeta$, we have also $h_1(-\zeta/2) = h_1(\zeta/2)$, whence $h_1(\zeta/2) = 0$. In the same way we would see that $h_1(\zeta'/2) = 0$. Since $\mathfrak{p}_{\zeta/2} \neq \mathfrak{p}_{\zeta'/2}$, we see that $h_1(z)$ is multiple of the divisor $\mathfrak{p}_0^{-1}\mathfrak{p}_{\zeta/2}\mathfrak{p}_{\zeta'/2}$, which is of degree 1. It follows immediately that $h_1(z) = 0$, whence $h(z) = C/2$. Thus we have proved that *a field of elliptic functions is of genus one.*

Since $f'(z)$ is an odd function, it does not belong to $K\langle f(z)\rangle$; since R is quadratic over $K\langle f(z)\rangle$, we have $R = K\langle f(z), f'(z)\rangle$. On the other hand, the function $(f'(z))^2$ is even and admits 0 as a pole of order 6. It follows that $(f'(z))^2$ is of the form $P(f(z))$, where P is a polynomial of degree 3. We rediscover in this way the generation of fields of genus one which was indicated in §3. In our case, it can be seen without difficulty that the polynomial $P(Z)$ is of the form $4Z^3 + aZ + b$.

CHAPTER III

THE $\mathfrak{p}$-ADIC COMPLETIONS

§1. Definition of the $\mathfrak{p}$-adic completion

Let K be the field of complex numbers and let x be transcendental over K. If $u(x)$ is a rational function of x and $a \in K$, then $u(x)$ can be represented on a neighbourhood of a by its Laurent series

$$u(x) = \sum_{k=r}^{\infty} c_k(x-a)^k. \tag{1}$$

However, not every function representable around a by a series of this type is rational. In some arguments, it is convenient to consider the set of all functions of x which are meromorphic at a, i.e., which are representable by a Laurent series around a. The existence of this set depends of course strongly on the topological properties of the field K, since a series like the one which occurs on the right side of (1) will only represent a function when it is convergent on some neighbourhood of a. However, for all algebraic purposes, only the sequence of coefficients of such series matters, not the fact that they converge in some region. This leads one to consider, instead of the set of meromorphic functions at a, the totality of all series of the form

$$\sum_{k=r}^{\infty} c_k(x-a)^k, \tag{2}$$

whether such series are convergent or not. The generalization of this set of formal power series to the abstract case of an arbitrary field R of algebraic functions of one variable (instead of the field $K\langle x\rangle$ considered above) is what we shall call a completion of the field R.

Let then $\mathfrak{p}$ be a place of the field R. We shall say that a sequence (x_n) of elements of R *converges* at $\mathfrak{p}$ to an element x of R if the condition

$$\lim_{n\to\infty} \nu_{\mathfrak{p}}(x - x_n) = \infty \tag{3}$$

is satisfied. If this is the case, we have, obviously,

$$\lim_{n\to\infty} \nu_{\mathfrak{p}}(x_{n+1} - x_n) = \infty. \tag{4}$$

However, a sequence (x_n) may satisfy the condition (4) without satisfying (3). Whenever a sequence (x_n) satisfies the condition (4), we shall say that it is a *Cauchy sequence* at $\mathfrak{p}$.

Let X be the set of all sequences $(x_n)_{1\leq n<\infty}$ of elements of R. We make X into a ring by defining addition and multiplication in the following way:

$$(x_n) + (y_n) = (x_n + y_n),$$

$$(x_n)(y_n) = (x_n y_n).$$

We shall see that the Cauchy sequences at $\mathfrak{p}$ form a subring of X. That the difference of two Cauchy sequences is again a Cauchy sequence follows immediately from the formula

$$\nu_\mathfrak{p}(x_{n+1} - y_{n+1} - (x_n - y_n)) \geqq \min \{\nu_\mathfrak{p}(x_{n+1} - x_n), \nu_\mathfrak{p}(y_{n+1} - y_n)\}.$$

In order to treat the case of the product, we establish first

LEMMA 1. *Let (x_n) be a Cauchy sequence at $\mathfrak{p}$. Then either $\nu_\mathfrak{p}(x_n)$ increases indefinitely with n (and then (x_n) converges to 0 at $\mathfrak{p}$), or the numbers $\nu_\mathfrak{p}(x_n)$ are all equal to each other from a certain value of n on.*

In fact, assume that $\nu_\mathfrak{p}(x_n)$ does not increase indefinitely with n. Then there exists an integer N such that $\nu_\mathfrak{p}(x_n) < N$ for infinitely many values of n. On the other hand, there exists an index n_0 such that $\nu_\mathfrak{p}(x_{n+1} - x_n) \geqq N$ for all $n \geqq n_0$; we may assume that n_0 is one of the indices n for which $\nu_\mathfrak{p}(x_n) < N$. Then we assert that $\nu_\mathfrak{p}(x_n) = \nu_\mathfrak{p}(x_{n_0})$ for all $n \geqq n_0$. This is true for $n = n_0$. Assume that it is true for some $n \geqq n_0$. Then we have $x_{n+1} = x_n + (x_{n+1} - x_n)$ and

$$\nu_\mathfrak{p}(x_{n+1} - x_n) \geqq N > \nu_\mathfrak{p}(x_{n_0}) = \nu_\mathfrak{p}(x_n),$$

from which it follows that $\nu_\mathfrak{p}(x_{n+1}) = \nu_\mathfrak{p}(x_n) = \nu_\mathfrak{p}(x_{n_0})$(cf. I, §3).

This being said, let (x_n) and (y_n) be Cauchy sequences. We write

$$x_{n+1}y_{n+1} - x_n y_n = y_n(x_{n+1} - x_n) + x_{n+1}(y_{n+1} - y_n).$$

It follows from Lemma 1 that there exist integers A and B such that

$$\nu_\mathfrak{p}(x_n) \geqq A, \qquad \nu_\mathfrak{p}(y_n) \geqq B$$

for all n's. Therefore we have

$$\nu_\mathfrak{p}(x_{n+1}y_{n+1} - x_n y_n) \geqq \min \{B + \nu_\mathfrak{p}(x_{n+1} - x_n), A + \nu_\mathfrak{p}(y_{n+1} - y_n)\}$$

which proves that $\nu_\mathfrak{p}(x_{n+1}y_{n+1} - x_n y_n)$ increases indefinitely with n and that $(x_n y_n)$ is a Cauchy sequence.

We shall denote by X_0 the ring of all Cauchy sequences at $\mathfrak{p}$, and by $\mathfrak{X}$ the set of all sequences which converge to 0 at $\mathfrak{p}$. Then $\mathfrak{X}$ is clearly a subset of X_0; we shall see that it is an ideal in X_0. It is clear that the conditions $(x_n) \in \mathfrak{X}$, $(y_n) \in \mathfrak{X}$ imply $(x_n - y_n) \in \mathfrak{X}$. Let (x_n) be a sequence belonging to $\mathfrak{X}$ and (z_n) a sequence belonging to X_0. It follows from Lemma 1 that there exists an integer C such that $\nu_\mathfrak{p}(z_n) \geqq C$ for all n, whence $\nu_\mathfrak{p}(x_n z_n) \geqq C + \nu_\mathfrak{p}(x_n)$, which shows that $\nu_\mathfrak{p}(x_n z_n)$ increases indefinitely with n, and that $(x_n z_n) \in \mathfrak{X}$. We shall denote by $\bar{R}_\mathfrak{p}$ the ring $X_0/\mathfrak{X}$, and we shall investigate the properties of this ring.

To every $x \in R$ we can associate the sequence (x_n) defined by $x_n = x$ for every n. We obtain in this way an isomorphic mapping of the field R into the ring of sequences of elements of R. If the sequence (x_n) corresponds to an element x of R, then it is clear that it is a Cauchy sequence, and that it converges to 0 only in the case where $x = 0$. It follows that the set of sequences which correspond to elements of R is in X_0 and is mapped isomorphically under the mapping of X_0 onto $\bar{R}_\mathfrak{p} = X_0/\mathfrak{X}$. This means that we have an isomorphic mapping of R onto a subfield of $\bar{R}_\mathfrak{p}$. We shall allow ourselves not to distinguish between the elements of R and the corresponding elements of $\bar{R}_\mathfrak{p}$; in other words, we shall speak as if R were a subring of $\bar{R}_\mathfrak{p}$. Moreover, if an element $\bar{x} \in \bar{R}_\mathfrak{p}$ is the residue class modulo $\mathfrak{X}$ of a Cauchy sequence (x_n), we shall indicate this fact by writing

$$\bar{x} = \lim_{n\to\infty} x_n .$$

If $x_n = x$ for every x, we clearly have $x = \lim_{n\to\infty} x_n$.

Now we shall prove that $\bar{R}_\mathfrak{p}$ is itself a field. Let $\bar{x}$ be an element $\neq 0$ of $\bar{R}_\mathfrak{p}$; set $\bar{x} = \lim_{n\to\infty} x_n$, $x_n \in R$. Since the sequence (x_n) does not converge to 0, but is a Cauchy sequence, it follows from Lemma 1 that $x_n \neq 0$ provided n is sufficiently large. Define x'_n to be x_n^{-1} if $x_n \neq 0$ and to be 0 if $x_n = 0$. Then, for n sufficiently large, we have $x'_{n+1} - x'_n = x_{n+1}^{-1} x_n^{-1}(x_n - x_{n+1})$. On the other hand, $\nu_\mathfrak{p}(x_n x_{n+1})$ remains constant for n large enough (by Lemma 1), while $\nu_\mathfrak{p}(x_{n+1} - x_n)$ increases indefinitely with n. It follows that $\nu_\mathfrak{p}(x'_{n+1} - x'_n)$ increases indefinitely with n, and that (x'_n) is a Cauchy sequence. Set $\bar{x}' = \lim_{n\to\infty} x'_n$; we shall see that $\bar{x}'$ is the inverse of $\bar{x}$ in $\bar{R}_\mathfrak{p}$. Set $u_n = 1$ for all n; then $x_n x'_n = u_n$ for n large enough, whence $\bar{x}\bar{x}' = \lim_{n\to\infty} x_n x'_n = \lim_{n\to\infty} u_n = 1$; but 1 is clearly the unit element of $\bar{R}_\mathfrak{p}$, and we see that $\bar{x}'$ is the inverse of $\bar{x}$. Thus we have proved that $\bar{R}_\mathfrak{p}$ is a field.

Now, we shall extend the domain of definition of the function $\nu_\mathfrak{p}$ from R to $\bar{R}_\mathfrak{p}$. Let $\bar{x}$ be an element $\neq 0$ of $\bar{R}_\mathfrak{p}$, and set $\bar{x} = \lim_{n\to\infty} x_n$, $x_n \in R$. Then the sequence (x_n) does not converge to 0; by Lemma 1, $\nu_\mathfrak{p}(x_n)$ is constant from a certain value of n on, and this constant value is not ∞. If (x'_n) is any other Cauchy sequence such that $\lim_{n\to\infty} x'_n = \bar{x}$, we have $\lim_{n\to\infty} \nu_\mathfrak{p}(x'_n - x_n) = \infty$, whence, from a certain value of n on, $\nu_\mathfrak{p}(x'_n - x_n) > \nu_\mathfrak{p}(x_n)$, and therefore, $\nu_\mathfrak{p}(x'_n) = \nu_\mathfrak{p}(x'_n - x_n + x_n) = \nu_\mathfrak{p}(x_n)$. We shall denote by $\nu_\mathfrak{p}(\bar{x})$ the constant value assumed for n large enough by $\nu_\mathfrak{p}(x_n)$ when (x_n) is a Cauchy sequence such that $\lim_{n\to\infty} x_n = \bar{x}$. If $\bar{x} = 0$, we set $\nu_\mathfrak{p}(\bar{x}) = \infty$. This definition does not lead to a contradiction in the case where $\bar{x} = x \in R$, because we may then take $x_n = x$ for every n.

It follows immediately from the definition that

$$\nu_\mathfrak{p}(\bar{x} + \bar{y}) \geqq \min\{\nu_\mathfrak{p}(\bar{x}), \nu_\mathfrak{p}(\bar{y})\},$$

$$\nu_\mathfrak{p}(\bar{x}\bar{y}) = \nu_\mathfrak{p}(\bar{x}) + \nu_\mathfrak{p}(\bar{y})$$

whenever $\bar{x}$ and $\bar{y}$ are in $\bar{R}_\mathfrak{p}$. Moreover, we see as in I, §3 that the first inequality may be replaced by an equality if $\nu_\mathfrak{p}(\bar{x}) \neq \nu_\mathfrak{p}(\bar{y})$.

Denote by $\bar{\mathfrak{o}}$ the set of elements $\bar{x} \in \bar{R}_\mathfrak{p}$ such that $\nu_\mathfrak{p}(\bar{x}) \geqq 0$. Then $\bar{\mathfrak{o}}$ is clearly a ring and $\bar{\mathfrak{o}} \cap R$ is the ring $\mathfrak{o}$ of the place $\mathfrak{p}$. We have $\bar{\mathfrak{o}} \neq \bar{R}_\mathfrak{p}$, and, if $\bar{x}$ is an element of $\bar{R}_\mathfrak{p}$ not in $\bar{\mathfrak{o}}$, then $\bar{x}^{-1}$ is in $\bar{\mathfrak{o}}$, because $\nu_\mathfrak{p}(\bar{x}^{-1}) = -\nu_\mathfrak{p}(\bar{x})$. This means that $\bar{\mathfrak{o}}$ is a V-ring in $\bar{R}_\mathfrak{p}$. The elements of $\bar{\mathfrak{o}}$ will be called the *integral elements* of $\bar{R}_\mathfrak{p}$. The ideal $\bar{\mathfrak{p}}$ of non-units in $\bar{\mathfrak{o}}$ is the set of elements $\bar{x}$ of $\bar{R}_\mathfrak{p}$ for which $\nu_\mathfrak{p}(\bar{x}) > 0$; it is clear that $\bar{\mathfrak{p}} \cap R = \mathfrak{p}$.

Let $\bar{x} = \lim_{n\to\infty} x_n$ be an element of $\bar{R}_\mathfrak{p}$, (x_n) being a Cauchy sequence of elements of R. Considering the elements of R as elements of $\bar{R}_\mathfrak{p}$, we shall see that $\lim_{n\to\infty} \nu_\mathfrak{p}(\bar{x} - x_n) = \infty$. We have $\bar{x} - x_m = \lim_{n\to\infty} (x_{m+n} - x_m)$ and $x_{m+n} - x_m = \sum_{k=1}^{n} (x_{m+k} - x_{m+k-1})$, whence $\nu_\mathfrak{p}(x_{m+n} - x_m) \geqq \min_{1 \leqq k \leqq n} \nu_\mathfrak{p}(x_{m+k} - x_{m+k-1})$. Since (x_n) is a Cauchy sequence, $\nu_\mathfrak{p}(x_{q+1} - x_q)$ increases indefinitely with q. It follows that, given any integer M, there exists an m_0 such that $q \geqq m_0$ implies $\nu_\mathfrak{p}(x_{q+1} - x_q) > M$. Therefore the condition $m \geqq m_0$ implies $\nu_\mathfrak{p}(x_{m+n} - x_m) > M$ for all $n > 0$. It follows that $\nu_\mathfrak{p}(\bar{x} - x_m) > M$ for all $m \geqq m_0$, which proves our assertion.

From this, we can first conclude that, if a Cauchy sequence (x_n) converges to an element x of R (i.e., if $\nu_\mathfrak{p}(x - x_n)$ increases indefinitely with n), then we have $x = \lim_{n\to\infty} x_n$. In fact, set $\bar{x} = \lim_{n\to\infty} x_n$; then $\nu_\mathfrak{p}(\bar{x} - x) \geqq \min \{\nu_\mathfrak{p}(\bar{x} - x_n), \nu_\mathfrak{p}(x - x_n)\}$. Since $\nu_\mathfrak{p}(\bar{x} - x_n)$ and $\nu_\mathfrak{p}(x - x_n)$ both increase indefinitely with n, we see that $\nu_\mathfrak{p}(\bar{x} - x) = \infty$, whence $\bar{x} = x$.

Next we see that, $\bar{x}$ being any element of $\bar{R}_\mathfrak{p}$ and a any integer, we can always find an $x \in R$ such that $\nu_\mathfrak{p}(\bar{x} - x) \geqq a$. If we take x in $\bar{\mathfrak{o}}$ and $a > 0$, then $\nu_\mathfrak{p}(x) \geqq \min \{a, \nu_\mathfrak{p}(\bar{x})\} \geqq 0$. It follows immediately that the elements of $\bar{\mathfrak{o}}$ are those which can be represented as limits of Cauchy sequences of elements of $\mathfrak{o}$. Taking $a = 1$, we see that every element of $\bar{\mathfrak{o}}$ is congruent modulo $\bar{\mathfrak{p}}$ to some element of $\mathfrak{o}$. This means that *the residue field* $\bar{\mathfrak{o}}/\bar{\mathfrak{p}}$ *may be identified with the residue field* $\mathfrak{o}/\mathfrak{p}$ *of the place* $\mathfrak{p}$. If $\bar{x} \in \bar{\mathfrak{o}}$, the residue class of $\bar{x}$ modulo $\bar{\mathfrak{p}}$ will also sometimes (although improperly) be called the residue class of $\bar{x}$ modulo $\mathfrak{p}$.

It is natural to say that a sequence $(\bar{x}_n)$ of elements of $\bar{R}_\mathfrak{p}$ is a Cauchy sequence if $\nu_\mathfrak{p}(\bar{x}_{n+1} - \bar{x}_n)$ increases indefinitely with n, and that $(\bar{x}_n)$ converges to an element $\bar{x}$ of $\bar{R}_\mathfrak{p}$ if $\nu_\mathfrak{p}(\bar{x} - \bar{x}_n)$ increases indefinitely with n. We shall see that *every Cauchy sequence of elements of* $\bar{R}_\mathfrak{p}$ *converges to some element of* $\bar{R}_\mathfrak{p}$. In fact, let $(\bar{x}_n)$ be a Cauchy sequence. For each n we can find an element x_n of R such that $\nu_\mathfrak{p}(\bar{x}_n - x_n) \geqq n$. We have $x_{n+1} - x_n = (x_{n+1} - \bar{x}_{n+1}) + (\bar{x}_{n+1} - \bar{x}_n) + (\bar{x}_n - x_n)$, whence $\nu_\mathfrak{p}(x_{n+1} - x_n) \geqq \min \{n, \nu_\mathfrak{p}(\bar{x}_{n+1} - \bar{x}_n)\}$. This proves that $\nu_\mathfrak{p}(x_{n+1} - x_n)$ increases indefinitely with n, i.e., that (x_n) is a Cauchy sequence of elements of R. Set $\bar{x} = \lim_{n\to\infty} x_n$, then $\bar{x} - \bar{x}_n = (\bar{x} - x_n) + (x_n - \bar{x}_n)$, and $\nu_\mathfrak{p}(\bar{x} - \bar{x}_n) \geqq \min \{\nu_\mathfrak{p}(\bar{x} - x_n), n\}$. We have seen that $\nu_\mathfrak{p}(\bar{x} - x_n)$ increases indefinitely with n. The same is therefore true of $\nu_\mathfrak{p}(\bar{x} - \bar{x}_n)$, which proves that the sequence $(\bar{x}_n)$ converges to $\bar{x}$.

We shall express the property which we have just proved by saying that $\bar{R}_\mathfrak{p}$ is a *complete field*; we shall also say that $\bar{R}_\mathfrak{p}$ is the $\mathfrak{p}$-*adic completion* of the field R.

If a sequence $(\bar{x}_n)$ of elements of $\bar{R}_\mathfrak{p}$ converges to the element $\bar{x}$, we shall write $\bar{x} = \lim_{n\to\infty} \bar{x}_n$. It is easily seen that the conditions $\bar{x} = \lim_{n\to\infty} \bar{x}_n$, $\bar{y} = \lim_{n\to\infty} \bar{y}_n$ imply $\bar{x} \pm \bar{y} = \lim_{n\to\infty} (\bar{x}_n \pm \bar{y}_n)$, $\bar{x}_n\bar{y}_n = \lim_{n\to\infty} (\bar{x}_n\bar{y}_n)$, and if $\bar{y} \neq 0$, $\bar{y}_n \neq 0$ for all n, $\bar{y}^{-1} = \lim_{n\to\infty} \bar{y}_n^{-1}$.

§2. Hensel's Lemma

Let $\bar{R}_{\mathfrak{p}}$ be the $\mathfrak{p}$-adic completion of a field R of algebraic functions of one variable, $\mathfrak{p}$ being a place of R. We denote by $\bar{\mathfrak{o}}$ the ring of integral elements of $\bar{R}_{\mathfrak{p}}$, by $\bar{\mathfrak{p}}$ the ideal of non-units of $\bar{\mathfrak{o}}$, and by Σ the field $\bar{\mathfrak{o}}/\bar{\mathfrak{p}}$. If $\bar{x} \in \bar{\mathfrak{o}}$, we denote by $\bar{x}^*$ the residue class of $\bar{x}$ modulo $\bar{\mathfrak{p}}$. We shall consider polynomials $f(X)$ in a letter X with coefficients in $\bar{R}_{\mathfrak{p}}$; if f is such a polynomial, we denote by $\nu_{\mathfrak{p}}(f)$ the smallest of the values assumed by the order function $\nu_{\mathfrak{p}}$ (extended to $\bar{R}_{\mathfrak{p}}$) on the set of coefficients of f. It is easy to see that, f and g being polynomials in X with coefficients in $\bar{R}_{\mathfrak{p}}$, we have

$$\nu_{\mathfrak{p}}(f \pm g) \geqq \min\{\nu_{\mathfrak{p}}(f), \nu_{\mathfrak{p}}(g)\}; \nu_{\mathfrak{p}}(fg) \geqq \nu_{\mathfrak{p}}(f) + \nu_{\mathfrak{p}}(g).$$

If f is a polynomial with coefficients in $\bar{\mathfrak{o}}$, we denote by f^* the polynomial with coefficients in Σ which is obtained from f by replacing its coefficients by their residue classes modulo $\bar{\mathfrak{p}}$; the mapping $f \to f^*$ is a homomorphism of $\bar{\mathfrak{o}}[X]$ onto $\Sigma[X]$.

Lemma 1 (Hensel's Lemma). *Let f be a polynomial with coefficients in $\bar{\mathfrak{o}}$. Assume that $f^* = \varphi\psi \neq 0$, where φ and ψ are polynomials with coefficients in Σ with no common factor. Then we may write $f = gh$, where g and h are polynomials with coefficients in $\bar{\mathfrak{o}}$ which satisfy the following conditions: $g^* = \varphi$, $h^* = \psi$, the degree of g is equal to the degree of φ.*

We shall denote by $\partial^\circ\theta$ the degree of a polynomial $\theta \neq 0$. We can find polynomials g_1, h_1 with coefficients in $\bar{\mathfrak{o}}$ which satisfy the following conditions: $g_1^* = \varphi$, $h_1^* = \psi$, $\partial^\circ g_1 = \partial^\circ\varphi$, $\nu_{\mathfrak{p}}(f - g_1h_1) \geqq 1$; for it is obviously possible to find polynomials g_1, h_1 which satisfy the three first conditions, and the last condition is then satisfied because $f^* = \varphi\psi$. Assume that, for some integer $n > 0$, we have already determined polynomials $g_1, \cdots, g_n, h_1, \cdots, h_n$ with coefficients in $\bar{\mathfrak{o}}$ which satisfy the following conditions:

a) $g_n^* = \varphi, \quad h_n^* = \psi;$
b) $\partial^\circ g_n = \partial^\circ\varphi; \quad \partial^\circ h_n \leqq \partial^\circ f - \partial^\circ g_n;$
c) $\nu_{\mathfrak{p}}(f - g_nh_n) \geqq n;$
d) $\nu_{\mathfrak{p}}(g_{k+1} - g_k) \geqq k; \quad \nu_{\mathfrak{p}}(h_{k+1} - h_k) \geqq k \quad (1 \leqq k \leqq n-1).$

Let t be an element of $\bar{\mathfrak{o}}$ such that $\nu_{\mathfrak{p}}(t) = 1$. Then $\zeta = t^{-n}(f - g_nh_n)$ is a polynomial with coefficients in $\bar{\mathfrak{o}}$. Since φ and ψ have no common factor, it is well known that there exist polynomials $\lambda(X)$ and $\mu(X)$ with coefficients in Σ such that $\zeta^* = \lambda\varphi + \mu\psi$. The polynomials λ and μ are not uniquely determined by this condition: they can be replaced by $\lambda - \chi\psi$ and $\mu + \chi\varphi$ respectively, where χ is an arbitrary polynomial with coefficients in Σ. We may dispose of this χ so as to make the degree of μ smaller than the degree of φ, (or, perhaps, to make $\mu = 0$). The polynomials λ and μ having been so selected, we determine polynomials $l(X)$ and $m(X)$ with coefficients in $\bar{\mathfrak{o}}$ such that $l^* = \lambda$, $m^* = \mu$; we may furthermore assume that m is either 0 or of degree $< \partial^\circ\varphi$ and that l is either 0 or of degree equal to that of λ. We set $g_{n+1} = g_n + t^n m$, $h_{n+1} = h_n + t^n l$. It is clear that $g_{n+1}^* = \varphi$, $h_{n+1}^* = \psi$, $\partial^\circ g_{n+1} = \partial^\circ \varphi$, $\nu_\mathfrak{p}(g_{n+1} - g_n) \geqq n$, $\nu_\mathfrak{p}(h_{n+1} - h_n) \geqq n$. We have $f - g_{n+1}h_{n+1} = t^n(\zeta - (lg_n + mh_n) - t^n lm)$. Since $\zeta^* = \lambda\varphi + \mu\psi$, we have $\nu_\mathfrak{p}(\zeta - (lg_n + mh_n)) \geqq 1$, whence $\nu_\mathfrak{p}(f - g_{n+1}h_{n+1}) \geqq n + 1$. It remains to check that $\partial^\circ h_{n+1} \leqq \partial^\circ f - \partial^\circ g_{n+1}$. Since $\partial^\circ h_n \leqq \partial^\circ f - \partial^\circ g_n$, we have $\partial^\circ \zeta \leqq \partial^\circ f$. On the other hand, the equality $\varphi\psi = f^*$ shows that $\partial^\circ \psi = \partial^\circ f^* - \partial^\circ \varphi \leqq \partial^\circ f - \partial^\circ \varphi$, whence (if $\mu \neq 0$) $\partial^\circ \mu\psi \leqq \partial^\circ f$. It follows that (if $\lambda \neq 0$) $\partial^\circ \lambda\varphi \leqq \partial^\circ f$, whence $\partial^\circ \lambda \leqq \partial^\circ f - \partial^\circ \varphi$ and $\partial^\circ l \leqq \partial^\circ f - \partial^\circ \varphi = \partial^\circ f - \partial^\circ g_n$. Now we have $h_{n+1} = h_n + t^n l$ and h_n and l are both of degrees $\leqq \partial^\circ f - \partial^\circ g_n$; therefore, h_{n+1} is of degree $\leqq \partial^\circ f - \partial^\circ g_n = \partial^\circ f - \partial^\circ g_{n+1}$.

Thus, our process can be continued indefinitely, and we obtain two sequences (g_n), (h_n) of polynomials for which the conditions a), b), c), d) are satisfied for all n.

Let a and b be the respective degrees of f and φ. Then we may write $g_n = \sum_{i=0}^{a} \gamma_{i,n} X^i$, $h_n = \sum_{j=0}^{a-b} \delta_{j,n} X^j$, where $\gamma_{i,n}$ and $\delta_{j,n}$ are elements of $\bar{\mathfrak{o}}$, and we have $\nu_\mathfrak{p}(\gamma_{i,n+1} - \gamma_{i,n}) \geqq n$ and $\nu_\mathfrak{p}(\delta_{j,n+1} - \delta_{j,n}) \geqq n$. This means that the sequences $(\gamma_{i,n})$, $(\delta_{j,n})$ (for $1 \leqq i \leqq b$, $1 \leqq j \leqq a - b$) are Cauchy sequences. Since $\bar{R}_\mathfrak{p}$ is complete, these sequences converge to elements γ_i, δ_j which clearly belong to $\bar{\mathfrak{o}}$. Set $g = \sum_{i=0}^{a} \gamma_i X^i$, $h = \sum_{j=0}^{a-b} \delta_j X^j$. Since $\nu_\mathfrak{p}(\gamma_i - \gamma_{i,1}) \geqq 1$, $\nu_\mathfrak{p}(\delta_j - \delta_{j,1}) \geqq 1$, we have $g^* = \varphi$, $h^* = \psi$. Since b is the degree of φ and g is of degree $\leqq b$, the equality $g^* = \varphi$ implies $\partial^\circ g = \partial^\circ \varphi$. We have $f - gh = f - g_n h_n + gh - g_n h_n$. We see easily that $\nu_\mathfrak{p}(gh - g_n h_n)$ increases indefinitely with n; on the other hand, we have $\nu_\mathfrak{p}(f - g_n h_n) \geqq n$. It follows that $\nu_p(f - gh) = \infty$, whence $f = gh$. Hensel's lemma is thereby proved.

§3. Structure of $\mathfrak{p}$-adic completions

Let R be a field of algebraic functions of one variable. We denote by $\mathfrak{p}$ a place of R, by $\bar{R}$ the $\mathfrak{p}$-adic completion of R, by Σ the residue field of $\mathfrak{p}$, by K the field of constants of R, by $\bar{\mathfrak{o}}$ the ring of integral elements of $\bar{R}$, by $\bar{\mathfrak{p}}$ the ideal of non-units of $\bar{\mathfrak{o}}$. We know that the field $\bar{\mathfrak{o}}/\bar{\mathfrak{p}}$ may be identified with Σ. If Σ' is any subset of Σ, we shall say that a subset $\bar{\Sigma}'$ of $\bar{R}$ is a *system of representatives for* Σ' if $\bar{\Sigma}'$ is in $\bar{\mathfrak{o}}$ and is mapped in a one-to-one way onto Σ' by the natural homomorphism of $\bar{\mathfrak{o}}$ onto $\Sigma = \bar{\mathfrak{o}}/\bar{\mathfrak{p}}$.

THEOREM 1. *The notation being as explained above, denote also by Σ_s the field of elements of Σ which are separable over K. Then $\bar{R}$ has a subfield $\bar{\Sigma}_s$ which contains K and which is a system of representatives for the elements of Σ_s. The field $\bar{\Sigma}_s$ is uniquely determined by these two conditions.*

Since Σ_s is separable of finite degree over K, we may write $\Sigma_s = K\langle\alpha^*\rangle$, where α^* is some element of Σ_s. This element satisfies an irreducible equation of the form $f(X) = 0$, where f is a polynomial with coefficients in K. We have $f(X) = (X - \alpha^*)g(X)$, where g is a polynomial with coefficients in Σ_s. Since Σ_s is separable over K, we have $g(\alpha^*) \neq 0$, which means that $X - \alpha^*$ and $g(X)$ have no common factor of degree >0.

We may consider f as a polynomial with coefficients in $\bar{\mathfrak{o}}$. Applying Hensel's lemma, we see that f admits a factor of the first degree, say $aX + b$, with coefficients in $\bar{\mathfrak{o}}$ such that the residue classes of a and b modulo $\bar{\mathfrak{p}}$ are 1 and $-\alpha^*$ respectively. It follows that $\alpha = -b/a$ is integral and belongs to the residue class α^*. Since $f(\alpha) = 0$, α is algebraic over K and every element of $K\langle\alpha\rangle$ is a linear combination of $1, \alpha, \cdots, \alpha^{d-1}$ with coefficients in K (where d is the degree of f). Since $1, \alpha^*, \cdots, (\alpha^*)^{d-1}$ are linearly independent over K, we see that $K\langle\alpha\rangle$ is a system of representatives for the elements of Σ_s. Let conversely $\bar{\Sigma}_s'$ be any subfield $\bar{R}$ containing K which is a system of representatives for the elements of Σ_s. Let α' be the element of $\bar{\Sigma}_s'$ which belongs to the residue class α^*. Since $\Sigma_s = K[\alpha^*]$, it is clear that $\bar{\Sigma}_s' = K[\alpha'] = K\langle\alpha'\rangle$. The element $f(\alpha')$ belongs to the residue class $f(\alpha^*) = 0$, whence $f(\alpha') = 0$. Write $f(X) = (aX + b)\bar{g}(X)$, where $\bar{g}$ is a polynomial with coefficients in $\bar{\mathfrak{o}}$. Then we have $(a\alpha' + b)\bar{g}(\alpha') = 0$. But $\bar{g}(\alpha')$ belongs to the residue class $g(\alpha^*)$ which is $\neq 0$, whence $\bar{g}(\alpha') \neq 0$ and therefore $a\alpha' + b = 0$, whence $\alpha = \alpha'$ and $\bar{\Sigma}_s' = K\langle\alpha\rangle$. Theorem 1 is thereby proved.

This being said, let x be any element $\neq 0$ of R such that $\nu_{\mathfrak{p}}(x) > 0$. Let there be given a sequence $(c_k)_{r \leqq k < \infty}$ of elements of $\bar{\Sigma}_s$ starting with a term c_r whose index r may be negative, null, or positive. If we set $u_n = \sum_{k=r}^{n} c_k x^k$, the sequence (u_n) is obviously a Cauchy sequence in $\bar{R}$. The limit of this sequence is represented by $\sum_{k=r}^{\infty} c_k x^k$. An equality of the form $\sum_{k=r}^{\infty} c_k x^k = \sum_{k=r}^{\infty} c_k' x^k$ implies $c_k = c_k'$ for every $k \geqq r$. In fact, assume for a moment that this were not the case. Then there would exist a least index, say m, for which $c_m \neq c_m'$, and we would have

$$\lim_{n \to \infty} \left((c_m' - c_m)x^m + \sum_{k=1}^{n} (c_{m+k}' - c_{m+k})x^{m+k} \right) = 0.$$

But we have $\nu_{\mathfrak{p}}((c_m' - c_m)x^m) = m\nu_{\mathfrak{p}}(x)$, while

$$\nu_{\mathfrak{p}} \left(\sum_{k=1}^{n} (c_{m+k}' - c_{m+k})x^{m+k} \right) \geqq (m + 1)\nu_{\mathfrak{p}}(x),$$

which proves that the equality written above cannot take place.

Any element of the form $\sum_{k=r}^{\infty} c_k x^k$ can also be written in the form $\sum_{k=s}^{\infty} c_k x^k$ if $s < r$, by simply defining c_k to be 0 if $s \leqq k < r$. It is clear that

$$\sum_{k=r}^{\infty} c_k x^k - \sum_{k=r}^{\infty} c_k' x^k = \sum_{k=r}^{\infty} (c_k - c_k')x^k,$$

$$\left(\sum_{k=r}^{\infty} c_k x^k \right) \left(\sum_{k=r'}^{\infty} c_k' x^k \right) = \sum_{l=r+r'}^{\infty} \left(\sum_{k+k'=l} c_k c_{k'}' \right) x^l.$$

This means that the elements of the form $\sum_{k=r}^{\infty} c_k x^k$ form a subfield of $\bar{R}$, isomorphic with the field of formal power series in x with coefficients in Σ_s.

Consider now the special case where the residue field Σ of $\mathfrak{p}$ is separable over

K (i.e. $\Sigma = \Sigma_s$). Let then $\bar{\Sigma}$ be the subfield of $\bar{R}$ containing K which is a system of representatives for the elements of Σ. Select a uniformizing variable x at $\mathfrak{p}$. *Then every element y of $\bar{R}$ can be represented in the form* $\sum_{k=r}^{\infty} c_k x^k$. We may of course assume that $y \neq 0$. Set then $r = \nu_{\mathfrak{p}}(y)$. We shall define a sequence $(c_k)_{r \leqq k < \infty}$ of elements of $\bar{\Sigma}$. To define c_r, we observe that $yx^{-r} \in \bar{\mathfrak{o}}$; c_r will then be the representative in $\bar{\Sigma}$ of the residue class of yx^{-r} modulo $\mathfrak{p}$. It is then clear that $\nu_{\mathfrak{p}}(y - c_r x^r) > r$. Suppose that $c_r, \cdots, c_n$ have already been defined and that $\nu_{\mathfrak{p}}(y - \sum_{k=r}^{n} c_k x^k) > n$. Then $x^{-n-1}(y - \sum_{k=r}^{n} c_k x^k)$ is integral and we define c_{n+1} to be the element of $\bar{\Sigma}$ which is congruent to $x^{-n-1}(y - \sum_{k=r}^{n} c_k x^k)$ modulo $\bar{\mathfrak{p}}$. It is then clear that $\nu_{\mathfrak{p}}(y - \sum_{k=r}^{n+1} c_k x^k) > n + 1$. The sequence (c_k) being defined in this way, it is clear that we have $y = \sum_{k=r}^{\infty} c_k x^k$. The formula $y = \sum_{k=r}^{\infty} c_k x^k$ (or $y = \sum_{k=s}^{\infty} c_k x^k$ with $s < r$, $c_k = 0$ for $s \leqq k < r$) is said to represent the *expansion of y in terms of x in the $\mathfrak{p}$-adic completion of R.*

REMARK. Even in the case where Σ is not separable over K, it can be proved that there always exists a subfield $\bar{\Sigma}$ of $\bar{R}$ which is a system of representatives for the elements of Σ. We shall not prove this result here. Besides the fact that the proof is rather difficult, the usefulness of the result is rather seriously impaired by the following circumstances: a) the field $\bar{\Sigma}$ is in general not unique; b) it is not always possible to take for $\bar{\Sigma}$ a field which contains the field of constants K or R, as we shall prove by an example.

Let K be a field of characteristic $p > 0$ which is not perfect, and let a be an element of K which is not a pth power in K. Set $R = K\langle x \rangle$, where x is transcendental over K. Then $x^p - a$ is irreducible in $K[x]$; this polynomial determines a place $\mathfrak{p}$ of $K\langle x \rangle$, and the residue field Σ of $\mathfrak{p}$ is of the form $K\langle \xi \rangle$, where ξ is such that $\xi^p = a$. We shall prove that a is not a pth power in the $\mathfrak{p}$-adic completion $\bar{R}$ of R. In fact, assume for a moment that $a = y^p$, $y \in \bar{R}$. Then we have $(y - x)^p = a - x^p$, whence $p\nu_{\mathfrak{p}}(y - x) = 1$, which is not possible. It follows immediately from this that no subfield of $\bar{R}$ containing K can be a system of representatives for the elements of Σ.

§4. GENERALIZATION OF THE NOTION OF REPARTITION

Let R be a field of algebraic functions of one variable, and let K be the field of constants of R. For each place $\mathfrak{p}$ of R we have a $\mathfrak{p}$-adic completion $\bar{R}_{\mathfrak{p}}$ of R. We have defined a repartition of R to be a mapping $\mathfrak{x}$ which assigns to every place $\mathfrak{p}$ of R an element $\mathfrak{x}(\mathfrak{p})$ of R in such a way that $\nu_{\mathfrak{p}}(\mathfrak{x}(\mathfrak{p})) \geqq 0$ for almost all places $\mathfrak{p}$. For technical reasons, it is convenient to generalize this notion by allowing each $\mathfrak{x}(\mathfrak{p})$ to lie in the $\mathfrak{p}$-adic completion $\bar{R}_{\mathfrak{p}}$ of R. We shall therefore, from now on, call *repartition* a mapping $\mathfrak{x}$ which assigns to every place $\mathfrak{p}$ of R an element $\mathfrak{x}(\mathfrak{p})$ of $\bar{R}_{\mathfrak{p}}$ in such a way that $\nu_{\mathfrak{p}}(\mathfrak{x}(\mathfrak{p})) \geqq 0$ for almost all $\mathfrak{p}$ ($\nu_{\mathfrak{p}}$ being now the extension to $\bar{R}_{\mathfrak{p}}$ of the order function at $\mathfrak{p}$). The element $\mathfrak{x}(\mathfrak{p})$ is called the *$\mathfrak{p}$-component* of the repartition $\mathfrak{x}$; we set $\nu_{\mathfrak{p}}(\mathfrak{x}) = \nu_{\mathfrak{p}}(\mathfrak{x}(\mathfrak{p}))$.

The repartitions in the new sense still form a ring, in which the operations are defined by: $(\mathfrak{x} + \mathfrak{y})(\mathfrak{p}) = \mathfrak{x}(\mathfrak{p}) + \mathfrak{y}(\mathfrak{p})$, $(\mathfrak{x}\mathfrak{y})(\mathfrak{p}) = \mathfrak{x}(\mathfrak{p})\mathfrak{y}(\mathfrak{p})$. This ring contains the set of repartitions in the old sense as a subring. In particular, it contains R as a subring, and it may be considered as a vector space over K.

Let $\mathfrak{a} = \prod_{\mathfrak{p}} \mathfrak{p}^{m(\mathfrak{p})}$ be a divisor. The notion of congruence (mod $\mathfrak{a}$) can be extended to our new repartitions: the repartitions $\mathfrak{x}$ and $\mathfrak{y}$ are said to be congruent to each other (mod $\mathfrak{a}$) if we have $\nu_{\mathfrak{p}}(\mathfrak{x} - \mathfrak{y}) \geqq m(\mathfrak{p})$ for every place $\mathfrak{p}$.

If $\mathfrak{x}$ is any repartition in our new sense and $\mathfrak{a}$ any divisor, there always exists a repartition $\mathfrak{x}'$ in the old sense such that $\mathfrak{x} \equiv \mathfrak{x}'$ (mod $\mathfrak{a}$). For, set $\mathfrak{a} = \Pi_{\mathfrak{p}} \mathfrak{p}^{m(\mathfrak{p})}$. For each place $\mathfrak{p}$ we can find an element $x(\mathfrak{p})$ of R such that $\nu_{\mathfrak{p}}(\mathfrak{x}(\mathfrak{p}) - x(\mathfrak{p})) \geqq m(\mathfrak{p})$. It follows that $\nu_{\mathfrak{p}}(x(\mathfrak{p})) \geqq \min \{\nu_{\mathfrak{p}}(\mathfrak{x}(\mathfrak{p})), m(\mathfrak{p})\}$, whence $\nu_{\mathfrak{p}}(x(\mathfrak{p})) \geqq 0$ for almost all $\mathfrak{p}$'s. This means that the mapping $\mathfrak{p} \rightarrow x(\mathfrak{p})$ is a repartition $\mathfrak{x}'$ in the old sense, and it is clear that $\mathfrak{x} \equiv \mathfrak{x}'$ (mod $\mathfrak{a}$).

Let $\mathfrak{a}$ and $\mathfrak{a}'$ be divisors such that $\mathfrak{a}'$ divides $\mathfrak{a}$. Denote by $\mathfrak{X}(\mathfrak{a})$ and $\mathfrak{X}(\mathfrak{a}')$ the spaces of repartitions in the old sense which are $\equiv 0$ modulo $\mathfrak{a}$ and modulo $\mathfrak{a}'$ respectively, and by $\mathfrak{X}^*(\mathfrak{a})$ and $\mathfrak{X}^*(\mathfrak{a}')$ the spaces of repartitions in the new sense satisfying respectively the same conditions. If $\mathfrak{x}^* \in \mathfrak{X}^*(\mathfrak{a}')$, then we can find a repartition $\mathfrak{x}$ in the old sense such that $\mathfrak{x} \equiv \mathfrak{x}^*$ (mod $\mathfrak{a}$). Since $\mathfrak{a}'$ divides $\mathfrak{a}$, it is clear that $\mathfrak{x} \in \mathfrak{X}(\mathfrak{a}')$; this means that $\mathfrak{X}^*(\mathfrak{a}') = \mathfrak{X}(\mathfrak{a}') + \mathfrak{X}^*(\mathfrak{a})$. By Noether's homomorphism theorem, $\mathfrak{X}^*(\mathfrak{a}')/\mathfrak{X}^*(\mathfrak{a})$ is isomorphic with $\mathfrak{X}(\mathfrak{a}')/(\mathfrak{X}(\mathfrak{a}') \cap \mathfrak{X}^*(\mathfrak{a})) = \mathfrak{X}(\mathfrak{a}')/\mathfrak{X}(\mathfrak{a})$. From this and from Lemma 1, II, §4, we conclude that $\mathfrak{X}^*(\mathfrak{a}')/\mathfrak{X}^*(\mathfrak{a})$ *is of finite dimension equal to* $d(\mathfrak{a}) - d(\mathfrak{a}')$ *over the field* K. In other words, we see that the result quoted just above remains valid for repartitions in the new sense.

Now, let ω be a differential of the field R. We shall see that ω may be extended in a natural way to a linear function on the space of repartitions in the new sense. If $\omega = 0$, we extend it of course by the null function. If not, let $\mathfrak{x}$ be any repartition in the new sense. Then we can find a repartition $\mathfrak{x}'$ in the old sense such that $\mathfrak{x} \equiv \mathfrak{x}'$ (mod $(\mathfrak{d}(\omega))^{-1}$); and, if $\mathfrak{x}'$ and $\mathfrak{x}''$ are repartitions in the old sense satisfying this condition, we have $\mathfrak{x}' \equiv \mathfrak{x}''$ (mod $(\mathfrak{d}(\omega))^{-1}$), whence $\omega(\mathfrak{x}') = \omega(\mathfrak{x}'')$. We may therefore define $\omega(\mathfrak{x})$ to be $\omega(\mathfrak{x}')$. The function ω, extended in this way to the space of repartitions in the new sense, is obviously linear. It is clear that Theorem 2, II, §5 remains true if $\mathfrak{x}$ is taken to represent a repartition in the new sense.

If $\mathfrak{p}$ is a place of R, we define (as we did in II, §7) the $\mathfrak{p}$-component $\omega^{\mathfrak{p}}$ of a differential ω by the formula

$$\omega^{\mathfrak{p}}(\mathfrak{x}) = \omega(\mathfrak{x}^{\mathfrak{p}})$$

where $\mathfrak{x}^{\mathfrak{p}}$ is the repartition which assigns $\mathfrak{x}(\mathfrak{p})$ to $\mathfrak{p}$ and 0 to every other place. The function $\omega^{\mathfrak{p}}$ defined in this way is an extension to the space of repartitions in the new sense of the $\mathfrak{p}$-component of ω, as defined in II, §7. On the other hand, it may also be considered as a linear function on $\bar{R}_{\mathfrak{p}}$. We see immediately that Lemmas 1 and 2, II, §7 remain true; in the statement of the second one of these lemmas, we may even replace the words "for every $x \in R$" by "for every $x \in \bar{R}_{\mathfrak{p}}$."

§5. Residues of a differential

Let R be a field of algebraic functions of one variable, and let K be the field of constants of R. We denote by ω a differential of R, by $\mathfrak{p}$ a place of R, by Σ the residue field of $\mathfrak{p}$, by Σ_s the field of elements of Σ which are separable over K, by $\bar{R}$ the $\mathfrak{p}$-adic completion of R, and by $\bar{\Sigma}_s$ the subfield of $\bar{R}$ containing K which is a system of representatives for the elements of Σ_s (Theorem 1, §3). The $\mathfrak{p}$-component $\omega^{\mathfrak{p}}$ of ω may be considered as a function on $\bar{R}$ with values in K; it induces a certain function on $\bar{\Sigma}_s$. If $\xi \in \Sigma_s$, denote by $\bar{\xi}$ the representative of ξ in $\bar{\Sigma}_s$, and by $\omega^{\mathfrak{p}}(\xi)$ the element $\omega^{\mathfrak{p}}(\bar{\xi})$. Thus we attach to $\omega^{\mathfrak{p}}$ a function on Σ_s with values in K. This function is obviously linear when Σ_s is regarded as a vector space over K.

Lemma 1. *Let K be a field and let L be an overfield of K which is of finite degree and separable over K. Then every linear function on L (considered as a vector space over K) can be represented in the form $\xi \to \mathrm{Sp}_{L/K}\xi\rho$, where ρ is a uniquely determined element of L.*

If $\rho \in L$, denote by λ_ρ the linear function defined by $\lambda_\rho(\xi) = \mathrm{Sp}_{L/K}\xi\rho$. Then the mapping $\rho \to \lambda_\rho$ is a linear mapping of L into the space L^* of linear functions on L. Since L is separable over K, there exists a $\gamma \in L$ such that $\mathrm{Sp}_{L/K}\gamma \neq 0$; if $\rho \neq 0$, then we have $\lambda_\rho(\gamma\rho^{-1}) \neq 0$, whence $\lambda_\rho \neq 0$. This means that the mapping $\rho \to \lambda_\rho$ is one-to-one. The spaces L and L^* having the same dimension, every element of L^* can be written in one and only one way in the form λ_ρ, which proves Lemma 1.

Returning to the notation introduced above, we see that there exists a uniquely determined element ρ of Σ_s such that $\omega^{\mathfrak{p}}(\xi) = \mathrm{Sp}_{\Sigma_s/K}\xi\rho$ for every $\xi \in \Sigma_s$. This element is called the *residue* of the differential ω at the place $\mathfrak{p}$, and is denoted by $\mathrm{res}_{\mathfrak{p}}\omega$.

It is clear that the mapping $\omega \to \mathrm{res}_{\mathfrak{p}}\omega$ is a linear mapping of the space of differentials (regarded as a vector space over K) into the field Σ_s. It follows immediately from the definition that we have

$$\omega^{\mathfrak{p}}(1) = \mathrm{Sp}_{\Sigma_s/K}(\mathrm{res}_{\mathfrak{p}}\omega). \tag{1}$$

In particular, if the residue field of $\mathfrak{p}$ is purely unseparable over K, we have $\mathrm{res}_{\mathfrak{p}}\omega = \omega^{\mathfrak{p}}(1)$.

Theorem 2. *Let ω be a differential of the field R of algebraic functions of one variable. If $\mathfrak{p}$ is a place of R which is not a pole of ω, then $\mathrm{res}_{\mathfrak{p}}\omega = 0$. If we know of a place $\mathfrak{p}$ that its residue field is separable over the field of constants of R, that $\mathrm{res}_{\mathfrak{p}}\omega = 0$, and that $\nu_{\mathfrak{p}}(\omega) \geqq -1$, then we may conclude that $\mathfrak{p}$ is not a pole of ω.*

If $\mathfrak{p}$ is not a pole of ω, then we have $\omega^{\mathfrak{p}}(x) = 0$ for every integral element x of $\bar{R}$ and, in particular, for every element of $\bar{\Sigma}_s$, from which it follows that $\mathrm{res}_{\mathfrak{p}}\omega = 0$. If $\mathrm{res}_{\mathfrak{p}}\omega = 0$, then we have $\omega^{\mathfrak{p}}(\bar{\xi}) = 0$ for every element $\bar{\xi} \in \bar{\Sigma}_s$. Now, if Σ is separable over K, then every integral element of $\bar{R}$ is congruent modulo $\bar{\mathfrak{p}}$ to an element of $\bar{\Sigma}_s$. If $\nu_{\mathfrak{p}}(\omega) \geqq -1$, we have $\omega^{\mathfrak{p}}(y) = 0$ for every $y \in \bar{R}$ such that $\nu_{\mathfrak{p}}(y) > 0$, which means that $\omega^{\mathfrak{p}}(x) = \omega^{\mathfrak{p}}(x')$ whenever x and x' are integral elements of $\bar{R}$ which are congruent to each other modulo $\bar{\mathfrak{p}}$. It follows that, under the assumptions made in the second part of Theorem 2, we have $\omega^{\mathfrak{p}}(x) = 0$ for every integral element x of $\bar{R}$, whence $\nu_{\mathfrak{p}}(\omega) \geqq 0$.

It follows in particular from Theorem 2 that any given differential has only a finite number of residues $\neq 0$.

THEOREM 3. *Let ω be a differential of a field R of algebraic functions of one variable and let K be the field of constants of R. For each place $\mathfrak{p}$ of R, denote by $\Sigma_s(\mathfrak{p})$ the field of elements of the residue field of $\mathfrak{p}$ which are separable over K. Then we have* $\sum_{\mathfrak{p}} \mathrm{Sp}_{\Sigma_s(\mathfrak{p})/K}\, \mathrm{res}_{\mathfrak{p}}\omega = 0$.

We have $0 = \omega(1) = \sum_{\mathfrak{p}}\omega^{\mathfrak{p}}(1)$, and Theorem 3 follows immediately from Formula (1) above.

COROLLARY. *If R is a field of algebraic functions of one variable over an algebraically closed field of constants, then the sum of the residues of a differential of R is* 0.

This follows immediately from Theorem 3.

LEMMA 2. *Let $\mathfrak{p}_1, \cdots, \mathfrak{p}_h$ be a finite number of distinct places of a field R of algebraic functions of one variable. For each i $(1 \leqq i \leqq h)$ let there be given a linear mapping μ_i of the residue field Σ_i of $\mathfrak{p}_i$ into the field of constants K of R. Assume that $\sum_{i=1}^h \mu_i(1) = 0$. Then there exists a differential ω of R which satisfies the following conditions*: a) *ω is multiple of $(\mathfrak{p}_1 \cdots \mathfrak{p}_h)^{-1}$*; b) *if an element x_i of R is integral at $\mathfrak{p}_i$, we have $\omega^{\mathfrak{p}_i}(x_i) = \mu_i(\xi_i)$, where ξ_i is the value taken by x_i at $\mathfrak{p}_i$.*

Let $\mathfrak{D}$ be the space of differentials which are $\equiv 0 \pmod{(\mathfrak{p}_1 \cdots \mathfrak{p}_h)^{-1}}$. If $h \geqq 1$ (which we assume), then no element $\neq 0$ of R is multiple of $\mathfrak{p}_1 \cdots \mathfrak{p}_h$, and it follows from the theorem of Riemann-Roch that $\mathfrak{D}$ is of dimension $\sum_{i=1}^h d(\mathfrak{p}_i) + g - 1$, where g is the genus of R. If $\omega \in \mathfrak{D}$ and $\nu_{\mathfrak{p}_i}(x_i) \geqq 0$, then $\omega^{\mathfrak{p}_i}(x_i)$ depends only on the value ξ_i taken by x_i at $\mathfrak{p}_i$; for, if $\nu_{\mathfrak{p}_i}(x_i' - x_i) > 0$, then we have $\omega^{\mathfrak{p}_i}(x_i' - x_i) = 0$. Set $\omega^{\mathfrak{p}_i}(x_i) = \lambda_{i,\omega}(\xi_i)$. Then $\lambda_{i,\omega}$ is a linear function on Σ_i (regarded as a vector space over K). Denote by Σ_i^* the set of linear functions on Σ_i, and form the vector space $\prod_{i=1}^h \Sigma_i^*$, product of the vector spaces Σ_i^* $(1 \leqq i \leqq h)$. If we assign to every $\omega \in \mathfrak{D}$ the element $\Lambda(\omega) = (\lambda_{1,\omega}, \cdots, \lambda_{h,\omega})$ of this direct product, we obtain a linear mapping Λ of $\mathfrak{D}$ into $\prod_{i=1}^h \Sigma_i^*$. If $\omega \in \mathfrak{D}$, we have $0 = \omega(1) = \sum_{i=1}^h \omega^{\mathfrak{p}_i}(1) = \sum_{i=1}^h \lambda_{i,\omega}(1)$; thus Λ maps $\mathfrak{D}$ into the subspace P of $\prod \Sigma_i^*$ composed of the elements $(\lambda_1, \cdots, \lambda_h)$ such that $\sum_{i=1}^h \lambda_i(1) = 0$. If $\lambda_{i,\omega} = 0$, then we have $\omega^{\mathfrak{p}_i}(x_i) = 0$ for every $x_i \in R$ which is integral at $\mathfrak{p}_i$, i.e., $\mathfrak{p}_i$ is not a pole of ω. It follows immediately that the kernel of the mapping Λ is the space of differentials of the first kind of R. This kernel is of dimension g, and $\Lambda(\mathfrak{D})$ is therefore of dimension

$$\left(\sum_{i=1}^h d(\mathfrak{p}_i) + g - 1\right) - g = \sum_{i=1}^h d(\mathfrak{p}_i) - 1.$$

But $\prod_{i=1}^{h} \Sigma_i^*$ is of dimension $\sum_{i=1}^{h} d(\mathfrak{p}_i)$, and therefore P of dimension $\sum_{i=1}^{h} d(\mathfrak{p}_i) - 1$. It follows that $\Lambda(\mathfrak{D}) = P$, which proves Lemma 1.

THEOREM 4. *Let $\mathfrak{p}_1, \cdots, \mathfrak{p}_h$ be a finite number of distinct places of a field R of algebraic functions of one variable. Let there be given for each i $(1 \leqq i \leqq h)$ an elemen ρ_i belonging to the field $\Sigma_{i,s}$ of elements of the residue field of $\mathfrak{p}_i$ which are separable over the field of constants K of R. Assume that $\sum_{i=1}^{h} \mathrm{Sp}_{\Sigma_{i,s}/K}\rho_i = 0$. Then there exists a differential ω of R which is $\equiv 0 \pmod{(\mathfrak{p}_1 \cdots \mathfrak{p}_h)^{-1}}$ such that* $\mathrm{res}_{\mathfrak{p}_i}\omega = \rho_i (1 \leqq i \leqq h)$

For each i we can construct a linear mapping μ_i of the residue field Σ_i of $\mathfrak{p}_i$ into K which extends the mapping $\xi_i \to \mathrm{Sp}_{\Sigma_{i,s}/K}\xi_i\rho_i$ of $\Sigma_{i,s}$ into K. Then we have $\sum_{i=1}^{h} \mu_i(1) = 0$. Let ω be a differential which satisfies the conditions of Lemma 1 relative to these linear functions μ_i. If $\xi_i \in \Sigma_{i,s}$, denote by x_i an element of R which takes the value ξ_i at $\mathfrak{p}_i$ and by $\bar{\xi}_i$ the element of the $\mathfrak{p}_i$-adic completion of R which belongs to the residue class ξ_i modulo $\mathfrak{p}_i$ and which is algebraic over K. Then $\nu_{\mathfrak{p}_i}(\bar{\xi}_i - x_i) > 0$, whence, since $\nu_{\mathfrak{p}_i}(\omega) \geqq -1$, $\omega^{\mathfrak{p}_i}(\bar{\xi}_i - x_i) = 0$ and therefore $\mathrm{Sp}_{\Sigma_{i,s}/K}\xi_i(\mathrm{res}_{\mathfrak{p}_i}\omega) = \omega^{\mathfrak{p}_i}(\bar{\xi}_i) = \omega^{\mathfrak{p}_i}(x_i) = \mu_i(\xi_i) = \mathrm{Sp}_{\Sigma_{i,s}/K}\xi_i\rho_i$. This being true for every $\xi_i \in \Sigma_{i,s}$, we have $\mathrm{res}_{\mathfrak{p}_i}\omega = \rho_i$ by Lemma 1.

A differential ω of a field of algebraic functions of one variable of characteristic 0 is said to be *of the second kind* if its residues are all equal to 0. It is clear that any differential of the first kind is also of the second kind.

LEMMA 3. *Let R be a field of algebraic functions of one variable of characteristic 0, and let $\mathfrak{a} = \Pi_{\mathfrak{p}}\mathfrak{p}^{a(\mathfrak{p})}$ be an integral divisor of R. Set $a^*(\mathfrak{p}) = \max\{0, a(\mathfrak{p}) - 1\}$. Then the differentials of the second kind of R which are $\equiv 0 \pmod{\mathfrak{a}^{-1}}$ form a vector space over the field of constants K of R, and the dimension of this space is $\sum_{\mathfrak{p}} a^*(\mathfrak{p})\, d(\mathfrak{p}) + g$, where g is the genus of R.*

This is obvious if $\mathfrak{a}$ is the unit divisor. Assume from now on that $\mathfrak{a}$ is not the unit divisor. Then no element $\neq 0$ of R is $\equiv 0 \pmod{\mathfrak{a}}$, and it follows from the theorem of Riemann-Roch that the dimension of the space $\mathfrak{D}$ of differentials which are $\equiv 0 \pmod{\mathfrak{a}^{-1}}$ is $d(\mathfrak{a}) + g - 1$. Let $\mathfrak{p}_1, \cdots, \mathfrak{p}_h$ be the distinct places for which $a(\mathfrak{p}) \neq 0$, and let Σ_i be the residue field of $\mathfrak{p}_i$. If we assign to every $\omega \in \mathfrak{D}$ the element $(\mathrm{res}_{\mathfrak{p}_1} \omega, \cdots, \mathrm{res}_{\mathfrak{p}_h} \omega)$ of the product $\prod_{i=1}^{h} \Sigma_i$, we obtain a linear mapping P of $\mathfrak{D}$ into $\prod_{i=1}^{h} \Sigma_i$. It follows from Theorem 4 that $\mathrm{P}(\mathfrak{D})$ is the space of elements $(\rho_1, \cdots, \rho_h)$ of $\prod_{i=1}^{h} \Sigma_i$ such that $\sum_{i=1}^{h} \mathrm{Sp}_{\Sigma_i/K}\, \rho_i = 0$; this space is clearly of dimension $\sum_{i=1}^{h} d(\mathfrak{p}_i) - 1$. The space of differentials of the second kind which are multiples of $\mathfrak{a}^{-1}$ is the kernel of the mapping P; the dimension of this space is therefore $d(\mathfrak{a}) + g - 1 - (\sum_{i=1}^{h} d(\mathfrak{p}_i) - 1)$, and this is easily seen to be equal to $\sum_{\mathfrak{p}} a^*(\mathfrak{p})\, d(\mathfrak{p}) + g$.

CHAPTER IV

EXTENSIONS OF FIELDS OF ALGEBRAIC FUNCTIONS OF ONE VARIABLE

§1. The relative degree and the ramification index

We shall consider in this chapter pairs (R, S) of fields of algebraic functions of one variable such that R is a subfield of S. *We shall always tacitly assume that the field of constants K of R is contained in the field of constants L of S and that any element of R which is transcendental over K is also transcendental over L.* This can also be expressed by saying that $K = L \cap R$.

Let $\mathfrak{P}$ be a place of S, and let $\mathfrak{O}$ be the ring of the place $\mathfrak{P}$. If $\mathfrak{O}$ contains R, then we say that $\mathfrak{P}$ is a *variable place* (with respect to R); if not, then we say that $\mathfrak{P}$ is a *fixed place* (with respect to R). In the latter case, the intersection $\mathfrak{O} \cap R = \mathfrak{o}$ is obviously a V-ring in R which contains K; it determines a place $\mathfrak{p}$ of R, of which we say that it *lies below* $\mathfrak{P}$, or that $\mathfrak{P}$ *lies above* it. If S is algebraic over R, then every place of S is fixed with respect to R. For, in this case, every element of S is algebraic and therefore also (since R is a field) integral over R; if R would be contained in $\mathfrak{O}$, then every element of S would be integral over $\mathfrak{O}$, which is impossible since $\mathfrak{O}$ is integrally closed in S (cf. I, §2). On the contrary, if S is transcendental over R (which can happen only if L is transcendental over K), then it is easily seen that there always exist places of S which are variable with respect to R.

Assume that the place $\mathfrak{P}$ of S lies above the place $\mathfrak{p}$ of R. The values taken by the order function $\nu_{\mathfrak{P}}$ at $\mathfrak{P}$ on the multiplicative group of elements $\neq 0$ in R form a subgroup of the additive group of integers, and this subgroup contains numbers $\neq 0$. It follows that the subgroup in question consists of all multiples of some integer $e > 0$, which is uniquely determined by this condition. The number e is called the *ramification index* of $\mathfrak{P}$ with respect to $\mathfrak{p}$ (or with respect to R); if $e > 1$, then we say that $\mathfrak{P}$ is *ramified* with respect to R, if not that it is *unramified*. If no place of S lying above $\mathfrak{p}$ is ramified with respect to R, then we say that $\mathfrak{p}$ is *unramified* in S.

In the general case, it is clear that

$$\nu_{\mathfrak{P}}(x) = e\nu_{\mathfrak{p}}(x) \qquad \text{for every } x \neq 0 \text{ in } R.$$

On the other hand, since the intersection of $\mathfrak{P}$ with the ring $\mathfrak{o}$ of the place $\mathfrak{p}$ is $\mathfrak{p}$, we may identify the residue field $\Sigma_{\mathfrak{p}} = \mathfrak{o}/\mathfrak{p}$ of $\mathfrak{p}$ with a subfield of the residue field $\Sigma_{\mathfrak{P}} = \mathfrak{O}/\mathfrak{P}$ of $\mathfrak{P}$. If $\Sigma_{\mathfrak{P}}$ is algebraic of finite degree over $\Sigma_{\mathfrak{p}}$, then we say that $[\Sigma_{\mathfrak{P}}:\Sigma_{\mathfrak{p}}]$ is the *relative degree* of $\mathfrak{P}$ with respect to $\mathfrak{p}$ (or with respect to R); if not, then we say that $\mathfrak{P}$ is of infinite relative degree with respect to R.

THEOREM 1. *Let R and S be fields of algebraic functions of one variable such that S contains R as a subfield. If $\mathfrak{p}$ is a place of R, then there exists at least one place of S which lies above $\mathfrak{p}$, and there exist only a finite number of such places, say $\mathfrak{P}_1, \cdots, \mathfrak{P}_h$. If S is of finite degree over R, the relative degrees $d_1, \cdots, d_h$ of $\mathfrak{P}_1, \cdots, \mathfrak{P}_h$ are finite, and we have $[S:R] = \sum_{i=1}^{h} d_i e_i$, where e_i is the ramification index of $\mathfrak{P}_i$ with respect to R.*

First we prove

LEMMA 1. *If $\mathfrak{p}$ is a place of R, there exists an element x of R which has $\mathfrak{p}$ as its only zero.*

Denote by g the genus of R and by d the degree of $\mathfrak{p}$. Then we have, by Riemann's theorem, $l(\mathfrak{p}^{-g-1}) \geqq d(g+1) - g + 1 \geqq 2$, which means that there exists a non constant element y of R which is $\equiv 0 \pmod{\mathfrak{p}^{-g-1}}$; the element $x = y^{-1}$ then has $\mathfrak{p}$ as its only zero.

LEMMA 2. *Let K, L, and T be subfields of some field; assume that K is contained in L and T and that L is of finite degree over K. Let $\{\lambda_1, \cdots, \lambda_m\}$ be a base of L over K; then every element of $T\langle L\rangle$ can be expressed as a linear combination of $\lambda_1, \cdots, \lambda_m$ with coefficients in T, whence $[T\langle L\rangle:T] \leqq [L:K]$.*

Let U be the set of linear combinations of $\lambda_1, \cdots, \lambda_m$ with coefficients in T. It is clear that the products $\lambda_i\lambda_j$ $(1 \leqq i, j \leqq m)$ belong to U and therefore that U is a ring. We have $L \subset U$, whence $1 \in U$, and $T \subset U$. If u_1 is an element $\neq 0$ in U, the mapping $u \to u_1 u$ of U into itself is linear (when U is considered as a vector space over T) and does not map any $u \neq 0$ upon 0. Since U is of finite dimension over T, our mapping maps U onto itself; since $1 \in U$, we have $u_1^{-1} \in U$, and U is a field. Since U contains T and L, we have $U = T\langle L\rangle$.

LEMMA 3. *Let L be an overfield of a field K, and let x be an element of an overfield of L which is transcendental over L. In order for $L\langle x\rangle$ to be of finite degree over $K\langle x\rangle$, it is necessary and sufficient that L be of finite degree over K; if this condition is satisfied, we have $[L\langle x\rangle:K\langle x\rangle] = [L:K]$.*

If L is of finite degree over K, then it follows from Lemma 2 that $L\langle x\rangle$ is of finite degree over $K\langle x\rangle$ and that $[L\langle x\rangle:K\langle x\rangle] \leqq [L:K]$. Now, assume that $L\langle x\rangle$ is of finite degree over $K\langle x\rangle$. Let $\lambda_1, \cdots, \lambda_m$ be elements of L which are linearly independent over K; then they are also linearly independent over $K\langle x\rangle$. In fact, assume that $\sum_{i=1}^{m} \lambda_i r_i = 0$, $r_i \in K\langle x\rangle$. We can find a $q \neq 0$ in $K[x]$ such that the elements $p_i = qr_i (1 \leqq i \leqq m)$ are in $K[x]$. Writing that the coefficients of the polynomial $\sum_{i=1}^{m} \lambda_i p_i$ are all zero, and remembering that $\lambda_1, \cdots, \lambda_m$ are linearly independent over K, we see immediately that $p_i = 0$ $(1 \leqq i \leqq m)$, which proves our assertion. It follows that $[L:K]$ is finite and at most equal to $[L\langle x\rangle:K\langle x\rangle]$. Lemma 3 is thereby proved.

We can now prove Theorem 1. We denote by K and L the fields of constants of R and S respectively, and we take an element $x \in R$ which has $\mathfrak{p}$ as its only zero. Then x, which is in R and transcendental over K, is not in L, which proves that x has at least one zero $\mathfrak{P}$ in S. If $\mathfrak{O}$ is the ring of $\mathfrak{P}$, then $\mathfrak{O}$ does not contain x^{-1}; it follows that $\mathfrak{O} \cap R$ is the ring of a place of R which contains x. This place must be $\mathfrak{p}$, and $\mathfrak{P}$ lies above $\mathfrak{p}$. Conversely, any place of S which lies above $\mathfrak{p}$ is a zero of x, which proves that there are only a finite number of such places. Assume now that S is of finite degree over R. Since S is of finite degree over $L\langle x\rangle$ and R over $K\langle x\rangle$, it follows easily that $L\langle x\rangle$ is of finite degree over $K\langle x\rangle$, and therefore that L is of finite degree over K and that $[L\langle x\rangle : K\langle x\rangle] = [L:K]$. If $a = \nu_{\mathfrak{p}}(x)$, the divisor of zeros of x in S is $\mathfrak{P}_1^{ae_1} \cdots \mathfrak{P}_h^{ae_h}$. Making use of Theorem 4, I, §8, we have $[S:L\langle x\rangle] = \sum_{i=1}^{h} ae_i[\mathfrak{O}_i/\mathfrak{P}_i:L]$, where $\mathfrak{O}_i$ is the ring of $\mathfrak{P}_i$. Since $[L\langle x\rangle:K\langle x\rangle] = [L:K]$, it follows that $[S:K\langle x\rangle] = a\sum_{i=1}^{h} e_i[\mathfrak{O}_i/\mathfrak{P}_i:K]$. Since $\mathfrak{O}_i/\mathfrak{P}_i$ is of finite degree over K, it is also of finite degree f_i over $\mathfrak{o}/\mathfrak{p}$, and we have $[S:R][R:K\langle x\rangle] = a[\mathfrak{o}/\mathfrak{p}:K]\sum_{i=1}^{h} e_if_i$. Applying the theorem quoted above to R instead of S, we obtain $[R:K\langle x\rangle] = a[\mathfrak{o}/\mathfrak{p}:K]$. Theorem 1 is thereby proved.

LEMMA 4. *Let R, S, and T be fields of algebraic functions of one variable; assume that R is a subfield of S and S of T. Let $\mathfrak{P}$ be a place of T which is fixed with respect to R; then $\mathfrak{P}$ is fixed with respect to S. If $\mathfrak{p}$ is the place of S which lies below $\mathfrak{P}$, $\mathfrak{p}$ is fixed with respect to R and the ramification index of $\mathfrak{P}$ with respect to R is the product of its ramification index with respect to S by the ramification index of $\mathfrak{p}$ with respect to R. If T is of finite degree over R, the relative degree of $\mathfrak{P}$ with respect to R is the product of its relative degree with respect to S by the relative degree of $\mathfrak{p}$ with respect to R.*

It is obvious that $\mathfrak{P}$ is fixed with respect to S and that $\mathfrak{p}$ is fixed with respect to R. Let $\mathfrak{q}$ be the place of R which lies below $\mathfrak{P}$, and let x be a uniformizing variable at $\mathfrak{q}$ in R. Then $e_0 = \nu_{\mathfrak{p}}(x)$ is the ramification degree of $\mathfrak{p}$ with respect to R, while the ramification index of $\mathfrak{P}$ with respect to R is $\nu_{\mathfrak{P}}(x) = e\nu_{\mathfrak{p}}(x)$, where e is the ramification index of $\mathfrak{P}$ with respect to S. Let $\Sigma(\mathfrak{q})$, $\Sigma(\mathfrak{p})$, and $\Sigma(\mathfrak{P})$ be the residue fields of $\mathfrak{q}$, $\mathfrak{p}$, and $\mathfrak{P}$ respectively. Then we have $\Sigma(\mathfrak{q}) \subset \Sigma(\mathfrak{p}) \subset \Sigma(\mathfrak{P})$, and, if T is of finite degree over R, we may write $[\Sigma(\mathfrak{P}):\Sigma(\mathfrak{q})] = [\Sigma(\mathfrak{P}):\Sigma(\mathfrak{p})][\Sigma(\mathfrak{p}):\Sigma(\mathfrak{q})]$. Lemma 4 is thereby proved.

§2. THE CASE OF NORMAL ALGEBRAIC EXTENSIONS

Let R and S be fields of algebraic functions of one variable such that S contains R as a subfield. If σ is an isomorphism of S with a field S' which maps every element of R upon itself, S' may be considered as a field of algebraic functions of one variable, whose field of constants is the image under σ of the field of constants of S. It is clear that, if $\mathfrak{P}$ is a place of S, then $\sigma\mathfrak{P}$ is a place of S', and that σ defines an isomorphism of the residue field of $\mathfrak{P}$ with the residue field of $\sigma\mathfrak{P}$, and also an isomorphism of the $\mathfrak{P}$-adic completion of S with the $\sigma\mathfrak{P}$-adic completion of S'. If x is an element of S, we have $\nu_{\sigma\mathfrak{P}}(\sigma x) = \nu_{\mathfrak{P}}(x)$. If $\mathfrak{P}$ is fixed with respect to R, the same is true of $\sigma\mathfrak{P}$, and $\sigma\mathfrak{P}$ lies above the same place of

R as $\mathfrak{P}$; the ramification index of $\sigma\mathfrak{P}$ with respect to R is the same as that of $\mathfrak{P}$. If S is of finite degree over R, then the relative degree of $\sigma\mathfrak{P}$ with respect to R is equal to that of $\mathfrak{P}$.

When places $\mathfrak{P}$ and $\sigma\mathfrak{P}$ lie in the relation described above, then we say that $\sigma\mathfrak{P}$ is *conjugate to* $\mathfrak{P}$ with respect to R. In particular, the places of the field S itself which are conjugate to $\mathfrak{P}$ are those deduced from $\mathfrak{P}$ by the automorphisms of S which leave the elements of R fixed.

THEOREM 2. *Let R and S be fields of algebraic functions of one variable such that S contains R as a subfield and is algebraic and normal over R. Then the places of S which lie above a given place $\mathfrak{p}$ of R are all conjugate to each other.*

Let $\mathfrak{P}_1, \cdots, \mathfrak{P}_h$ be these places. Then we can find an element z of S such that $\nu_{\mathfrak{P}_1}(z) > 0$, $\nu_{\mathfrak{P}_i}(z) = 0$ for $i > 1$ (Theorem 3, I, §6). Let S' be the field generated by adjunction to R of z and of its conjugates with respect to R; then S' is algebraic, normal, and of finite degree over R. Let $\sigma_1', \cdots, \sigma_m'$ be the distinct automorphisms of S' over R; then it is well known that each σ_k' can be extended to an automorphism σ_k of S. On the other hand, we know that the relative norm $N_{S'/R}z$ of z from S' to R can be written in the form $x = \prod_{k=1}^{m} (\sigma_k z)^q$ where q is some exponent >0. We have $\nu_{\sigma_k\mathfrak{P}_1}(\sigma_k z) > 0$, but $\nu_{\mathfrak{P}_i}(\sigma_k z) = 0$ if $\mathfrak{P}_i \neq \sigma_k\mathfrak{P}_1$. This shows that $\nu_{\mathfrak{P}_1}(\sigma_k z) \geqq 0$ for all k's; since z is one of the elements $\sigma_k z$, we have $\nu_{\mathfrak{P}_1}(x) > 0$, whence $\nu_{\mathfrak{p}}(x) > 0$ and $\nu_{\mathfrak{P}_i}(x) > 0$ for $1 \leqq i \leqq h$. From this and from what we have seen above, it follows immediately that each $\mathfrak{P}_i$ must occur among the places $\sigma_k\mathfrak{P}_1$. Theorem 2 is thereby proved.

§3. INTEGRAL BASES

Let R and S be fields of algebraic functions of one variable, of which we assume that S contains R as a subfield and is of finite degree over R. Let $\mathfrak{p}$ be a place of R, and let $\mathfrak{P}_1, \cdots, \mathfrak{P}_h$ be the places of S which lie above $\mathfrak{p}$. We shall consider the following rings: 1) $\mathfrak{o}$, the ring of the place $\mathfrak{p}$; 2) $\mathfrak{r}$, the ring of elements of R which do not have any pole outside $\{\mathfrak{p}\}$; 3) $\mathfrak{O}$, the intersection of the rings of the places $\mathfrak{P}_1, \cdots, \mathfrak{P}_h$; $\mathfrak{O}$ is therefore the ring of elements of S which do not have any of the places $\mathfrak{P}_1, \cdots, \mathfrak{P}_h$ as a pole; 4) $\mathfrak{R}$, the ring of elements of S which do not have any pole outside the set $\{\mathfrak{P}_1, \cdots, \mathfrak{P}_h\}$. It is clear that $\mathfrak{o} \cap \mathfrak{r}$ is the field of constants of R, while $\mathfrak{O} \cap \mathfrak{R}$ is the field of constants of S. Moreover, we have $\mathfrak{O} \cap R = \mathfrak{o}$ and $\mathfrak{R} \cap R = \mathfrak{r}$.

Let $\{y_1, \cdots, y_n\}$ be a base of S with respect to R. Then any $y \in S$ can be expressed in the form $\sum_{i=1}^{n} x_i y_i$, $x_i \in R$. We propose to investigate the properties of the coefficients $x_1, \cdots, x_n$ when y is assumed to belong to either $\mathfrak{R}$ or $\mathfrak{O}$.

LEMMA 1. *There exists an element $u \neq 0$ in R, which depends only on $\mathfrak{p}$ and on the base $\{y_1, \cdots, y_n\}$, which has the following property: if $y = \sum_{i=1}^{n} x_i y_i$ (with $x_i \in R$, $1 \leqq i \leqq n$) is any element of $\mathfrak{R}$, then $ux_i \in \mathfrak{r}$ $(1 \leqq i \leqq n)$.*

Let K and L be the fields of constants of R and S respectively, and let x be an element of R which admits $\mathfrak{p}$ as its only zero (Lemma 1, §1). Then it follows from

Lemma 3, I, §8 that there exists a finite subset $\{w_1, \cdots, w_N\}$ of R such that $\mathfrak{R} = \sum_{\nu=1}^{N} L[x^{-1}]w_\nu$. Since S is of finite degree over R, $L\langle x\rangle$ is of finite degree over $K\langle x\rangle$ and L is of finite degree over K (Lemma 3, §1). If $\{\lambda_1, \cdots, \lambda_t\}$ is a base of L over K, every element of $L[x^{-1}]$ can be expressed as a linear combination of $\lambda_1, \cdots, \lambda_t$ with coefficients in $K[x^{-1}]$, whence $\mathfrak{R} = \sum_{\tau=1}^{t} \sum_{\nu=1}^{N} K[x^{-1}]\lambda_\tau w_\nu$. Set $\lambda_\tau w_\nu = \sum_{i=1}^{n} a_{\tau\nu i} y_i$, $a_{\tau\nu i} \in R$, and let $\mathfrak{q}_1, \cdots, \mathfrak{q}_r$ be all places of R which are poles of at least one of the coefficients $a_{\tau\nu i}$. Set $m_k = \min_{\tau,\nu,i} \nu_{\mathfrak{q}_k}(a_{\tau\nu i})$; the degree of the divisor $\mathfrak{p}^m \prod_{k=1}^{r} \mathfrak{q}_k^{m_k}$ increasing indefinitely with m, it follows from Riemann's theorem that, for m large enough, there exists an element $u \neq 0$ of R which is $\equiv 0 \pmod{\mathfrak{p}^{-m} \prod_{k=1}^{r} \mathfrak{q}_k^{-m_k}}$. We then have $ua_{\tau\nu i} \in \mathfrak{r}$ $(1 \leqq \tau \leqq t$, $1 \leqq \nu \leqq N$, $1 \leqq i \leqq n)$; since $K[x^{-1}] \subset \mathfrak{r}$, it follows immediately that u has the property required in Lemma 1.

LEMMA 2. *There exists an element $v \neq 0$ in R, which depends only on $\mathfrak{p}$ and on the base $\{y_1, \cdots, y_n\}$, which has the following property: if $y = \sum_{i=1}^{n} x_i y_i$ $(x_i \in R, 1 \leqq i \leqq n)$ is any element of $\mathfrak{O}$, then $vx_i \in \mathfrak{o}$ $(1 \leqq i \leqq n)$.*

Let $\mathfrak{p}^*$ be a place of R different from $\mathfrak{p}$; let $\mathfrak{r}^*$ and $\mathfrak{R}^*$ be the rings constructed with respect to $\mathfrak{p}^*$ in the same way as $\mathfrak{r}$ and $\mathfrak{R}$ have been constructed with respect to $\mathfrak{p}$. Let y be an element of $\mathfrak{O}$; we shall prove the existence of a unit ξ in $\mathfrak{o}$ such that $\xi y \in \mathfrak{R}^*$. Let $\mathfrak{Q}_1, \cdots, \mathfrak{Q}_r$ be those poles of y which lie above places of R distinct from $\mathfrak{p}^*$ (if any), and let $\mathfrak{q}_\rho$ be the place of R below $\mathfrak{Q}_\rho$. We can find for each ρ an integer $m_\rho \geqq 0$ such that the conditions $\xi \in R$, $\nu_{\mathfrak{q}_\rho}(\xi) \geqq m_\rho$ imply $\nu_{\mathfrak{Q}_\rho}(\xi y) \geqq 0$. Let μ be an integer >0; if ξ is an element of R which is $\equiv 0 \pmod{\mathfrak{p}^{*-\mu} \prod_{\rho=1}^{r} \mathfrak{q}_\rho^{m_\rho}}$, then it is clear that $\xi y \in \mathfrak{R}^*$. Now, we shall take μ to be so large that $d(\mathfrak{p}^{*\mu} \prod_{\rho=1}^{r} \mathfrak{q}_\rho^{-m_\rho}) \geqq 2g - 2 + d(\mathfrak{p})$, where g is the genus of R. Then the dimension l_μ of the space Ξ of elements of R which are $\equiv 0 \pmod{\mathfrak{p}^{*-\mu} \prod_{\rho=1}^{r} \mathfrak{q}_\rho^{m_\rho}}$ is $d(\mathfrak{p}^{*\mu} \prod_{\rho=1}^{r} \mathfrak{q}_\rho^{-m_\rho}) - g + 1$, while the dimension l'_μ of the space Ξ' of elements of R which are $\equiv 0 \pmod{\mathfrak{p}\mathfrak{p}^{*-\mu} \prod_{\rho=1}^{r} \mathfrak{q}_\rho^{m_\rho}}$ is $l_\mu - d(\mathfrak{p})$. It follows that Ξ then contains an element ξ not contained in Ξ'. We have $\xi y \in R^*$; on the other hand, since $y \in \mathfrak{O}$, the places $\mathfrak{q}_\rho$ are $\neq \mathfrak{p}$, and, since $\xi \notin \Xi'$, we have $\nu_\mathfrak{p}(\xi) = 0$, and ξ is a unit in $\mathfrak{o}$.

Now we determine an element $u^* \neq 0$ of R which satisfies with respect to $\mathfrak{p}^*$ (and to the base $\{y_1, \cdots, y_n\}$) the same condition as the element u of Lemma 1 with respect to $\mathfrak{p}$. If $y = \sum_{i=1}^{n} x_i y_i \in \mathfrak{O}$ (with $x_i \in R$, $1 \leqq i \leqq n$), then we have $\xi u^* x_i \in \mathfrak{r}^*$ $(1 \leqq i \leqq n)$, and, *a fortiori*, $\xi u^* x_i \in \mathfrak{o}$. Since ξ is a unit in $\mathfrak{o}$, we see that $v = u^*$ satisfies the condition of Lemma 2.

Moreover, we see that, the base $\{y_1, \cdots, y_n\}$ and the place $\mathfrak{p}^*$ being kept fixed, the element v may be taken the same for all places $\mathfrak{p} \neq \mathfrak{p}^*$. Now, there are only a finite number of places $\mathfrak{p}$ of R which fail to satisfy simultaneously all the following conditions: 1) $\mathfrak{p} \neq \mathfrak{p}^*$; 2) $\nu_\mathfrak{P}(y_i) \geqq 0$ $(1 \leqq i \leqq n)$ for all places $\mathfrak{P}$ of S which lie above $\mathfrak{p}$; 3) $\nu_\mathfrak{p}(u^*) = 0$. Let us say that the base $\{y_1, \cdots, y_n\}$ is an *integral base* at the place $\mathfrak{p}$ if the following conditions are satisfied: 1) each y_i belongs to the intersection $\mathfrak{O}$ of the rings of the places of S which lie above $\mathfrak{p}$; 2) if $y = \sum_{i=1}^{n} x_i y_i$ $(x_i \in R, 1 \leqq i \leqq n)$ is any element of $\mathfrak{O}$, then $x_1, \cdots, x_n$ belong to the ring of the place $\mathfrak{p}$. Then we see that we have proved

LEMMA 3. *Any base of S with respect to R is an integral base at almost all places of R.*

Now we prove

THEOREM 3. *Let R and S be fields of algebraic functions of one variable such that S is an overfield of finite degree of R. If $\mathfrak{p}$ is any place of R, then there exists a base of S with respect to R which is integral at $\mathfrak{p}$.*

Denote by $\mathfrak{o}$ the ring of $\mathfrak{p}$, by $\mathfrak{O}$ the intersection of the rings of the places of S which lie above $\mathfrak{p}$, and by $\{y_1', \cdots, y_n'\}$ any base of S with respect to R. If $1 \leqq m \leqq n$, let $\mathfrak{M}_m$ be the set of elements of $\mathfrak{O}$ which are linear combinations of $y_1', \cdots, y_m'$ with coefficients in R. If $y \in \mathfrak{M}_m$, set $y = \sum_{i=1}^{m} u_i' y_i'$, $u_i' \in R$ $(1 \leqq i \leqq m)$; then it follows immediately from Lemma 2 that the numbers $\nu_{\mathfrak{p}}(u_m')$ for all $y \in \mathfrak{M}_m$ have a finite lower bound. If t is any element of R such that $\nu_{\mathfrak{p}}(t) > 0$, then $t^{\mu} y_m' \in \mathfrak{M}_m$ for μ large enough, which shows that there exist elements $y \in \mathfrak{M}_m$ for which $u_m' \neq 0$. Let $-q_m$ be the smallest of the numbers $\nu_{\mathfrak{p}}(u_m')$ for all $y \in \mathfrak{M}_m$, and let $y_m = \sum_{i=1}^{m} u_{i,m}' y_i'$ be an element of $\mathfrak{M}_m$ such that $\nu_{\mathfrak{p}}(u_{m,m}') = -q_m$. Since $u_{m,m}' \neq 0$, it is clear that the elements $y_1, \cdots, y_n$ form a base of S with respect to R. Since $y_m \in \mathfrak{M}_m$, $y_1, \cdots, y_n$ belong to $\mathfrak{O}$, and the same is true of the linear combinations of $y_1, \cdots, y_n$ with coefficients in $\mathfrak{o}$. Let conversely $y = \sum_{i=1}^{n} u_i y_i$ be any element of $\mathfrak{O}$, with $u_i \in R$ $(1 \leqq i \leqq n)$. We shall see that $u_i \in \mathfrak{o}$ $(1 \leqq i \leqq n)$. In fact, assume for a moment that $u_i \in \mathfrak{o}$ for $i > m$, but that $u_m \notin \mathfrak{o}$. Set $y' = y - \sum_{i>m} u_i y_i$; then y' belongs to $\mathfrak{O}$, and therefore also to $\mathfrak{M}_m$, since $y_1, \cdots, y_m$ are linear combinations of $y_1', \cdots, y_m'$ with coefficients in R. If we set $y' = \sum_{i=1}^{m} u_i' y_i'$, we have $u_m' = u_m u_{m,m}'$, whence $\nu_{\mathfrak{p}}(u_m') = -q_m + \nu_{\mathfrak{p}}(u_m) < -q_m$, which is impossible. Theorem 3 is thereby proved.

Let $\{y_1, \cdots, y_n\}$ be a base of S with respect to R which is integral at a place $\mathfrak{p}$ of R. Let $\mathfrak{O}$ be the intersection of the rings of the places of S which lie above $\mathfrak{p}$, and let y be any element of $\mathfrak{O}$. Since $yy_i \in \mathfrak{O}$, we have $yy_i = \sum_{j=1}^{n} a_{ij} y_j$, where the elements a_{ij} belong to the ring $\mathfrak{o}$ of $\mathfrak{p}$. The characteristic polynomial $F(X)$ of y with respect to R is the determinant of the matrix $XE - (a_{ij})$, where E is the unit matrix of degree n; the coefficients of $F(X)$ belong therefore to $\mathfrak{o}$. In particular, we see that the trace and the norm of y, taken from S to R, belong to $\mathfrak{o}$, and that y is integral with respect to $\mathfrak{o}$. Conversely, any element of S which is integral with respect to $\mathfrak{o}$ belongs to $\mathfrak{O}$, as follows from the fact that the ring of any place is integrally closed (I, §2). Thus we have proved

LEMMA 4. *Let $\mathfrak{p}$ be a place of a field R of algebraic functions of one variable, and let S be an overfield of R which is of finite degree over R. Then the ring $\mathfrak{O}$, intersection of the rings of the places of S which lie above $\mathfrak{p}$, is also the ring of elements of S which are integral over the ring $\mathfrak{o}$ of $\mathfrak{p}$. If $y \in \mathfrak{O}$, the coefficients of the characteristic polynomial of y with respect to R belong to $\mathfrak{o}$.*

Now we prove

LEMMA 5. *Let the notation be as in Lemma 4, and let* $\{y_1, \cdots, y_n\}$ *be a base of* S *with respect to* R *which is integral at* $\mathfrak{p}$. *Let* $\mathfrak{P}$ *be a place of* S *which lies above* $\mathfrak{p}$, *and let* y_i^* *be the residue class of* y_i *modulo* $\mathfrak{P}$ $(1 \leqq i \leqq n)$. *Then every element of the residue field of* $\mathfrak{P}$ *can be written as a linear combination of* $y_1^*, \cdots, y_n^*$ *with coefficients in the residue field of* $\mathfrak{p}$.

Let η^* be an element of the residue field of $\mathfrak{P}$, and let η_0 be an element of S belonging to the class η^*. Let $\mathfrak{P}_1 = \mathfrak{P}, \mathfrak{P}_2, \cdots, \mathfrak{P}_h$ be all the distinct places of S lying above $\mathfrak{p}$. Making use of Theorem 3, I, §6, we see that there exists an $\eta \in S$ such that $\nu_{\mathfrak{P}}(\eta - \eta_0) > 0$, $\nu_{\mathfrak{P}_i}(\eta) \geqq 0$ $(2 \leqq i \leqq h)$. Then η belongs to $\mathfrak{O}$, whence $\eta = \sum_{i=1}^n x_i y_i$ with $x_i \in \mathfrak{o}$ $(1 \leqq i \leqq n)$. If x_i^* is the value taken by x_i at $\mathfrak{p}$, we have $\eta^* = \sum_{i=1}^n x_i^* y_i^*$.

§4. Kronecker products of commutative algebras

Let $\mathfrak{L}$ and $\mathfrak{M}$ be commutative algebras (of finite or infinite dimensions) over a field K. We shall associate with $\mathfrak{L}$ and $\mathfrak{M}$ a commutative algebra $\mathfrak{L} \otimes \mathfrak{M}$ over K which will have the following properties:

1) to every pair (x, y) formed of an $x \in \mathfrak{L}$ and a $y \in \mathfrak{M}$ there is associated an element $xy \in \mathfrak{L} \otimes \mathfrak{M}$, and we have $(x + x')y = xy + x'y$, $x(y + y') = xy + xy'$, $(xx')(yy') = (xy)(x'y')$, $(\alpha x)y = x(\alpha y) = \alpha(xy)$ (where x, x' are any elements of $\mathfrak{L}$, y, y' are any elements of $\mathfrak{M}$, and α is any element of K);

2) every element of $\mathfrak{L} \otimes \mathfrak{M}$ can be written as a finite sum of elements of the form xy, $x \in \mathfrak{L}$, $y \in \mathfrak{M}$;

3) if $x_1, \cdots, x_m$ are elements of $\mathfrak{L}$ which are linearly independent over K, and $y_1, \cdots, y_n$ elements of $\mathfrak{M}$ which are linearly independent over K, then the elements $x_i y_j$ $(1 \leqq i \leqq m, 1 \leqq j \leqq n)$ are linearly independent over K.

To prove that there exists an algebra with these properties, we select bases $(\xi_\alpha)_{\alpha \in A}$ and $(\eta_\beta)_{\beta \in B}$ of $\mathfrak{L}$ and $\mathfrak{M}$ respectively (A and B being sets of indices which may be finite or infinite). We introduce a set of symbols $\zeta_{\alpha,\beta}$ indexed by means of the pairs $(\alpha, \beta) \in A \times B$, and we form the vector space of all formal linear combinations $\Sigma_{\alpha,\beta} a_{\alpha\beta} \zeta_{\alpha\beta}$ of these symbols with coefficients $a_{\alpha\beta} \in K$ (it being understood that, in any such linear combination, there are only a finite number of pairs (α, β) for which $a_{\alpha\beta} \neq 0$). If $x = \Sigma_{\alpha \in A} c_\alpha \xi_\alpha \in \mathfrak{L}$ and $y = \Sigma_{\beta \in B} d_\beta \eta_\beta \in \mathfrak{M}$, we define xy to be $\Sigma_{\alpha,\beta} c_\alpha d_\beta \zeta_{\alpha\beta}$; then the conditions stated under 1) and 2) are satisfied, except the condition $(xx')(yy') = (xy)(x'y')$, which involves the multiplication in $\mathfrak{L} \otimes \mathfrak{M}$, which has not been defined yet. Let us now check that 3) is satisfied. We can express $x_1, \cdots, x_m$ as linear combinations of a finite number of the ξ_α's, say of $\xi_{\alpha_1}, \cdots, \xi_{\alpha_p}$. We have $p \geqq m$ and, by adjoining to $x_1, \cdots, x_m$ $p - m$ suitably selected elements ξ_{α_i}, we obtain p elements $x_1, \cdots, x_p$ which span the same subspace of $\mathfrak{L}$ as $\xi_{\alpha_1}, \cdots, \xi_{\alpha_p}$. Similarly, we can adjoin to $y_1, \cdots, y_n$ a certain number of elements $y_{n+1}, \cdots, y_q$ of the base $(\eta_\beta)_{\beta \in B}$ such that $y_1, \cdots, y_q$ form a base of a subspace of $\mathfrak{M}$ which has also a base composed of q elements $\eta_{\beta_1}, \cdots, \eta_{\beta_q}$ of the base $(\eta_\beta)_{\beta \in B}$. The elements $x_i y_j$ $(1 \leqq i \leqq p, 1 \leqq j \leqq q)$ are linear combinations of the elements

$\zeta_{\alpha_i\beta_j}$ $(1 \leqq i \leqq p, 1 \leqq j \leqq q)$ and conversely, from which it follows that these elements are linearly independent over K.

To define a bilinear multiplication in our vector space, it is sufficient to define the products of the basic elements $\zeta_{\alpha\beta}$. We set $\zeta_{\alpha\beta}\zeta_{\alpha'\beta'} = (\xi_\alpha\xi_{\alpha'})(\eta_\beta\eta_{\beta'})$; it is easy to see that the multiplication defined in this way is associative and commutative, and that it defines a structure of algebra on our vector space. The formula $(xx')(yy') = (xy)(x'y')$ holds whenever x and x' are among the ξ_α's and y and y' among the η_β's; by making use of the linearity, it is easy to check that our formula is generally valid.

Thus we have proved the existence of an algebra $\mathfrak{L} \otimes \mathfrak{M}$ with the required properties. Furthermore, the structure of $\mathfrak{L} \otimes \mathfrak{M}$ is uniquely determined by the conditions 1), 2), and 3). In fact, using the bases $(\xi_\alpha)_{\alpha\epsilon A}$ and $(\eta_\beta)_{\beta\epsilon B}$, it follows from 1) and 2) that every element of $\mathfrak{L} \otimes \mathfrak{M}$ is a linear combination of the products $\xi_\alpha\eta_\beta$; making use of 3), we see that these products form a base of $\mathfrak{L} \otimes \mathfrak{M}$. In virtue of 1), we see that $(\xi_\alpha\eta_\beta)(\xi_{\alpha'}\eta_{\beta'}) = (\xi_\alpha\xi_{\alpha'})(\eta_\beta\eta_{\beta'})$, which shows that the multiplication in $\mathfrak{L} \otimes \mathfrak{M}$ is uniquely determined.

The algebra $\mathfrak{L} \otimes \mathfrak{M}$ is called the *Kronecker product* of $\mathfrak{L}$ and $\mathfrak{M}$. It is clear that, if $\mathfrak{L}'$ and $\mathfrak{M}'$ are subalgebras of $\mathfrak{L}$ and $\mathfrak{M}$ respectively, then $\mathfrak{L}' \otimes \mathfrak{M}'$ is a subalgebra of $\mathfrak{L} \otimes \mathfrak{M}$.

From now on, we shall assume that $\mathfrak{L}$ and $\mathfrak{M}$ have unit elements $1_\mathfrak{L}$ and $1_\mathfrak{M}$ respectively. Then it is clear that $1_\mathfrak{L}1_\mathfrak{M}$ is the unit element of $\mathfrak{L} \otimes \mathfrak{M}$. Moreover the elements $x1_\mathfrak{M}$ $(x \epsilon \mathfrak{L})$ form a subalgebra of $\mathfrak{L} \otimes \mathfrak{M}$ isomorphic with $\mathfrak{L}$; we shall identify this subalgebra with $\mathfrak{L}$ itself. Similarly, we shall treat $\mathfrak{M}$ as a subalgebra of $\mathfrak{L} \otimes \mathfrak{M}$.

If $\mathfrak{L} = L$ is a field, then the fact that L is a subalgebra of $L \otimes \mathfrak{M}$ (and that the unit element of L is also the unit element of $L \otimes \mathfrak{M}$) shows that we may consider the elements of $L \otimes \mathfrak{M}$ as forming an algebra over L. This algebra will be denoted by $\mathfrak{M}_L$, and we shall say that it is the algebra deduced from $\mathfrak{M}$ by extending the basic field from K to L. Any base $(\eta_\beta)_{\beta\epsilon B}$ of $\mathfrak{M}$ with respect to K is also a base of $\mathfrak{M}_L$ with respect to L. For, first it is clear that every element of M_L can be written as a linear combination of the elements η_β with coefficients in L. On the other hand, assume that $\Sigma_{\beta\epsilon B}x_\beta\eta_\beta = 0$, the x_β's being elements of L of which only a finite number are $\neq 0$. If $(\xi_\alpha)_{\alpha\epsilon A}$ is a base of L with respect to K, we may write each x_β in the form $\Sigma_{\alpha\epsilon A}c_{\alpha\beta}\xi_\alpha$, $c_{\alpha\beta} \epsilon K$, and we have $\Sigma_{\alpha\beta}c_{\alpha\beta}\xi_\alpha\eta_\beta = 0$, whence $c_{\alpha\beta} = 0$ for all α's and β's, and therefore $x_\beta = 0$ for every β.

If L' is a subfield of L containing K, $\mathfrak{M}_{L'}$ is an algebra over L', and it follows immediately from what was said above that $\mathfrak{M}_L = (\mathfrak{M}_{L'})_L$ is the algebra deduced from $\mathfrak{M}_{L'}$ by extending the basic field from L' to L.

We shall be mainly interested in the case where $\mathfrak{L} = L$ and $\mathfrak{M} = M$ are both fields. It may then happen that L and M are both subfields of some field Ω. In that case, there arises a difficulty in the notation, because expressions such as $x + y$ or xy (where $x \epsilon L$, $y \epsilon M$) do not have the same meaning according as to whether we consider the operations of addition and multiplication as performed in $L \otimes M$ or in Ω. Whenever confusions are possible, we shall endeavour to clarify the situation by giving supplementary indications in the context.

LEMMA 1. *Let L and M be overfields of a field K. Assume that we are given isomorphisms λ and μ of L and M respectively with subfields of an overfield Ω of K, and that each of λ and μ maps every element of K upon itself. Then there exists a homomorphism φ of $L \otimes M$ into Ω which coincides with λ on L and with μ on M; $\varphi(L \otimes M)$ is the subring of Ω generated by $\lambda(L)$ and $\mu(M)$.*

Let $(\xi_\alpha)_{\alpha \in A}$ and $(\eta_\beta)_{\beta \in B}$ be bases of L and M respectively with respect to K. Then the products $\xi_\alpha \eta_\beta$ form a base of $L \otimes M$. We can define a linear mapping φ of $L \otimes M$ into Ω which assigns to $\xi_\alpha \eta_\beta$ the element $\lambda(\xi_\alpha)\mu(\eta_\beta)$ of Ω. We may assume that the unit element of L is one of the elements ξ_α, say ξ_{α_0}, and that the unit element of M is one of the elements η_β, say η_{β_0}. Since $\xi_\alpha = \xi_\alpha \eta_{\beta_0}$, $\eta_\beta = \xi_{\alpha_0}\eta_\beta$, we see that φ coincides with λ on L and with μ on M. It follows immediately from our definition of the multiplication in $L \otimes M$ that φ is a homomorphism of the structure of algebra of $L \otimes M$ into that of Ω. The image $\varphi(L \otimes M)$ is therefore a subring of Ω; it is obviously the ring generated by $\lambda(L)$ and $\mu(M)$.

§5. EXTENSION OF THE $\mathfrak{p}$-ADIC COMPLETION

Let R and S be fields of algebraic functions of one variable such that S is an overfield of R (with the usual assumptions on the fields of constants; cf. §1). Denote by $\mathfrak{p}$ a place of R and by $\mathfrak{P}$ a place of S which lies above $\mathfrak{p}$. Let $\bar{R}$ be the $\mathfrak{p}$-adic completion of R and $\bar{S}$ the $\mathfrak{P}$-adic completion of S. If an element $\bar{u} \in \bar{R}$ is the limit of a sequence (u_n) of elements of R, then the sequence (u_n), considered as a sequence of elements of S, converges to an element of $\bar{S}$. For, if e is the ramification index of $\mathfrak{P}$ with respect to R, the number $\nu_\mathfrak{P}(u_{n+1} - u_n) = e\nu_\mathfrak{p}(u_{n+1} - u_n)$ increases indefinitely with n. Moreover, if $\bar{u}$ is the limit of another sequence (u_n') of elements of R, then $\nu_\mathfrak{P}(u_n' - u_n) = e\nu_\mathfrak{p}(u_n' - u_n)$ increases indefinitely with n, which proves that the sequences (u_n) and (u_n') have the same limit in $\bar{S}$. If we assign this limit to $\bar{u}$, we obtain a mapping of $\bar{R}$ into $\bar{S}$ which maps every element of R upon itself. We see immediately that this mapping is a ring homomorphism. Since $\bar{R}$ is a field, our mapping is actually an isomorphism. We shall often identify $\bar{R}$ with its image under the isomorphism which we have just defined; i.e. we shall treat $\bar{R}$ as a subfield of $\bar{S}$.

From now on, we shall assume that S is of finite degree over R. We shall denote by $\mathfrak{P}_1, \cdots, \mathfrak{P}_h$ the distinct places of S which lie above $\mathfrak{p}$, and by $\bar{S}_\lambda$ the $\mathfrak{P}_\lambda$-adic completion of S ($1 \leqq \lambda \leqq h$). We construct the product $\prod_{\lambda=1}^{h} \bar{S}_\lambda$; this is an algebra over $\bar{R}$ whose elements are the h-uples $(y_1, \cdots, y_h)$ with $y_\lambda \in \bar{S}_\lambda$ ($1 \leqq \lambda \leqq h$) and in which the operations are defined by the formulas

$$(y_1, \cdots, y_h) + (z_1, \cdots, z_h) = (y_1 + z_1, \cdots, y_h + z_h),$$

$$(y_1, \cdots, y_h)(z_1, \cdots, z_h) = (y_1 z_1, \cdots, y_h z_h),$$

$$x(y_1, \cdots, y_h) = (xy_1, \cdots, xy_h)$$

(where $y_\lambda, z_\lambda \in \bar{S}_\lambda$, $1 \leqq \lambda \leqq h$, and $x \in \bar{R}$).

THEOREM 4. *Let R and S be fields of algebraic functions of one variable such that S is an overfield of finite degree of R. Let $\mathfrak{p}$ be a place of R, $\bar{R}$ the $\mathfrak{p}$-adic completion of R, and $\mathfrak{P}_1, \cdots, \mathfrak{P}_h$ the distinct places of S which lie above $\mathfrak{p}$. Then the product of the $\mathfrak{P}_\lambda$-adic completions of S $(1 \leqq \lambda \leqq h)$ is an algebra over $\bar{R}$ which is isomorphic with the algebra $S_{\bar{R}}$ deduced from S (considered as an algebra over R) by extending the basic field from R to $\bar{R}$.*

We shall construct a homomorphism Φ of $S_{\bar{R}}$ into $\prod_{\lambda=1}^h \bar{S}_\lambda$ and a mapping Ψ of $\prod_{\lambda=1}^h \bar{S}_\lambda$ into $S_{\bar{R}}$, and we shall prove that $\Phi\Psi$ is the identity mapping of $\prod_{\lambda=1}^h \bar{S}_\lambda$ into itself, while $\Psi\Phi$ is the identity mapping of $S_{\bar{R}}$ into itself.

Each $\bar{S}_\lambda$ contains the fields S and R; it follows from Lemma 1, §4 that there exists a homomorphism Φ_λ of $S_{\bar{R}}$ into $\bar{S}_\lambda$ which maps every element of either S or R upon itself. If we set $\Phi(u) = (\Phi_1(u), \cdots, \Phi_h(u))$ (for $u \in S_{\bar{R}}$), we clearly obtain a homomorphism Φ of $S_{\bar{R}}$ into $\prod_{\lambda=1}^h \bar{S}_\lambda$.

Let $(y_1, \cdots, y_h)$ be any element of $\prod_{\lambda=1}^h \bar{S}_\lambda$. If m is any integer > 0, we can find for each λ $(1 \leqq \lambda \leqq h)$ an element $y'_{m,\lambda} \in S$ such that

$$\nu_{\mathfrak{P}_\lambda}(y_\lambda - y'_{m,\lambda}) \geqq m e_\lambda$$

(where e_λ is the ramification index of $\mathfrak{P}_\lambda$ with respect to R). By Theorem 3, I, §6, we can find a $y'_m \in S$ such that $\nu_{\mathfrak{P}_\lambda}(y'_m - y'_{m,\lambda}) \geqq m e_\lambda$ $(1 \leqq \lambda \leqq h)$. It follows that

$$(1) \qquad \nu_{\mathfrak{P}_\lambda}(y_\lambda - y'_m) \geqq m e_\lambda \qquad (1 \leqq \lambda \leqq h),$$

whence

$$(2) \qquad \nu_{\mathfrak{P}_\lambda}(y'_{m+1} - y'_m) \geqq m e_\lambda \qquad (1 \leqq \lambda \leqq h).$$

Now we introduce a base $\{z_1, \cdots, z_n\}$ of S with respect to R which we assume to be integral at $\mathfrak{p}$. Let t be a uniformizing variable at $\mathfrak{p}$ in R. Write

$$y'_m = \sum_{i=1}^n x_{im} z_i, \qquad x_{im} \in R;$$

it follows from (2) that $\nu_{\mathfrak{P}_\lambda}(t^{-m}(y'_{m+1} - y'_m)) \geqq 0$, whence, since the base

$$\{z_1, \cdots, z_n\}$$

is integral at $\mathfrak{p}$, $\nu_{\mathfrak{p}}(t^{-m}(x_{i,m+1} - x_{im})) \geqq 0$ and therefore $\nu_{\mathfrak{p}}(x_{i,m+1} - x_{im}) \geqq m$. We conclude that each of the sequences $(x_{im})_{1 \leqq m < \infty}$ converges to an x_i in $\bar{R}$. Moreover, these elements x_i do not depend on the way in which we select the approximating elements y'_m, provided the conditions (2) are satisfied. For, if (y^*_m) is another sequence of elements of S such that $\nu_{\mathfrak{P}_\lambda}(y_\lambda - y^*_m) \geqq m e_\lambda$ $(1 \leqq \lambda \leqq h)$, then we have $\nu_{\mathfrak{P}_\lambda}(y^*_m - y'_m) \geqq m e_\lambda$, and, if we set $y^*_m = \sum_{i=1}^n x^*_{im} z_i$, then the same argument as above shows that $\nu_{\mathfrak{p}}(x^*_{im} - x_{im}) \geqq m$, whence $\lim_{m \to \infty} x^*_{im} = x_i$. Thus we can define a mapping Ψ of $\prod_{\lambda=1}^h \bar{S}_\lambda$ into $S_{\bar{R}}$ by the formula $\Psi(y_1, \cdots, y_h) = \sum_{i=1}^n x_i z_i$.

The image of an element $\sum_{i=1}^n x_i z_i$ of $S_{\bar{R}}$ $(x_i \in \bar{R}, 1 \leqq i \leqq n)$ by the homomorphism Φ_λ of $S_{\bar{R}}$ into $\bar{S}_\lambda$ is the element of $\bar{S}_\lambda$ which is represented by the same expression $\sum_{i=1}^n x_iz_i$, but in which the operations of addition and multiplication are now assumed to be performed in $\bar{S}_\lambda$.

Since $\nu_{\mathfrak{P}_\lambda}(x_i - x_{im}) \geqq me_\lambda$, $\nu_{\mathfrak{P}_\lambda}(z_i) \geqq 0$, we have $\nu_{\mathfrak{P}_\lambda}(\Phi_\lambda\Psi(y_1, \cdots, y_h) - y'_m) \geqq me_\lambda$. Comparing with (1), we see that $\Phi_\lambda\Psi(y_1, \cdots, y_h) = y_\lambda$ $(1 \leqq \lambda \leqq h)$, which means that the mapping $\Phi\Psi$ is the identity mapping of $\prod_{\lambda=1}^h \bar{S}_\lambda$ onto itself.

Let $x_1, \cdots, x_n$ be any elements of $\bar{R}$. For each integer $m > 0$, let

$$x_{1m}, \cdots, x_{nm}$$

be elements of R such that $\nu_\mathfrak{p}(x_i - x_{im}) \geqq m$ $(1 \leqq i \leqq n)$, and set

$$y'_m = \sum_{i=1}^n x_{im}z_i .$$

Then, for each λ, we have $\nu_{\mathfrak{P}_\lambda}(y'_{m+1} - y'_m) \geqq me_\lambda$, which shows that the sequence (y'_m) converges in $\bar{S}_\lambda$ to some y_λ, and that $y_\lambda = \sum_{i=1}^n x_iz_i$, the operations being performed in $\bar{S}_\lambda$. If $u = \sum_{i=1}^n x_iz_i$ (operations performed in $S_{\bar{R}}$), then $\Phi(u) = (y_1, \cdots, y_h)$, and $\Psi\Phi(u) = u$. This proves that $\Psi\Phi$ is the identity mapping of $S_{\bar{R}}$ onto itself.

It follows that Φ is a one to one mapping of $S_{\bar{R}}$ onto $\prod_{\lambda=1}^h \bar{S}_\lambda$. Since we know already that Φ is a homomorphism, we see that Φ is actually an isomorphism. Theorem 4 is thereby proved.

REMARK. Call *integral* an element $(y_1, \cdots, y_h)$ of $\prod_{\lambda=1}^h \bar{S}_\lambda$ when, for each λ, y_λ is integral in $\bar{S}_\lambda$. Using the notation of the proof of Theorem 4, we see that the elements y'_m of S are then integral at each place $\mathfrak{P}_\lambda$; the coefficients x_{im} are therefore integral at $\mathfrak{p}$, and so are their limits x_i. Thus we see that *the isomorphism which we have established between $S_{\bar{R}}$ and $\prod_{\lambda=1}^h \bar{S}_\lambda$ has the property that the integral elements of $\prod_{\lambda=1}^h \bar{S}_\lambda$ are those which correspond to the linear combinations of the elements of a base of S with respect to R which is integral at $\mathfrak{p}$ with coefficients which are integral in $\bar{R}$.*

It follows in particular from Theorem 4 that each $\bar{S}_\lambda$ is of finite degree n_λ over $\bar{R}$, and that $\sum_{\lambda=1}^h n_\lambda = n = [S:R]$. Let e_λ and f_λ be the ramification index and the relative degree of $\mathfrak{P}_\lambda$ with respect to R. Then we know (Theorem 1, §1) that $\sum_{\lambda=1}^h e_\lambda f_\lambda = n$. We shall prove that n_λ is in fact equal to $e_\lambda f_\lambda$.

THEOREM 5. *Let R and S be fields of algebraic functions of one variable; assume that S is an overfield of finite degree of R. Denote by $\mathfrak{p}$ a place of R, by $\mathfrak{P}$ a place of S which lies above $\mathfrak{p}$, by $\bar{R}$ the $\mathfrak{p}$-adic completion of R, by $\bar{S}$ the $\mathfrak{P}$-adic completion of S, by e and f respectively the ramification index and the relative degree of $\mathfrak{P}$ with respect to R, by $\Sigma(\mathfrak{p})$ and $\Sigma(\mathfrak{P})$ the residue fields of $\mathfrak{p}$ and $\mathfrak{P}$ respectively. Let $v_1, \cdots, v_f$ be integral elements of $\bar{S}$ whose residue classes modulo $\mathfrak{P}$ form a base of $\Sigma(\mathfrak{P})$ with respect to $\Sigma(\mathfrak{p})$, and let u be an element of $\bar{S}$ such that $\nu_\mathfrak{P}(u) = 1$. Then the ef elements u^iv_j $(0 \leqq i \leqq e - 1, 1 \leqq j \leqq f)$ form a base of $\bar{S}$ with respect to $\bar{R}$, and every integral element of $\bar{S}$ can be written in the form $\sum_{i=0}^{e-1}\sum_{j=1}^f x_{ij}u^iv_j$ with coefficients x_{ij} which are integral elements of $\bar{R}$.*

We shall prove that, if the x_{ij}'s are elements of $\bar{R}$ such that

$$\nu_{\mathfrak{P}}\left(\sum_{i=0}^{e-1}\sum_{j=1}^{f} x_{ij}u^i v_j\right) \geqq em,$$

where m is an integer $\geqq 0$, then $\nu_{\mathfrak{p}}(x_{ij}) \geqq m$ for all i and j. Let m_0 be the smallest of the numbers $\nu_{\mathfrak{p}}(x_{ij})$; we may assume that $m_0 = \nu_{\mathfrak{p}}(x_{i_0j_0})$ and that $\nu_{\mathfrak{p}}(x_{ij}) > m_0$ if $i < i_0$, $1 \leqq j \leqq f$. Let t be a uniformizing variable at $\mathfrak{p}$ in R. Then the elements $t^{-m_0}x_{i_0j}$ $(1 \leqq j \leqq f)$ are integral in $\bar{R}$. Let ξ_j be the residue class of $t^{-m_0}x_{i_0j}$ modulo $\mathfrak{p}$; since the residue classes $v_1^*, \cdots, v_f^*$ of $v_1, \cdots, v_f$ form a base of $\Sigma(\mathfrak{P})$ with respect to $\Sigma(\mathfrak{p})$, and $\xi_{j_0} \neq 0$, we have $\sum_{j=1}^{f} \xi_j v_j^* \neq 0$, whence $\nu_{\mathfrak{P}}(\sum_{j=1}^{f} t^{-m_0}x_{i_0j}v_j) = 0$ and

$$\nu_{\mathfrak{P}}\left(\sum_{j=1}^{f} x_{i_0j}u^{i_0}v_j\right) = i_0 + m_0e. \tag{3}$$

If $i < i_0$, we have

$$\nu_{\mathfrak{P}}(x_{ij}u^iv_j) = i + e\nu_{\mathfrak{p}}(x_{ij}) \geqq i + e(m_0 + 1) > i_0 + m_0e \tag{4}$$

because $i + e \geqq e > i_0$. If $i > i_0$, we have

$$\nu_{\mathfrak{p}}(x_{ij}u^iv_j) \geqq i + m_0e > i_0 + m_0e. \tag{5}$$

Comparing (3), (4), and (5), we obtain $\nu_{\mathfrak{P}}(\sum_{i=0}^{e-1}\sum_{j=1}^{f} x_{ij}u^iv_j) = i_0 + m_0e \geqq me$. Since $i_0 < e$, it follows that $m_0 \geqq m$.

In particular, if $\sum_{i=0}^{e-1}\sum_{j=1}^{f} x_{ij}u^iv_j = 0$, we must have $x_{ij} = 0$ $(0 \leqq i \leqq e - 1$, $1 \leqq j \leqq f)$. This proves that the ef elements u^iv_j are linearly independent over $\bar{R}$, whence $[\bar{S}:\bar{R}] \geqq ef$. Now let $\mathfrak{P}_1 = \mathfrak{P}, \cdots, \mathfrak{P}_h$ be all the distinct places of S which lie above $\mathfrak{p}$, and let e_λ and f_λ be respectively the ramification index and the relative degree of $\mathfrak{P}_\lambda$ with respect to R. If $\bar{S}_\lambda$ is the $\mathfrak{P}_\lambda$-adic completion of S, we have just proved that $[\bar{S}_\lambda:\bar{R}] \geqq e_\lambda f_\lambda$. On the other hand, we know from Theorem 4 that $\sum_{\lambda=1}^{h} [\bar{S}_\lambda:\bar{R}] = [S:R] = \sum_{\lambda=1}^{h} e_\lambda f_\lambda$ (Theorem 1, §1); it follows that $[\bar{S}_\lambda:\bar{R}] = e_\lambda f_\lambda$ for $1 \leqq \lambda \leqq h$, and, in particular, $[\bar{S}:\bar{R}] = ef$. The ef elements u^iv_j, which are linearly independent over $\bar{R}$, form a base of $\bar{S}$ with respect to $\bar{R}$. Applying the result which was established at the beginning of this proof to the case where $m = 0$, we see that our base has the property stated at the end of Theorem 5.

COROLLARY 1. *The notation being as in Theorem 5, the characteristic polynomial with respect to $\bar{R}$ of an integral element of $\bar{S}$ has its coefficients in the ring of integral elements of $\bar{R}$.*

Let y be an integral element of $\bar{S}$, and let $z_1, \cdots, z_{\bar{n}}$ be the elements u^iv_j arranged in some order (with $\bar{n} = ef$). Then we have $yz_i = \sum_{j=1}^{\bar{n}} \eta_{ij}z_j$ with elements η_{ij} which are integral in $\bar{R}$. The characteristic polynomial $F(X)$ of y with respect to $\bar{R}$ being the determinant of the matrix $XE - (\eta_{ij})$, where E is the unit matrix of degree n, the corollary 1 is established.

COROLLARY 2. *If y is an element $\neq 0$ of $\bar{S}$, then we have $\nu_{\mathfrak{p}}(\mathrm{N}_{\bar{S}/\bar{R}}y) = f\nu_{\mathfrak{P}}(y)$.*

It follows from Corollary 1 that the norm (from $\bar{S}$ to $\bar{R}$) of an integral element of $\bar{S}$ is integral in $\bar{R}$, from which we conclude that the norm of any unit in the ring of integral elements of $\bar{S}$ is a unit in the ring of integral elements of $\bar{R}$. Now, we may write $y = wu^m$, where w is a unit in the ring of integral elements of $\bar{S}$ and $m = \nu_{\mathfrak{P}}(y)$. It follows that $\nu_{\mathfrak{p}}(\mathrm{N}_{\bar{S}/\bar{R}}y) = a\nu_{\mathfrak{P}}(y)$, where $a = \nu_{\mathfrak{p}}(\mathrm{N}_{\bar{S}/\bar{R}}u)$ does not depend on y. Take for y an element t of $\bar{R}$ such that $\nu_{\mathfrak{p}}(t) = 1$; then $\mathrm{N}_{\bar{S}/\bar{R}}t = t^{ef}$, whence $ef = a\nu_{\mathfrak{P}}(t) = ae$, and therefore $a = f$.

COROLLARY 3. *The notation being as in Theorem 4, we have $\bar{S} = \bar{R}\langle S\rangle$.*

This follows immediately from the fact that the elements $u, v_1, \cdots, v_f$ of $\bar{S}$ may be taken in S.

This being said, let us return to the notation of Theorem 4. Let y be any element of $S_{\bar{R}}$; then the mapping $z \to yz$ is an endomorphism of the structure of vector space of $S_{\bar{R}}$ over $\bar{R}$. Let us for a moment identify the algebra $S_{\bar{R}}$ with the algebra $\prod_{\lambda=1}^{h} \bar{S}_\lambda$ by means of the isomorphism which was constructed in the proof of Theorem 4. For each λ, let $\{z_{1\lambda}, \cdots, z_{n_\lambda\lambda}\}$ be a base of $\bar{S}_\lambda$ with respect to $\bar{R}$; denote by $z'_{i\lambda}$ the element of $S_{\bar{R}}$ which, considered as element of $\prod_{\lambda=1}^{h} \bar{S}_\lambda$, has its λth coordinate equal to $z_{i\lambda}$ and its other coordinates equal to 0. Then all the elements $z'_{i\lambda}$ $(1 \leqq i \leqq n_\lambda, 1 \leqq \lambda \leqq h)$ taken together form a base of $S_{\bar{R}}$ with respect to $\bar{R}$. If we denote by y_λ the λth coordinate of y, considered as an element of $\prod_{\lambda=1}^{h} \bar{S}_\lambda$, then we have $y_\lambda z_{i\lambda} = \sum_{j=1}^{n_\lambda} x_{ij\lambda} z_{j\lambda}$ with $x_{ij\lambda} \in \bar{R}$, whence $yz'_{i\lambda} = \sum_{j=1}^{n_\lambda} x_{ij\lambda} z'_{j\lambda}$. Denote by Ξ_λ the matrix $(x_{ij\lambda})_{1 \leqq i,j \leqq n_\lambda}$; then we see that the characteristic polynomial of the endomorphism $z \to yz$ is the product of the characteristic polynomials of the matrices Ξ_λ, and that the characteristic polynomial of Ξ_λ is the characteristic polynomial of y_λ with respect to $\bar{R}$. In particular, if $y \in S$, the elements y_λ are all equal to y, and the characteristic polynomial of the endomorphism $z \to yz$ of $S_{\bar{R}}$ is clearly the characteristic polynomial of y with respect to R. Thus we obtain

THEOREM 6. *Let R and S be fields of algebraic functions of one variable, S being an overfield of finite degree of R. Let $\mathfrak{p}$ be a place of R; denote by $\mathfrak{P}_1, \cdots, \mathfrak{P}_h$ the distinct places of S which lie above $\mathfrak{p}$, by $\bar{R}$ the $\mathfrak{p}$-adic completion of R, and by $\bar{S}_\lambda$ the $\mathfrak{P}_\lambda$-adic completion of S $(1 \leqq \lambda \leqq h)$. If $y \in S$, then the characteristic polynomial of y with respect to R is the product of the characteristic polynomials with respect to $\bar{R}$ of y, considered successively as element of $\bar{S}_1, \cdots, \bar{S}_h$.*

If $\mathfrak{P}$ is any one of the places $\mathfrak{P}_\lambda$, and $\bar{S}$ the corresponding completion of S, we shall denote (for $y \in \bar{S}$) by $\mathrm{Sp}^{\mathfrak{P}} y$ and $\mathrm{N}^{\mathfrak{P}} y$ relatively the relative trace and the relative norm of y from $\bar{S}$ to $\bar{R}$. Then the following corollary follows immediately from Theorem 6:

COROLLARY 1. *The notation being as in Theorem* 6, *we have* $\mathrm{N}_{S/R}y = \prod_{\lambda=1}^{h} \mathrm{N}^{\mathfrak{P}_\lambda} y$, $\mathrm{Sp}_{S/R}y = \sum_{\lambda=1}^{h} \mathrm{Sp}^{\mathfrak{P}_\lambda} y$.

COROLLARY 2. *The notation being as in Theorem* 6, *we have*

$$\nu_{\mathfrak{p}}(\mathrm{N}_{S/R}y) = \sum_{\lambda=1}^{h} f_\lambda \nu_{\mathfrak{P}_\lambda}(y),$$

where f_λ *is the relative degree of* $\mathfrak{P}_\lambda$ *with respect to* R.

This follows immediately from Corollary 1 and from Corollary 2 to Theorem 5.

§6. THE PUISEUX EXPANSIONS

Let R be a field of algebraic functions of one variable over an algebraically closed field of constants K, and let S be an overfield of finite degree of R. Denote by $\mathfrak{p}$ a place of R and by $\mathfrak{P}$ a place of S which lies above $\mathfrak{p}$. Since K is algebraically closed, $\mathfrak{p}$ and $\mathfrak{P}$ are of degree 1. Let t be a uniformizing variable at $\mathfrak{p}$; then the $\mathfrak{p}$-adic completion $\bar{R}$ of R may be identified with the field of formal power series in t with coefficients in K (III, §3).

Let e be the ramification index of $\mathfrak{P}$ with respect to R. *We shall assume that e is not divisible by the characteristic p of K.* This being the case, we shall prove that t is the e-th power of some element of the $\mathfrak{P}$-adic completion $\bar{S}$ of S. Let u be a uniformizing variable at $\mathfrak{P}$ in S. We know that $\nu_{\mathfrak{P}}(t) = e$; it follows that $t = u^e v$, with $\nu_{\mathfrak{P}}(v) = 0$. Since K is algebraically closed, there exists an $a \in K$ such that $\nu_{\mathfrak{P}}(v - a) > 0$, and we may write a in the form b^e, $b \in K$. Consider the polynomial $W^e - v = f(W)$ in a letter W. Its coefficients are in the ring of integral elements of $\bar{S}$; if we replace these coefficients by their residue classes modulo $\mathfrak{P}$, we obtain a polynomial $f^*(W) = W^e - b^e$, which admits a linear factor $W - b$: $f^*(W) = (W - b)h^*(W)$. Moreover, b is a simple zero of $f^*(W)$; in fact, we have $(df^*(W)/dW)(b) = eb^{e-1} \neq 0$, because $b \neq 0$ and e is not divisible by p. Making use of Hensel's lemma (III, §2), we conclude that $f(W)$ has a linear factor with coefficients in $\bar{S}$; this means that $v = w^e$, $w \in \bar{S}$, whence $t = (uw)^e$.

Denote by $t^{1/e}$ any one of the e-th roots of t in $\bar{S}$; then it is clear that

$$\nu_{\mathfrak{P}}(t^{1/e}) = 1.$$

It follows that $\bar{S}$ may be considered to be the field of formal power series in $t^{1/e}$ with coefficients in K. The power series expansions in terms of $t^{1/e}$ of the elements of S are called the *Puiseux expansions.*

If we consider now a place $\mathfrak{P}$ of S whose ramification index e is divisible by the characteristic p of K, then it is not true in general that t has an e-th root in $\bar{S}$. Consider for instance the case where $R = K\langle x\rangle$ (x transcendental over K) and $S = R\langle x, y\rangle$ where y satisfies the equation $y^p - y - x^{-1} = 0$. Let $\mathfrak{p}$ be the zero of x in R, and let $\mathfrak{P}$ be a place of S which lies above $\mathfrak{p}$. We have $\nu_{\mathfrak{P}}(1/x) = -e$, where e is the ramification index of $\mathfrak{P}$ with respect to R. It follows that $\nu_{\mathfrak{P}}(y) < 0$, whence $\nu_{\mathfrak{P}}(y^p) < \nu_{\mathfrak{P}}(y)$ and $\nu_{\mathfrak{P}}(y^p) = \nu_{\mathfrak{P}}(y^p - y) = -e$, which proves that e is divisible by p. On the other hand, we have $e \leqq [S:R] \leqq p$; it follows that $e = [S:R] = p$, and that $Y^p - Y - x^{-1} = 0$ is an irreducible equation in R. We have $[\bar{S}:\bar{R}] = e = p$, which proves that $Y^p - Y - x^{-1}$ remains irreducible in $\bar{R}$. Now, $y + 1$ is also a root of the equation $Y^p - Y - x^{-1} = 0$. It follows easily that $\bar{S}$ has an automorphism of order p which leaves the elements of $\bar{R}$ fixed, and therefore that $\bar{S}$ is separable over $\bar{R}$. Since t is obviously not a p-th power in $\bar{R}$, it is also not a p-th power in $\bar{S}$.

§7. Norm and conorm; trace and cotrace

Let R and S be fields of algebraic functions of one variable of which we assume that R is a subfield of S, with the usual assumptions about the fields of constants (cf. §1). We propose to set up certain correspondences between divisors and repartitions in R and in S.

If x is an element $\neq 0$ of R, we denote by $\mathfrak{d}_R(x)$ and $\mathfrak{d}_S(x)$ the divisors of x in R and S respectively. Let $\mathfrak{p}$ be any place of R and let $\mathfrak{P}$ be a place of S which lies above $\mathfrak{p}$; the exponent $a(\mathfrak{p})$ of $\mathfrak{p}$ in $\mathfrak{d}_R(x)$ is $\nu_\mathfrak{p}(x)$, while the exponent of $\mathfrak{P}$ in $\mathfrak{d}_S(x)$ is $\nu_\mathfrak{P}(x) = e(\mathfrak{P})\nu_\mathfrak{p}(x) = e(\mathfrak{P})a(\mathfrak{p})$, where $e(\mathfrak{P})$ is the ramification index of $\mathfrak{P}$ with respect to R. On the other hand, if $\mathfrak{P}$ is a variable place of S with respect to R, then clearly $\nu_\mathfrak{P}(x) = 0$. This leads to the following generalization. Let $\mathfrak{a} = \prod_\mathfrak{p} \mathfrak{p}^{a(\mathfrak{p})}$ be any divisor in R. If $\mathfrak{P}$ is a place of S which lies above a place $\mathfrak{p}$ of R, we set $A(\mathfrak{P}) = e(\mathfrak{P})a(\mathfrak{p})$, where $e(\mathfrak{P})$ is the ramification index of $\mathfrak{P}$ with respect to R; if $\mathfrak{P}$ is a variable place of S, we set $A(\mathfrak{P}) = 0$. Since above any given place of R there lie only a finite number of places of S, there are only a finite number of places $\mathfrak{P}$ of S for which $A(\mathfrak{P}) \neq 0$, and the symbol $\prod_\mathfrak{P} \mathfrak{P}^{A(\mathfrak{P})}$ represents a divisor of S. We shall call this divisor the *conorm* of $\mathfrak{a}$ (from R to S), and we shall denote it by $\mathrm{Con}_{R/S}\mathfrak{a}$. It is clear that, if $\mathfrak{a}$ and $\mathfrak{b}$ are divisors of R, then

$$\mathrm{Con}_{R/S}\mathfrak{a}\mathfrak{b} = (\mathrm{Con}_{R/S}\mathfrak{a})(\mathrm{Con}_{R/S}\mathfrak{b})$$

and that, if x is an element $\neq 0$ in R, then

$$\mathrm{Con}_{R/S}\mathfrak{d}_R(x) = \mathfrak{d}_S(x).$$

Lemma 1. *Assume that S is of finite degree over R, and denote by K and L the fields of constants of R and S respectively. Then, if $\mathfrak{a}$ is any divisor of R, we have*

$$d(\mathrm{Con}_{R/S}\mathfrak{a}) = \frac{[S:R]}{[L:K]}\, d(\mathfrak{a})$$

It will be sufficient to consider the case where $\mathfrak{a}$ is a place $\mathfrak{p}$ of R. Let then $\mathfrak{P}_1, \cdots, \mathfrak{P}_h$ be the distinct places of S which lie above $\mathfrak{p}$, whence $\mathrm{Con}_{R/S}\mathfrak{p} = \mathfrak{P}_1^{e_1} \cdots \mathfrak{P}_h^{e_h}$, where e_i is the ramification index of $\mathfrak{P}_i$ with respect to R. Let Σ be the residue field of $\mathfrak{p}$, and Σ_i that of $\mathfrak{P}_i$; then $d(\mathrm{Con}_{R/S}\mathfrak{p}) = \sum_{i=1}^h e_i[\Sigma_i:L]$. Now, we have

$$[\Sigma_i:L] = \frac{[\Sigma_i:K]}{[L:K]} = \frac{[\Sigma_i:\Sigma][\Sigma:K]}{[L:K]}.$$

The number $[\Sigma:K]$ is equal to $d(\mathfrak{p})$, while $[\Sigma_i:\Sigma]$ is the relative degree f_i of $\mathfrak{P}_i$ with respect to R. By Theorem 1, §1, we have $\sum_{i=1}^{h} e_i f_i = [S:R]$; Lemma 1 is thereby proved.

Now, assuming S to be of finite degree over R, we shall generalize to divisors the notion of the norm from S to R of an element of S. Let y be any element $\neq 0$ in S. Let $\mathfrak{p}$ be a place of R, and let $\mathfrak{P}_1, \cdots, \mathfrak{P}_h$ be the places of S which lie above $\mathfrak{p}$. Then we have $\mathrm{N}_{S/R}y = \prod_{i=1}^{h} \mathrm{N}^{\mathfrak{P}_i}y$ (Corollary 1 to Theorem 6, §5). On the other hand, we have $\nu_{\mathfrak{p}}(\mathrm{N}^{\mathfrak{P}_i}y) = f_i\nu_{\mathfrak{P}_i}(y)$ (Corollary 2 to Theorem 5, §5). It follows that $\nu_{\mathfrak{p}}(\mathrm{N}_{S/R}y) = \sum_{i=1}^{h} f_i\nu_{\mathfrak{P}_i}(y)$. This being said, let $\mathfrak{A} = \prod_{\mathfrak{P}} \mathfrak{P}^{A(\mathfrak{P})}$ be any divisor of S. If $\mathfrak{p}$ is any place of R, set $a(\mathfrak{p}) = \sum_{i=1}^{h} f_i A(\mathfrak{P}_i)$, where $\mathfrak{P}_1, \cdots, \mathfrak{P}_h$ are the distinct places of S which lie above $\mathfrak{p}$, while f_i is the degree of $\mathfrak{P}_i$ with respect to R. Then it is clear that there are only a finite number of places $\mathfrak{p}$ of R for which $a(\mathfrak{p}) \neq 0$, and therefore that the symbol $\prod_{\mathfrak{p}} \mathfrak{p}^{a(\mathfrak{p})}$ represents a divisor of R. We shall call this divisor the *norm* of $\mathfrak{A}$ (from S to R) and we shall denote it by $\mathrm{N}_{S/R}\mathfrak{A}$. It is clear that, if $\mathfrak{A}$ and $\mathfrak{B}$ are divisors of S, then

$$\mathrm{N}_{S/R}\mathfrak{A}\mathfrak{B} = (\mathrm{N}_{S/R}\mathfrak{A})(\mathrm{N}_{S/R}\mathfrak{B})$$

and that, if y is an element $\neq 0$ of S, then

$$\mathrm{N}_{S/R}(\mathfrak{d}_S(y)) = \mathfrak{d}_R(\mathrm{N}_{S/R}y).$$

The operations of norm and conorm are related to each other by the formula

$$\mathrm{N}_{S/R}(\mathrm{Con}_{R/S}\mathfrak{a}) = \mathfrak{a}^{[S:R]}$$

in which $\mathfrak{a}$ stands for an arbitrary divisor of R. It will be sufficient to prove this formula in the case where $\mathfrak{a}$ is a place $\mathfrak{p}$. Let then $\mathfrak{P}_1, \cdots, \mathfrak{P}_h$ be the distinct places of S which lie above $\mathfrak{p}$; denote by e_i the ramification index of $\mathfrak{P}_i$ with respect to R and by f_i its relative degree. Then

$$\mathrm{N}_{S/R}(\mathrm{Con}_{R/S}\mathfrak{p}) = \mathrm{N}_{S/R} \prod_{i=1}^{h} \mathfrak{P}_i^{e_i} = \prod_{i=1}^{h} \mathfrak{p}^{e_i f_i} = \mathfrak{p}^{\sum_{i=1}^{h} e_i f_i}$$

and our formula follows from Theorem 1, §1.

LEMMA 2. *Assume that S is of finite degree over R, and let K and L be the fields of constants of R and S respectively. Then we have, for any divisor $\mathfrak{A}$ of S, $d(N_{S/R}\mathfrak{A}) = [L:K]d(\mathfrak{A})$.*

It will be sufficient to prove this formula in the case where $\mathfrak{A}$ is a place $\mathfrak{P}$. Denote by $\mathfrak{p}$ the place of R which lies below $\mathfrak{P}$, by $\Sigma(\mathfrak{P})$ and $\Sigma(\mathfrak{p})$ the residue fields of $\mathfrak{P}$ and $\mathfrak{p}$ respectively, and by f the relative degree of $\mathfrak{P}$ with respect to R. Then we have

$$d(\mathrm{N}_{S/R}\mathfrak{P}) = d(\mathfrak{p}^f) = f[\Sigma(\mathfrak{p}):K] = [\Sigma(\mathfrak{P}):\Sigma(\mathfrak{p})][\Sigma(\mathfrak{p}):K] = [\Sigma(\mathfrak{P}):K]$$
$$= [\Sigma(\mathfrak{P}):L][L:K] = [L:K]\, d(\mathfrak{P})$$

and Lemma 2 is proved.

Now, assuming that S is algebraic over R, we propose to associate with every repartition in R a repartition in S in such a way that to the principal repartition attached to an element x of R be associated the principal repartition attached to x in S.

Let $\mathfrak{P}$ be any place of S. Since S is algebraic over R, $\mathfrak{P}$ lies above some place $\mathfrak{p}$ of R. The $\mathfrak{P}$-adic completion $\bar{S}$ of S may therefore be assumed to contain the $\mathfrak{p}$-adic completion $\bar{R}$ of R as a subfield. Let $\mathfrak{x}$ be a repartition in R; if we assign to every place $\mathfrak{P}$ of S the element $\mathfrak{x}(\mathfrak{p}) \in \bar{R} \subset \bar{S}$, we obtain a repartition of S. For, since only a finite number of places of S lie above any given place of R, it is clear that there are only a finite number of places $\mathfrak{P}$ of S for which $\nu_{\mathfrak{P}}(\mathfrak{x}(\mathfrak{p})) < 0$. The repartition of S defined in this way is called the *cotrace* of $\mathfrak{x}$, *taken from R to S*; we shall denote this repartition by the symbol $\mathrm{Cosp}_{R/S}\mathfrak{x}$. It is clear that, if $\mathfrak{x}$ and $\mathfrak{x}'$ are repartitions in R and λ, λ' constants in R (and therefore also in S), then

$$\mathrm{Cosp}_{R/S}(\lambda\mathfrak{x} + \lambda'\mathfrak{x}') = \lambda\,\mathrm{Cosp}_{R/S}\mathfrak{x} + \lambda'\,\mathrm{Cosp}_{R/S}\mathfrak{x}'.$$

If $x \in R$, then we can attach to x a repartition in R and a repartition in S. Until now, we have used notations which did not differentiate between x and the repartitions attached to it; however, this convention would lead to ambiguities now that we have two fields to deal with simultaneously. For this reason, we shall denote by x_R the repartition attached to x in R, while, if $y \in S$, we shall denote by y_S the repartition attached to y in S. This being said, it is clear that, if $x \in R$, then

$$\mathrm{Cosp}_{R/S}x_R = x_S .$$

Now, assume that S is of finite degree over R. In this case, we shall generalize to repartitions the notion of the trace of an element of S with respect to R. Denote by y an element of S, by $\mathfrak{p}$ a place of R, and by $\mathfrak{P}_1, \cdots, \mathfrak{P}_h$ the distinct places of S which lie above $\mathfrak{p}$. Then we know that $\mathrm{Sp}_{S/R}y = \sum_{i=1}^{h} \mathrm{Sp}^{\mathfrak{P}_i}y$ (by Corollary 1 to Theorem 6, §5). Now, let $\mathfrak{y}$ be any repartition in S; for each place $\mathfrak{p}$ of R, set $\mathfrak{x}(\mathfrak{p}) = \sum_{i=1}^{h} \mathrm{Sp}^{\mathfrak{P}_i}\mathfrak{y}(\mathfrak{P}_i)$. Then the mapping $\mathfrak{p} \rightarrow \mathfrak{x}(\mathfrak{p})$ is a repartition in R. For, there are only a finite number of places $\mathfrak{P}$ of S for which $\nu_{\mathfrak{P}}(\mathfrak{y}(\mathfrak{P})) < 0$, and, if $\mathfrak{p}$ is a place of R which does not lie below any of these, then $\nu_{\mathfrak{p}}(\mathfrak{x}(\mathfrak{p})) \geqq 0$ (Corollary 1 to Theorem 5, §5). The repartition $\mathfrak{x}$ defined in this way is called the *trace* of $\mathfrak{y}$, *from S to R*, and is denoted by the symbol $\mathrm{Sp}_{S/R}\mathfrak{y}$. It is clear that, if $\mathfrak{y}$ and $\mathfrak{y}'$ are repartitions in S and λ, λ' constants in R, then

$$\mathrm{Sp}_{S/R}(\lambda\mathfrak{y} + \lambda'\mathfrak{y}') = \lambda\,\mathrm{Sp}_{S/R}\mathfrak{y} + \lambda'\,\mathrm{Sp}_{S/R}\mathfrak{y}'.$$

On the other hand, if $y \in S$, then

$$\mathrm{Sp}_{S/R}y_S = (\mathrm{Sp}_{S/R}y)_R .$$

REMARK. Even if we are dealing in S with a repartition $\mathfrak{y}$ such that $\mathfrak{y}(\mathfrak{P}) \in S$ for every place $\mathfrak{P}$, it will not be true in general that $(\mathrm{Sp}_{S/R}\mathfrak{y})(\mathfrak{p}) \in R$ for every place $\mathfrak{p}$ of R. This shows that the definition of the trace could not have been given in terms of our old definition of repartitions; the generalization which was accomplished in III, §4 is here necessary.

If the field S is not separable over R, then $\mathrm{Sp}_{S/R}\mathfrak{y} = 0$ *for every repartition* $\mathfrak{y}$ *in* S. To show this, it will be sufficient to prove that, if we denote by $\mathfrak{P}$ a place of S, by $\mathfrak{p}$ the place of R which lies below $\mathfrak{P}$, by $\bar{R}$ and $\bar{S}$ respectively the $\mathfrak{p}$-adic completion of R and the $\mathfrak{P}$-adic completion of S, then $\bar{S}$ is not separable over $\bar{R}$. Let R' be the subfield of S composed of the elements of S which are separable with respect to R. Denote by $\mathfrak{p}'$ the place of R' which lies below $\mathfrak{P}$ and by $\bar{R}'$ the $\mathfrak{p}'$-adic completion of R'. If p is the characteristic of S, there exists an exponent m such that $y^{p^m} \in R'$ for every $y \in S$. It is clear that the conditions $\nu_{\mathfrak{P}}(y) > 0$ and $\nu_{\mathfrak{p}'}(y^{p^m}) > 0$ are equivalent to each other, which proves that $\mathfrak{P}$ is the only place of S to lie above $\mathfrak{p}'$. It follows that $[\bar{S}:\bar{R}'] = [S:R'] > 1$ (Theorem 4, §5). On the other hand, we have $\bar{S} = \bar{R}'\langle S\rangle$ (Corollary 3 to Theorem 5, §5), which proves that $\bar{S}$ is purely unseparable over $\bar{R}'$; our assertion is thereby proved.

Let $\mathfrak{x}$ be a repartition in R and z an element of S. We shall prove the formula

$$\mathrm{Sp}_{S/R}(z\ \mathrm{Cosp}_{R/S}\mathfrak{x}) = (\mathrm{Sp}_{S/R}z)\mathfrak{x}. \tag{1}$$

Denote by $\mathfrak{p}$ a place of R and by $\mathfrak{P}_1, \cdots, \mathfrak{P}_h$ the distinct places of S above $\mathfrak{p}$. Then we have $(\mathrm{Sp}_{S/R}(z\ \mathrm{Cosp}_{R/S}\mathfrak{x}))(\mathfrak{p}) = \sum_{i=1}^{h} \mathrm{Sp}^{\mathfrak{P}_i}(z\ \mathrm{Cosp}_{R/S}\mathfrak{x})(\mathfrak{P}_i)$. Since $(\mathrm{Cosp}_{R/S}\mathfrak{x})(\mathfrak{P}_i) = \mathfrak{x}(\mathfrak{p})$ is in the $\mathfrak{p}$-adic completion of R, this is equal to

$$\left(\sum_{i=1}^{h} \mathrm{Sp}^{\mathfrak{P}_i}z\right)\mathfrak{x}(\mathfrak{p}) = (\mathrm{Sp}_{S/R}z)\mathfrak{x}(\mathfrak{p}),$$

which proves our formula.

LEMMA 3. *Let R, S, and T be fields of algebraic functions of one variable such that R is a subfield of S and S of T. If $\mathfrak{a}$ is any divisor of R, then we have* $\mathrm{Con}_{R/T}\mathfrak{a} = \mathrm{Con}_{S/T}(\mathrm{Con}_{R/S}\mathfrak{a})$. *If T is algebraic over R, and if $\mathfrak{x}$ is any repartition in R, then we have* $\mathrm{Cosp}_{R/T}\,\mathfrak{x} = \mathrm{Cosp}_{S/T}(\mathrm{Cosp}_{R/S}\mathfrak{x})$. *Assume now that T is of finite degree over R. Then, if $\mathfrak{A}$ is any divisor in T, we have* $\mathrm{N}_{T/R}\mathfrak{A} = \mathrm{N}_{S/R}(\mathrm{N}_{T/S}\mathfrak{A})$, *and, if $\mathfrak{y}$ is any repartition in T, we have* $\mathrm{Sp}_{T/R}\mathfrak{y} = \mathrm{Sp}_{S/R}(\mathrm{Sp}_{T/S}\mathfrak{y})$.

1. It will be sufficient to prove the first formula in the case where $\mathfrak{a}$ is a place $\mathfrak{p}_0$. In this case, the only places of T which actually occur in either one of the divisors $\mathrm{Con}_{R/T}\mathfrak{a}$ or $\mathrm{Con}_{S/T}(\mathrm{Con}_{R/S}\mathfrak{a})$ are those which lie above $\mathfrak{p}_0$. If $\mathfrak{P}$ is any one of these, then $\mathfrak{P}$ occurs with the same exponent in our two divisors in virtue of Lemma 4, §1.

2. If T is algebraic over R, the formula $\mathrm{Cosp}_{R/T}\mathfrak{x} = \mathrm{Cosp}_{S/T}(\mathrm{Cosp}_{R/S}\mathfrak{x})$ is obvious.

3. Assume from now on that T is of finite degree over R. It will be sufficient to prove the formula $\mathrm{N}_{T/R}\mathfrak{A} = \mathrm{N}_{S/R}(\mathrm{N}_{T/S}\mathfrak{A})$ in the case where $\mathfrak{A}$ is a place of T. In this case, the formula follows immediately from the last assertion of Lemma 4, §1.

4. We shall now prove the formula $\mathrm{Sp}_{T/R}\mathfrak{y} = \mathrm{Sp}_{S/R}(\mathrm{Sp}_{T/S}\mathfrak{y})$. Let $\mathfrak{p}_0$ be any place of R; then $(\mathrm{Sp}_{T/R}\mathfrak{y})(\mathfrak{p}_0) = \sum_{i=1}^{h} \mathrm{Sp}_{T/R}^{\mathfrak{P}_i}\mathfrak{y}(\mathfrak{P}_i)$, where $\mathfrak{P}_1, \cdots, \mathfrak{P}_h$ are the distinct places of T which lie above $\mathfrak{p}_0$, while $\mathrm{Sp}_{T/R}^{\mathfrak{P}_i}$ represents the trace from the $\mathfrak{P}_i$-adic completion of T to the $\mathfrak{p}_0$-adic completion of R. If $\mathfrak{p}_i$ is the place of S which lies below $\mathfrak{P}_i$, then $\mathrm{Sp}_{T/R}^{\mathfrak{P}_i}\mathfrak{y}(\mathfrak{P}_i) = \mathrm{Sp}_{S/R}^{\mathfrak{P}_i}(\mathrm{Sp}_{T/S}^{\mathfrak{P}_i}\mathfrak{y}(\mathfrak{P}_i))$, where $\mathrm{Sp}_{S/R}^{\mathfrak{P}_i}$ is the

trace from the $\mathfrak{p}_i$-adic completion of S to the $\mathfrak{p}_0$-adic completion of R, while $\mathrm{Sp}^{\mathfrak{P}_i}_{T/R}$ is the trace from the $\mathfrak{P}_i$-adic completion of T to the $\mathfrak{p}_i$-adic completion of S. Let $\mathfrak{p}$ be any one of the places of S which lie above $\mathfrak{p}_0$. If we denote by $U_\mathfrak{p}$ the sum of the terms $\mathrm{Sp}^{\mathfrak{P}_i}_{T/R}\mathfrak{y}(\mathfrak{P}_i)$ which are relative to the indices i for which $\mathfrak{p}_i = \mathfrak{p}$, then $U_\mathfrak{p}$ is equal to $(\mathrm{Sp}_{T/S}\mathfrak{y})(\mathfrak{p})$, and $(\mathrm{Sp}_{T/R}\mathfrak{y})(\mathfrak{p}_0)$ is equal to the sum of the quantities $\mathrm{Sp}^{\mathfrak{p}}_{S/R}U_\mathfrak{p}$ for the distinct places $\mathfrak{p}$ of S which lie above $\mathfrak{p}_0$ (where $\mathrm{Sp}^{\mathfrak{p}}_{S/R}$ is the trace from the $\mathfrak{p}$-adic completion of S to the $\mathfrak{p}_0$-adic completion of R). Our formula is thereby proved.

§8. The different

In this section, we shall assume that R and S are fields of algebraic functions of one variable such that S is a separable overfield of finite degree of R. Let $\mathfrak{P}$ be any place of S, and let $\mathfrak{p}$ be the place of R which lies below $\mathfrak{P}$. Then the $\mathfrak{P}$-adic completion $\bar{S}$ of S is separable over the $\mathfrak{p}$-adic completion $\bar{R}$ of R, as follows immediately from Corollary 3 to Theorem 5, §5. If $y \in \bar{S}$, we know that the condition $\nu_\mathfrak{P}(y) \geqq 0$ implies $\nu_\mathfrak{p}(\mathrm{Sp}^\mathfrak{P} y) \geqq 0$ (Corollary 1 to Theorem 5, §5). But the converse is in general not true: we may have elements $y \in \bar{S}$ such that $\nu_\mathfrak{P}(y) < 0$, but $\nu_\mathfrak{p}(\mathrm{Sp}^\mathfrak{P} y) \geqq 0$. However, we cannot have $\nu_\mathfrak{p}(\mathrm{Sp}^\mathfrak{P} y) \geqq 0$ for every $y \in \bar{S}$. For, since $\bar{S}$ is separable over $\bar{R}$, there exists an element $y_1 \in \bar{S}$ such that $\mathrm{Sp}^\mathfrak{P} y_1 \neq 0$. If x is an element of $\bar{R}$ such that $\nu_\mathfrak{p}(x) < -\nu_\mathfrak{p}(\mathrm{Sp}^\mathfrak{P} y_1)$, then we have $\nu_\mathfrak{p}(\mathrm{Sp}^\mathfrak{P} xy_1) < 0$. It follows that there exists a largest integer $m = m(\mathfrak{P})$ with the following property: if y is any element of S such that $\nu_\mathfrak{P}(y) \geqq -m(\mathfrak{P})$, then $\nu_\mathfrak{p}(\mathrm{Sp}^\mathfrak{P} y) \geqq 0$. The integer $m(\mathfrak{P})$, which is of course $\geqq 0$, is called the *differential exponent of* $\mathfrak{P}$ (with respect to R).

Theorem 7. *Let R and S be fields of algebraic functions of one variable; assume that R is a subfield of S and that S is of finite degree and separable over R. Let $\mathfrak{P}$ be a place of S. Then the differential exponent $m(\mathfrak{P})$ of $\mathfrak{P}$ with respect to R is at least equal to $e - 1$, where e is the ramification index of $\mathfrak{P}$. In order for $m(\mathfrak{P})$ to be equal to $e - 1$, it is necessary and sufficient that the following conditions be satisfied:* 1) *the residue field of $\mathfrak{P}$ is separable over the residue field of the place $\mathfrak{p}$ of R which lies below $\mathfrak{P}$;* 2) *e is not divisible by the characteristic of R.*

Let $\Sigma(\mathfrak{P})$ and $\Sigma(\mathfrak{p})$ be the residue fields of $\mathfrak{P}$ and $\mathfrak{p}$ respectively. If z is any integral element of the $\mathfrak{P}$-adic completion $\bar{S}$ of S, we shall denote by z^* its residue class modulo $\mathfrak{P}$. We shall prove the formula

$$(1) \qquad (\mathrm{Sp}^\mathfrak{P} z)^* = e\, \mathrm{Sp}_{\Sigma(\mathfrak{P})/\Sigma(\mathfrak{p})}\, z^*.$$

Let u be a uniformizing variable at $\mathfrak{P}$ in S, and let $\theta_1, \cdots, \theta_f$ be elements of the ring of $\mathfrak{P}$ whose residue classes $\theta_1^*, \cdots, \theta_f^*$ modulo $\mathfrak{P}$ form a base of $\Sigma(\mathfrak{P})$ over $\Sigma(\mathfrak{p})$. Then we know that the elements $u^i\theta_j$ ($0 \leqq i \leqq e - 1$, $1 \leqq j \leqq f$) form a base of $\bar{S}$ with respect to the $\mathfrak{p}$-adic completion $\bar{R}$ of R, and that, if we set

$$(2) \qquad zu^i\theta_j = \sum_{k=0}^{e-1} \sum_{l=1}^{f} a_{ijkl}\, u^k \theta_l, \qquad a_{ijkl} \in \bar{R},$$

the elements a_{ijkl} of $\bar{R}$ are integral (Theorem 5, §5). We have $\mathrm{Sp}^{\mathfrak{P}} z = \sum_{i=0}^{e-1} \sum_{j=1}^{f} a_{ijij}$. Multiplying the formulas (2) relative to the case $i = 0$ by u^i, we obtain

$$(3) \qquad zu^i \theta_j = \sum_{k=0}^{e-1} \sum_{l=1}^{f} a_{0jkl} u^{k+i} \theta_l .$$

Let t be a uniformizing variable at $\mathfrak{p}$ in R, whence $\nu_{\mathfrak{P}}(t) = e$. If $k + i \geqq e$, we have $u^{k+i}\theta_l = tv_{ikl}$ with v_{ikl} integral in $\bar{S}$. We may express each v_{ikl} as a linear combination of the products $u^{k'}\theta_{l'}$ ($0 \leqq k' \leqq e - 1, 1 \leqq l' \leqq f$) with coefficients which are integral in $\bar{R}$. Substituting these expressions in (3) and grouping like terms, we must find the expressions (2). From this, we conclude easily that $\nu_{\mathfrak{p}}(a_{ijkl}) > 0$ if $k < i$, $\nu_{\mathfrak{p}}(a_{ijkl} - a_{0j,k-i,l}) > 0$ if $i \leqq k \leqq e - 1$. This implies that $a_{ijij}^* = b_j^*$ does not depend on i and that $(\mathrm{Sp}^{\mathfrak{P}} z)^* = e \sum_{j=1}^{f} b_j^*$. On the other hand, the formulas (2) for $i = 0$ give $z^*\theta_j^* = \sum_{l=1}^{f} a_{0j0l}^* \theta_l^*$; since $b_j^* = a_{0j0j}^*$, we find that $\mathrm{Sp}_{\Sigma(\mathfrak{P})/\Sigma(\mathfrak{p})} z^* = \sum_{j=1}^{f} b_j^*$. Formula (1) is thereby proved.

Now, let y be an element of $\bar{S}$ such that $\nu_{\mathfrak{P}}(y) \geqq -(e - 1)$. Since $\nu_{\mathfrak{P}}(t) = e$, we have $\nu_{\mathfrak{P}}(ty) > 0$, whence, by (1), $\nu_{\mathfrak{p}}(\mathrm{Sp}^{\mathfrak{P}} ty) = 1 + \nu_{\mathfrak{p}}(\mathrm{Sp}^{\mathfrak{P}} y) \geqq 1$ and therefore $\nu_{\mathfrak{p}}(\mathrm{Sp}^{\mathfrak{P}} y) \geqq 0$. This proves that $m(\mathfrak{P}) \geqq e - 1$. If either one of the conditions stated in Theorem 7 fails to be satisfied, then it follows from (1) that $\nu_{\mathfrak{p}}(\mathrm{Sp}^{\mathfrak{P}} z) > 0$ for every integral element z of $\bar{S}$ (we know that, if $\Sigma(\mathfrak{P})$ is not separable over $\Sigma(\mathfrak{p})$, then the relative trace with respect to $\Sigma(\mathfrak{p})$ of every element of $\Sigma(\mathfrak{P})$ is 0). Applying the same argument as above to an element y such that $\nu_{\mathfrak{P}}(y) \geqq -e$, we see that $m(\mathfrak{P}) \geqq e$. If, however, the two conditions of Theorem 7 are satisfied, then $\Sigma(\mathfrak{P})$ contains an element z^* whose trace with respect to $\Sigma(\mathfrak{P})$ is $\neq 0$. If z is an integral element of $\bar{S}$ belonging to the class z^*, then we see that $\nu_{\mathfrak{p}}(\mathrm{Sp}^{\mathfrak{P}} z) = 0$, whence $\nu_{\mathfrak{p}}(\mathrm{Sp}^{\mathfrak{P}} t^{-1} z) = -1$, which shows that $m(\mathfrak{P}) < e$, whence $m(\mathfrak{P}) = e - 1$.

Now, we shall connect the differential exponent with the discriminant of a base. Let U be a field, and let V be an algebraic extension of U which is separable and of finite degree over U. If $\{v_1, \cdots, v_n\}$ is any base of V with respect to U, the determinant of the matrix $(\mathrm{Sp}_{V/U} v_i v_j)_{1 \leqq i,j \leqq n}$ is called the *discriminant* of the base $\{v_1, \cdots, v_n\}$; we shall denote this element by $D(v_1, \cdots, v_n)$. We have $D(v_1, \cdots, v_n) \neq 0$. In fact, the matrix $(\mathrm{Sp}_{V/U} v_i v_j)$ is the representative matrix (with respect to the base $\{v_1, \cdots, v_n\}$) of the bilinear function B defined by $B(v, v') = \mathrm{Sp}_{V/U} vv'$. Since V is separable over U, there exists a $v_0 \in V$ such that $\mathrm{Sp}_{V/U} v_0 \neq 0$; if v is any element $\neq 0$ of V, then $\mathrm{Sp}_{V/U} v(v^{-1} v_0) \neq 0$, which proves that the bilinear form B is non degenerate.

LEMMA 1. *The notation being as in Theorem 7, let also $\{y_1, \cdots, y_n\}$ be a base of the $\mathfrak{P}$-adic completion $\bar{S}$ of S with respect to the $\mathfrak{p}$-adic completion $\bar{R}$ of R whose elements are integral in $\bar{S}$. Then we have $\nu_{\mathfrak{p}}(D(y_1, \cdots, y_n)) \geqq fm(\mathfrak{P})$, where f is the relative degree of $\mathfrak{P}$ with respect to R. A necessary and sufficient condition for the equality $\nu_{\mathfrak{p}}(D(y_1, \cdots, y_n)) = fm(\mathfrak{P})$ to hold is that our base be an integral base* (i.e., *that every integral element of $\bar{S}$ be expressible as a linear combination of $y_1, \cdots, y_n$ with integral coefficients in $\bar{R}$*).

We know that there exists an integral base $\{z_1, \cdots, z_n\}$ of $\bar{S}$ with respect to $\bar{R}$ (Theorem 5, §5). We can then write

$$z_i = \sum_{j=1}^{n} b_{ij} y_j, \qquad y_i = \sum_{j=1}^{n} c_{ij} z_j$$

with coefficients b_{ij} and c_{ij} in $\bar{R}$; furthermore, the elements c_{ij} are integral, and the matrices (b_{ij}) and (c_{ij}) are inverse to each other. We see immediately that $D(y_1, \cdots, y_n) = \gamma^2 D(z_1, \cdots, z_n)$, where $\gamma = \det(c_{ij})$. The elements c_{ij} being integral, we have $\nu_\mathfrak{p}(\gamma) \geqq 0$, whence $\nu_\mathfrak{p}(D(y_1, \cdots, y_n)) \geqq \nu_\mathfrak{p}(D(z_1, \cdots, z_n))$. If $\nu_\mathfrak{p}(\gamma) = 0$, then γ^{-1} is integral, from which it follows easily that the elements b_{ij} of the inverse matrix of (c_{ij}) are integral. In this case, every integral element of $\bar{S}$, being expressible as a linear combination of $z_1, \cdots, z_n$ with integral coefficients in $\bar{R}$, is also expressible as a linear combination of $y_1, \cdots, y_n$ with integral coefficients, which shows that $\{y_1, \cdots, y_n\}$ is an integral base. Conversely, if the base $\{y_1, \cdots, y_n\}$ is integral, then the coefficients b_{ij} are integral, and γ^{-1}, which is $\det(b_{ij})$, is integral, whence $\nu_\mathfrak{p}(\gamma) = 0$. Thus we see that a necessary and sufficient condition for the base $\{y_1, \cdots, y_n\}$ to be integral is that $\nu_\mathfrak{p}(D(y_1, \cdots, y_n)) = \nu_\mathfrak{p}(D(z_1, \cdots, z_n))$. Now, we have to compute $\nu_\mathfrak{p}(D(z_1, \cdots, z_n))$. Let $\mathfrak{A}$ be the set of elements $\zeta \in \bar{S}$ such that $\mathrm{Sp}^\mathfrak{P}\zeta y$ is integral whenever y is integral. If u is a uniformizing variable at $\mathfrak{P}$ in S, we clearly have $\mathfrak{A} = u^{-m(\mathfrak{P})}\mathfrak{O}$, where $\mathfrak{O}$ is the ring of integral elements in $\bar{S}$. We shall determine the set $\mathfrak{A}$ in another manner. Since $D(z_1, \cdots, z_n) \neq 0$, we can find elements $z_1', \cdots, z_n'$ of $\bar{S}$ such that $\mathrm{Sp}^\mathfrak{P} z_i z_j'$ is 1 if $i = j$ and 0 otherwise. If $x_1', \cdots, x_n'$ are in $\bar{R}$, then $\mathrm{Sp}^\mathfrak{P}(\sum_{i=1}^n x_i' z_i') z_j = x_j'$, whence $\mathrm{Sp}^\mathfrak{P}(\sum_{i=1}^n x_i' z_i') z = \sum_{j=1}^n x_j x_j'$ if $z = \sum_{j=1}^n x_j z_j$, $x_j \in \bar{R}$. It follows that a necessary and sufficient condition for $\sum_{i=1}^n x_i' z_i'$ to be in $\mathfrak{A}$ is that $x_1', \cdots, x_n'$ be integral. This shows in particular that $\{z_1', \cdots, z_n'\}$ is a base, and that $\{u^{m(\mathfrak{P})} z_1', \cdots, u^{m(\mathfrak{P})} z_n'\}$ is an integral base. Write $u^{m(\mathfrak{P})} z_i' = \sum_{j=1}^n \alpha_{ij} z_j'$, with $\alpha_{ij} \in \bar{R}$. Then it is clear that $\det(\alpha_{ij}) = \mathrm{N}^\mathfrak{P} u^{m(\mathfrak{P})}$, whence $\nu_\mathfrak{p}(\det(\alpha_{ij})) = fm(\mathfrak{P})$ (Corollary 2 to Theorem 5, §5). On the other hand, if we set $z_i = \sum_{j=1}^n \beta_{ij} u^{m(\mathfrak{P})} z_j'$, we know that $\det(\beta_{ij})$ is a unit. If we denote by (γ_{ij}) the product of the matrices (α_{ij}) and (β_{ij}), then we have $z_i = \sum_{j=1}^n \gamma_{ij} z_j'$, whence $\gamma_{ij} = \mathrm{Sp}^\mathfrak{P} z_i z_j$. It follows that $\nu_\mathfrak{p}(D(z_1, \cdots, z_n)) = \nu_\mathfrak{p}(\det(\beta_{ij})) + \nu_\mathfrak{p}(\det(\alpha_{ij})) = fm(\mathfrak{P})$. Lemma 1 is thereby proved.

LEMMA 2. *Let R and S be fields of algebraic functions of one variable such that S is a separable overfield of finite degree of R. Let $\mathfrak{p}$ be a place of R, and let $\mathfrak{P}_1, \cdots, \mathfrak{P}_h$ be the distinct places of S which lie above $\mathfrak{p}$. Denote by f_λ the relative degree of $\mathfrak{P}_\lambda$ with respect to R and by m_λ its differential exponent. Let $\{y_1, \cdots, y_n\}$ be a base of S with respect to R whose elements are integral over the ring of $\mathfrak{p}$. Then we have $\nu_\mathfrak{p}(D(y_1, \cdots, y_n)) \geqq \sum_{\lambda=1}^h m_\lambda f_\lambda$; and a necessary and sufficient condition for the equality to take place is for $\{y_1, \cdots, y_n\}$ to be integral at $\mathfrak{p}$.*

Let $\bar{R}$ be the $\mathfrak{p}$-adic completion of R. We construct the algebra $S_{\bar{R}}$ deduced from S by extending the basic field from R to $\bar{R}$, and we identify (by means of the isomorphism constructed in the proof of Theorem 4, §5) the algebra $S_{\bar{R}}$ with the product $\prod_{\lambda=1}^{h} \bar{S}_\lambda$ of the $\mathfrak{P}_\lambda$-adic completions $\bar{S}_\lambda$ of S $(1 \leqq \lambda \leqq h)$. For each λ we determine an integral base $\{z'_{1\lambda}, \cdots, z'_{n_\lambda\lambda}\}$ of $\bar{S}_\lambda$ with respect to $\bar{R}$, and we denote by $z_{i\lambda}$ the element of $S_{\bar{R}}$ whose λ-th coordinate is $z'_{i\lambda}$, while its other coordinates are 0. Then the elements $z_{i\lambda}$ $(1 \leqq i \leqq n_\lambda, 1 \leqq \lambda \leqq h)$ form a base of $S_{\bar{R}}$, and the integral elements of $S_{\bar{R}}$ are expressible as linear combinations of the elements $z_{i\lambda}$ with coefficients which are integral in $\bar{R}$. Let $\zeta_1, \cdots, \zeta_n$ be the elements $z_{i\lambda}$ arranged in some order. Arguing as in the proof of Lemma 2, we see immediately that, if $\{z_1, \cdots, z_n\}$ is any integral base at $\mathfrak{p}$ of S with respect to R, and if $z_i = \sum_{j=1}^{n} \alpha_{ij}\zeta_j$, then $\nu_\mathfrak{p}(\det(\alpha_{ij})) = 0$ (cf. the remark which follows the proof of Theorem 4, §5). Now, if $u \in S$, we have $\mathrm{Sp}_{S/R}u = \sum_{\lambda=1}^{h} \mathrm{Sp}^{\mathfrak{P}_\lambda}u$ (Corollary 1 to Theorem 6, §5); it is therefore natural to define for every $u = (u_1, \cdots, u_h) \in S_{\bar{R}}$ (with $u_\lambda \in \bar{S}_\lambda$, $1 \leqq \lambda \leqq h$) the element $\mathrm{Sp}_{S_{\bar{R}}/\bar{R}}u$ to be $\sum_{\lambda=1}^{h} \mathrm{Sp}^{\mathfrak{P}_\lambda}u$. Then we have

$$D(z_1, \cdots, z_n) = (\det(\alpha_{ij}))^2 \, (\det(\mathrm{Sp}_{S_{\bar{R}}/\bar{R}}\zeta_i\zeta_j)).$$

Now we have $z_{i\lambda}z_{i'\lambda'} = 0$ if $\lambda \neq \lambda'$ and $z_{i\lambda}z_{i'\lambda}$ is the element of $\prod_{\lambda=1}^{h} \bar{S}_\lambda$ whose λ-coordinate is $z'_{i\lambda}z'_{i'\lambda}$, the other coordinates being zero. It follows immediately that

$$\det(Sp_{S_{\bar{R}}/\bar{R}}\, \zeta_i\zeta_j) = \prod_{\lambda=1}^{h} D(z'_{1\lambda}, \cdots, z'_{n_\lambda\lambda})$$

whence, by Lemma 2, $\nu_\mathfrak{p}(D(z_1, \cdots, z_n)) = \sum_{\lambda=1}^{h} f_\lambda m_\lambda$. Now, we may write $y_i = \sum_{j=1}^{n} b_{ij}z_j$, $b_{ij} \in R$; the elements b_{ij} are in the ring of $\mathfrak{p}$ whence $\nu_\mathfrak{p}(D(y_1, \cdots, y_n)) \geqq \nu_\mathfrak{p}(D(z_1, \cdots, z_n))$, and we see as in the proof of Lemma 2 that a necessary and sufficient condition for $\{y_1, \cdots, y_n\}$ to be an integral base at $\mathfrak{p}$ is that $\nu_\mathfrak{p}(\det(b_{ij})) = 0$, i.e., that $\nu_\mathfrak{p}(D(y_1, \cdots, y_n)) = \nu_\mathfrak{p}(D(z_1, \cdots, z_n))$.

LEMMA 3. *Let R and S be fields of algebraic functions of one variable; assume that S is an overfield of R and is of finite degree and separable over R. Then there are only a finite number of places $\mathfrak{P}$ of S which fail to satisfy simultaneously the following two conditions: $\mathfrak{P}$ is unramified with respect to R; the residue field of $\mathfrak{P}$ is separable over the residue field of the place $\mathfrak{p}$ of R which lies below $\mathfrak{P}$.*

Let $\{y_1, \cdots, y_n\}$ be any base of S with respect to R. Then we know that $\{y_1, \cdots, y_n\}$ is integral at almost every place of R (Lemma 3, §3). Since $D(y_1, \cdots, y_n) \neq 0$, we have $\nu_\mathfrak{p}(D(y_1, \cdots, y_n)) = 0$ for almost every place $\mathfrak{p}$ of R. If $\mathfrak{p}$ is a place of R at which the base $\{y_1, \cdots, y_n\}$ is integral and for which $\nu_\mathfrak{p}(D(y_1, \cdots y_n)) = 0$, then it follows from Lemma 2 that the differential exponent of every place of S which lies above $\mathfrak{p}$ is 0. Making use of Theorem 7, we see that any such place is unramified with respect to R and that its residue field is separable over the residue field of $\mathfrak{p}$. Lemma 3 is thereby proved.

The situation being as described in Lemma 3, let, for any place $\mathfrak{P}$ of S, $m(\mathfrak{P})$ be the differential exponent of $\mathfrak{P}$ with respect to R. Then we have $m(\mathfrak{P}) = 0$ for almost all places $\mathfrak{P}$ of S, from which it follows that the symbol $\prod_\mathfrak{P} \mathfrak{P}^{m(\mathfrak{P})}$ represents an integral divisor of S. This divisor is called the *different* of S with respect to R.

Lemma 4. *Let R and S be fields of algebraic functions of one variable; assume that R is a subfield of S and that S is of finite degree and separable over R. Let F be a finite set of places of R, and let F^* be the set of places of S which lie above the places of F; let there be given for each $\mathfrak{p} \in F$ an integer $a(\mathfrak{p})$. Then the following two conditions on an arbitrary element y of S are equivalent to each other:* 1) *we have $\nu_{\mathfrak{P}}(y) \geqq -m(\mathfrak{P}) + a(\mathfrak{p})e(\mathfrak{P})$ for every $\mathfrak{P} \in F^*$, where $e(\mathfrak{P})$ and $m(\mathfrak{P})$ are respectively the ramification index and the differential exponent of $\mathfrak{P}$ with respect to R, while $\mathfrak{p}$ is the place of R below $\mathfrak{P}$;* 2) *if z is any element of S such that $\nu_{\mathfrak{P}}(z) \geqq 0$ for all $\mathfrak{P} \in F^*$, then $\nu_{\mathfrak{p}}(\mathrm{Sp}_{S/R}yz) \geqq a(\mathfrak{p})$ for all $\mathfrak{p} \in F$.*

Let t be an element of R such that $\nu_{\mathfrak{p}}(t) = -a(\mathfrak{p})$ for all $\mathfrak{p} \in F$ (Corollary to Theorem 3, I, §6). Then we have, for $\mathfrak{P} \in F^*$, $\nu_{\mathfrak{P}}(ty) = \nu_{\mathfrak{P}}(y) - a(\mathfrak{p})e(\mathfrak{P})$, where $\mathfrak{p}$ is the place of R below $\mathfrak{P}$. On the other hand, we have $\mathrm{Sp}_{S/R}tyz = t \cdot \mathrm{Sp}_{S/R}yz$, whence $\nu_{\mathfrak{p}}(\mathrm{Sp}_{S/R}tyz) = \nu_{\mathfrak{p}}(\mathrm{Sp}_{S/R}yz) - a(\mathfrak{p})$ for $\mathfrak{p} \in F$. It follows that it will be sufficient to prove Lemma 4 in the case where the integers $a(\mathfrak{p})$ are 0. We shall assume from now on that this is the case.

Assume first that 1) is satisfied. Denote by $\mathfrak{p}$ a place in F, and by $\mathfrak{P}_1, \cdots, \mathfrak{P}_h$ the distinct places of S which lie above $\mathfrak{p}$. Then we have, for any $z \in S$, $\mathrm{Sp}_{S/R}yz = \sum_{i=1}^{h} \mathrm{Sp}^{\mathfrak{P}_i}yz$ (Corollary 1 to Theorem 6, §5). If $\nu_{\mathfrak{P}}(z) \geqq 0$ for all $\mathfrak{P} \in F^*$, the $\nu_{\mathfrak{P}_i}(yz) \geqq -m(\mathfrak{P}_i)$ $(1 \leqq i \leqq h)$, whence $\nu_{\mathfrak{p}}(\mathrm{Sp}^{\mathfrak{P}_i}yz) \geqq 0$, and therefore $\nu_{\mathfrak{p}}(\mathrm{Sp}_{S/R}yz) \geqq 0$.

Now, assume that condition 2) is satisfied and that $y \neq 0$. Let $\mathfrak{P}_0$ be any place in F^*, and let $\bar{z}$ be any integral element of the $\mathfrak{P}_0$-adic completion $\bar{S}$ of S. Let z be an element of S which satisfies the following conditions: $\nu_{\mathfrak{P}_0}(z - \bar{z}) \geqq \max\{0, -\nu_{\mathfrak{P}_0}(y)\}$, $\nu_{\mathfrak{P}}(z) \geqq \max\{0, -\nu_{\mathfrak{P}}(y)\}$ for every place $\mathfrak{P}$ of F^* other than $\mathfrak{P}_0$ (cf. Theorem 3, I, §6). Then it is clear that z is integral at every place in F^*, whence $\nu_{\mathfrak{p}}(\mathrm{Sp}_{S/R}yz) \geqq 0$ for every $\mathfrak{p} \in F$. Let $\mathfrak{p}_0$ be the place of R which lies below $\mathfrak{P}_0$, and let $\mathfrak{P}_1 = \mathfrak{P}_0, \mathfrak{P}_2, \cdots, \mathfrak{P}_h$ be the distinct places of S above $\mathfrak{p}_0$. Then $\mathrm{Sp}_{S/R}yz = \sum_{i=1}^{h} \mathrm{Sp}^{\mathfrak{P}_i}yz$. If $i > 1$, we have $\nu_{\mathfrak{P}_i}(yz) \geqq 0$, whence $\nu_{\mathfrak{p}_0}(\mathrm{Sp}^{\mathfrak{P}_i}yz) \geqq 0$; it follows that $\nu_{\mathfrak{p}_0}(\mathrm{Sp}^{\mathfrak{P}_0}yz) \geqq 0$. Since $\nu_{\mathfrak{P}_0}(yz - y\bar{z}) \geqq 0$, this implies that $\nu_{\mathfrak{p}_0}(\mathrm{Sp}^{\mathfrak{P}_0}y\bar{z}) \geqq 0$. Every element $\bar{y}$ of $\bar{S}$ which is such that $\nu_{\mathfrak{P}_0}(\bar{y}) \geqq \nu_{\mathfrak{P}_0}(y)$ can be written in the form $y\bar{z}$, with $\bar{z}$ integral in $\bar{S}$; thus, for any such $\bar{y}$, we have $\nu_{\mathfrak{p}_0}(\mathrm{Sp}^{\mathfrak{P}_0}\bar{y}) \geqq 0$. From this, it follows that $\nu_{\mathfrak{P}_0}(\bar{y}) \geqq -m(\mathfrak{P}_0)$; Lemma 4 is thereby proved.

Theorem 8. *Let R, S, and T be fields of algebraic functions of one variable. Assume that R is a subfield of T, that T is of finite degree and separable over R, and that S is an intermediary field between R and T. Denote by $\mathfrak{D}_{T/R}$, $\mathfrak{D}_{T/S}$, and $\mathfrak{D}_{S/R}$ respectively the differents of T with respect to R, of T with respect to S, and of S with respect to R. Then we have $\mathfrak{D}_{T/R} = \mathfrak{D}_{T/S} \cdot \mathrm{Con}_{S/T}\mathfrak{D}_{S/R}$.*

Denote by $\mathfrak{p}$ a place of R, by F the set of places of S which lie above $\mathfrak{p}$, and by F^* the set of places of T which lie above $\mathfrak{p}$. If $\mathfrak{Q} \in F^*$, denote by $m(\mathfrak{Q})$, $\mu(\mathfrak{Q})$, and $e(\mathfrak{Q})$ respectively the differential exponent of $\mathfrak{Q}$ with respect to R, the differential exponent of $\mathfrak{Q}$ with respect to S, and the ramification index of $\mathfrak{Q}$ with respect to S. If $\mathfrak{q} \in F$, denote by $m(\mathfrak{q})$ the differential exponent of $\mathfrak{q}$ with respect to R. We have to prove that $m(\mathfrak{Q}) = \mu(\mathfrak{Q}) + e(\mathfrak{Q})m(\mathfrak{q})$, where $\mathfrak{q}$ is the place of S which lies below $\mathfrak{Q}$. We denote by $m'(\mathfrak{Q})$ the right side of this formula. Let y be an element of T such that $\nu_{\mathfrak{Q}}(y) = -m(\mathfrak{Q})$ for all $\mathfrak{Q} \in F^*$ (cf. Corollary to Theorem 3, I, §6). Then, by Lemma 4, we have $\nu_{\mathfrak{p}}(\mathrm{Sp}_{T/R}yz) \geqq 0$ for all $z \in T$ such that $\nu_{\mathfrak{Q}}(z) \geqq 0$ for all $\mathfrak{Q} \in F^*$. We know that $\mathrm{Sp}_{T/R}yz = \mathrm{Sp}_{S/R}(\mathrm{Sp}_{T/S}yz)$. Let z_1 be an element of S such that $\nu_{\mathfrak{q}}(z_1) \geqq 0$ for all $\mathfrak{q} \in F$. Then $\nu_{\mathfrak{p}}(\mathrm{Sp}_{S/R}z_1\mathrm{Sp}_{T/S}yz) = \nu_{\mathfrak{p}}(\mathrm{Sp}_{T/R}yzz_1) \geqq 0$ since $\nu_{\mathfrak{q}}(zz_1) \geqq 0$ for all $\mathfrak{Q} \in F^*$. Making use of Lemma 4, we conclude first that $\nu_{\mathfrak{Q}}(\mathrm{Sp}_{T/S}yz) \geqq -m(\mathfrak{q})$ for all $\mathfrak{q} \in F$, and then that $\nu_{\mathfrak{Q}}(y) \geqq -m'(\mathfrak{Q})$ for all $\mathfrak{Q} \in F^*$, whence $m(\mathfrak{Q}) \leqq m'(\mathfrak{Q})$. Now, let y' be any element of T such that $\nu_{\mathfrak{Q}}(y') = -m'(\mathfrak{Q})$ for all $\mathfrak{Q} \in F^*$ (cf. Corollary to Theorem 3, I, §6). Making use of Lemma 4, we see first that $\nu_{\mathfrak{q}}(\mathrm{Sp}_{T/S}y'z) \geqq -m(\mathfrak{q})$ for all $z \in T$ such that $\nu_{\mathfrak{Q}}(z) \geqq 0$ for all $\mathfrak{Q} \in F^*$, and then that $\nu_{\mathfrak{p}}(\mathrm{Sp}_{T/R}y'z) \geqq 0$, whence $\nu_{\mathfrak{Q}}(y') \geqq -m(\mathfrak{Q})$ for all $\mathfrak{Q} \in F^*$. This proves that $m'(\mathfrak{Q}) \leqq m(\mathfrak{Q})$. Theorem 8 is thereby proved.

§9. Structure of hyperelliptic fields

Let R be a field of algebraic functions of one variable, and let ω be a differential $\neq 0$ in R. Then any differential ω' of R may be written in the form $u\omega$, where u is an element of R which is uniquely determined when ω and ω' are given. The element u is called the *ratio of the differentials* ω' *and* ω, and is often denoted by ω'/ω.

Let us assume that R is of genus $g > 0$. Then it is clear that every constant can be represented as a ratio of differentials of the first kind. If $g = 1$, then conversely every ratio of differentials of the first kind is a constant. However, if $g > 1$, then there are non constant elements which may be represented as ratios of differentials of the first kind. The field R is said to be *hyperelliptic* if the subfield of R generated by the ratios of differentials of the first kind is different from R itself. We propose to prove the following result:

THEOREM 9. *Let R be a field of algebraic functions of one variable which is assumed to be of genus >1. Then a necessary and sufficient condition for R to be hyperelliptic is for R to have a subfield S of genus* 0, *containing the field of constants of R and over which R is of degree* 2. *If this condition is satisfied, then S is unique and is the field generated by the ratios of differentials of the first kind of R.*

We shall first prove that the condition is necessary.

LEMMA 1. *Let $\mathfrak{p}$ be a place of a field R of algebraic functions of one variable of genus $g > 0$. Then there exists a differential of the first kind of R which does not have $\mathfrak{p}$ as a zero.*

Assume for a moment that this were not the case. If $\mathfrak{a}$ is any divisor, we denote by $\delta(\mathfrak{a})$ (as we did in II) the dimension of the space of differentials which are $\equiv 0 \pmod{\mathfrak{a}}$. Then we would have $\delta(\mathfrak{p}) = g$, whence, by the theorem of Riemann-Roch, $l(\mathfrak{p}^{-1}) = d(\mathfrak{p}) - g + 1 + g = d(\mathfrak{p}) + 1$, and there would exist a non constant $x \in R$ which is multiple of $\mathfrak{p}^{-1}$; we would obviously have $\nu_{\mathfrak{p}}(x) = -1$.

Let ω be any differential $\neq 0$ of the first kind; then $x^{-\nu_{\mathfrak{p}}(\omega)}\omega$ would be of the first kind but would not have $\mathfrak{p}$ as a zero.

COROLLARY. *Let $\mathfrak{a}$ be an integral divisor of a field R of algebraic functions of one variable. If we have $l(\mathfrak{a}^{-1}) = d(\mathfrak{a}) + 1$, $d(\mathfrak{a}) > 0$, then R is of genus 0.*

For the equality $l(\mathfrak{a}^{-1}) = d(\mathfrak{a}) + 1$ implies, in virtue of the theorem of Riemann-Roch, $\delta(\mathfrak{a}) = g$, where g is the genus of R, and this means that every differential of the first kind of R is $\equiv 0 \pmod{\mathfrak{a}}$.

This being said, assume that R is of genus $g > 1$ and hyperelliptic, and denote by S the field generated by the ratios of differentials of the first kind of R. Since $2g - 2 > 0$, we can find a place $\mathfrak{p}$ of R which is a zero of some differential of the first kind of R (i.e., we have $\delta(\mathfrak{p}) > 0$). We shall denote by $\mathfrak{p}_0$ the place of S below $\mathfrak{p}$, and by $\mathfrak{s}$ the divisor $\mathrm{Con}_{S/R}\mathfrak{p}_0$. Let ω_0 be a differential of the first kind of R of which $\mathfrak{p}$ is not a zero. If ω is a differential which is $\equiv 0 \pmod{\mathfrak{p}}$, then $\omega = u\omega_0$, with $u \in R$, and, since ω and ω_0 are of the first kind, $u \in S$. It is clear that $\mathfrak{p}$ is a zero of u in R; therefore, $\mathfrak{p}_0$ is a zero of u in S, which implies that the divisor of zeros of u in R is divisible by $\mathfrak{s}$. Thus we see that every differential which is multiple of $\mathfrak{p}$ is also multiple of $\mathfrak{s}$.

LEMMA 2. *Let $\mathfrak{p}$ be a place of a field R of algebraic functions of one variable. Assume that every differential of R which is multiple of $\mathfrak{p}$ is also multiple of an integral divisor $\mathfrak{s}$ which is divisible by $\mathfrak{p}$, and that there exists at least one such differential which is $\neq 0$. Then we have $l(\mathfrak{s}^{-1}) = l(\mathfrak{p}^{-1}) + d(\mathfrak{s}\mathfrak{p}^{-1})$ and $d(\mathfrak{s}\mathfrak{p}^{-1}) \leqq d(\mathfrak{p})$.*

We have by assumption $\delta(\mathfrak{s}) = \delta(\mathfrak{p})$. On the other hand, we have $l(\mathfrak{p}^{-1}) = d(\mathfrak{p}) - g + 1 + \delta(\mathfrak{p})$, where g is the genus of R, and $l(\mathfrak{s}^{-1}) = d(\mathfrak{s}) - g + 1 + \delta(\mathfrak{s})$. The first assertion of Lemma 2 follows immediately from these formulas. Let u be any element of R which is multiple of $\mathfrak{s}^{-1}\mathfrak{p}$. Then, if ω is any differential which is multiple of $\mathfrak{p}$, and therefore also of $\mathfrak{s}$, $u\omega$ is again multiple of $\mathfrak{p}$. We shall see that u must be a constant. If not, u would have a pole $\mathfrak{q}$; let then m be the smallest value assumed by $\nu_{\mathfrak{q}}(\omega)$ when ω runs over the set of differentials which are multiples of $\mathfrak{p}$; then $\nu_{\mathfrak{q}}(\omega) \neq \infty$ since our set contains at least one differential $\neq 0$. There would exist a differential ω_1, multiple of $\mathfrak{p}$, for which $\nu_{\mathfrak{q}}(\omega_1) = m$; but then we would have $\nu_{\mathfrak{q}}(u\omega_1) < m$, which brings a contradiction. Thus we have $l(\mathfrak{s}^{-1}\mathfrak{p}) = 1$. Making use of Formula (4), II, §1, we see that $l(\mathfrak{s}^{-1}) - l(\mathfrak{s}^{-1}\mathfrak{p}) \leqq d(\mathfrak{p})$, whence, in virtue of the formula proved above, $l(\mathfrak{s}^{-1}\mathfrak{p}) \geqq l(\mathfrak{p}^{-1}) + d(\mathfrak{s}\mathfrak{p}^{-1}) - d(\mathfrak{p}) \geqq 1 + d(\mathfrak{s}\mathfrak{p}^{-1}) - d(\mathfrak{p})$; it follows that $d(\mathfrak{s}\mathfrak{p}^{-1}) \leqq d(\mathfrak{p})$.

This being said, let us return to the proof of Theorem 9 and to the notation introduced above. Among all places of R which lie above $\mathfrak{p}_0$, let $\mathfrak{p}'$ be one of smallest relative degree with respect to R. Then $\delta(\mathfrak{p}') > 0$ and every differential which is multiple of $\mathfrak{p}'$ is multiple of $\mathfrak{s}$. Thus we have $d(\mathfrak{s}\mathfrak{p}'^{-1}) \leqq d(\mathfrak{p}')$; it follows that either $\mathfrak{s} = \mathfrak{p} = \mathfrak{p}'$ or $\mathfrak{s}$ is the product of two places (distinct or not) which have the same relative degree f with respect to R. Assume that we are in the second case. Observe that any element of R which is multiple of $\mathfrak{s}^{-1}$ is in S (because, if y is such an element, and ω a differential $\neq 0$ which is multiple of $\mathfrak{s}$, then $y\omega$ is of the first kind). It follows immediately that $l(\mathfrak{s}^{-1}) = l(\mathfrak{p}_0^{-1})$ whence, by Lemma 2, $l(\mathfrak{p}_0^{-1}) \geqq 1 + d(\mathfrak{p}) = 1 + fd(\mathfrak{p}_0)$. But it is clear that we have $l(\mathfrak{p}_0^{-1}) \leqq 1 + d(\mathfrak{p}_0)$ (because $\delta(\mathfrak{p}_0)$ is at most equal to the genus of S). Thus we have $f = 1$ and $l(\mathfrak{p}_0^{-1}) = 1 + d(\mathfrak{p}_0)$. Making use of Theorem 1, §1, the equality $f = 1$ implies that $[R:S] = 2$. Making use of the corollary to Lemma 1, we see that S is of genus 0.

Now, we have to consider the case where $\mathfrak{s} = \mathfrak{p}$. Let $f = [R:S]$ be the relative degree of $\mathfrak{p}$ with respect to S. Denote by $\mathfrak{d}$ the divisor of some differential $\omega_0 \neq 0$ of the first kind which is not multiple of $\mathfrak{p}$, and by $\Sigma(\mathfrak{p})$ and $\Sigma(\mathfrak{p}_0)$ the residue fields of $\mathfrak{p}$ and $\mathfrak{p}_0$ respectively. Let ξ be any element of $\Sigma(\mathfrak{p})$ and η a differential of R which is multiple of $\mathfrak{p}^{-1}\mathfrak{d}$. Let $\mathfrak{x}$ be a repartition of R which assigns to $\mathfrak{p}$ some element of R which takes the value ξ at $\mathfrak{p}$ and 0 to every other place. Then $\eta(\mathfrak{x})$ depends only on ξ and η, for, if $\mathfrak{x}$ and $\mathfrak{x}'$ are two repartitions which satisfy our conditions, then $\mathfrak{x}' - \mathfrak{x} \equiv 0 \pmod{\mathfrak{p}}$ and, *a fortiori*, $\mathfrak{x}' - \mathfrak{x} \equiv 0 \pmod{\mathfrak{d}^{-1}\mathfrak{p}}$ Set $L_\eta(\xi) = \eta(\mathfrak{x})$. Then $\eta \rightarrow L_\eta$ is a linear mapping of the space $\mathfrak{D}$ of differentials $\equiv 0 \pmod{\mathfrak{p}^{-1}\mathfrak{d}}$ into the space of linear functions on $\Sigma(\mathfrak{p})$ (considered as a vector space over the field of constants K of R). The differential ω_0 is of the first kind and contained in $\mathfrak{D}$; we have $L_{\omega_0} = 0$, which shows that the dimension of the space formed by the functions L_η is $\leqq \dim \mathfrak{D} - 1$. Thus the subspace X of $\Sigma(\mathfrak{p})$ formed by the elements ξ such that $L_\eta(\xi) = 0$ for every $\eta \in \mathfrak{D}$ is of dimension $\geqq d(\mathfrak{p}) - \dim \mathfrak{D} + 1$. The differentials $\eta \in \mathfrak{D}$ are those which can be written in the form $y\omega_0$, with $y \in R$, $y \equiv 0 \pmod{\mathfrak{p}^{-1}}$. Thus we have $\dim \mathfrak{D} = l(\mathfrak{p}^{-1})$, and this is equal (by an argument used above) to $l(\mathfrak{p}_0^{-1})$. On the other hand, let ξ be an element of X, and let the repartition $\mathfrak{x}$ be defined as above. Making use of Theorem 2, II, §5, we see that there exists an element $u \in R$ such that $u - \mathfrak{x} \equiv 0 \pmod{\mathfrak{p}\mathfrak{d}^{-1}}$. Then u is multiple of $\mathfrak{d}^{-1}$ and takes the value ξ at $\mathfrak{p}$. The differential $u\omega_0$ being of the first kind, we have $u \in S$, whence $\xi \in \Sigma(\mathfrak{p}_0)$. Thus we have $\dim X \leqq d(\mathfrak{p}_0)$, whence $d(\mathfrak{p}_0) \geqq d(\mathfrak{p}) - l(\mathfrak{p}^{-1}) + 1 = fd(\mathfrak{p}_0) - l(\mathfrak{p}_0^{-1}) + 1$, and $l(\mathfrak{p}_0^{-1}) \geqq (f - 1)\, d(\mathfrak{p}_0) + 1$. But we know that $l(\mathfrak{p}_0^{-1}) \leqq d(\mathfrak{p}_0) + 1$; thus, we have $[R:S] = f = 2$ and $l(\mathfrak{p}_0^{-1}) = d(\mathfrak{p}_0) + 1$, which implies that S is of genus 0 (Corollary to Lemma 1). This concludes the proof of the fact that any hyperelliptic field of genus $g > 1$ is quadratic over a field of genus 0, viz. the field generated by the ratios of the differentials of the first kind.

Let conversely R be a field of algebraic functions of one variable which has a subfield S, containing the field of constants K of R, and which has the following properties: S is of genus 0 and $[R:S]$ is equal to 2. Assume furthermore that R is of genus $g > 1$. Let $\mathfrak{q}_0$ be a place of S of smallest possible degree; then we know that $d(\mathfrak{q}_0)$ is either 1 or 2 (II, §6). Let us first consider the case where either $d(\mathfrak{q}_0) = 1$ or $d(\mathfrak{q}_0) = 2$, $g \equiv 1 \pmod{2}$. In that case, we set $\mathfrak{a}_0 = \mathfrak{q}_0^{g-1}$ if $d(\mathfrak{q}_0) = 1$ or $\mathfrak{a}_0 = \mathfrak{q}_0^{(g-1)/2}$ if $d(\mathfrak{q}_0) = 2$; thus we have $d(\mathfrak{a}_0) = g - 1$. Set $\mathfrak{a} = \mathrm{Con}_{S/R}\mathfrak{a}_0$, whence $d(\mathfrak{a}) = 2g - 2$ (Lemma 1, §7). Since S is of genus 0, we have $l(\mathfrak{a}_0^{-1}) = d(\mathfrak{a}_0) + 1 = g$. Denote by $\mathfrak{L}$ the space of elements $z \in S$ which are multiples of $\mathfrak{a}_0^{-1}$, whence $\dim \mathfrak{L} = g$. We have $g = l(\mathfrak{a}_0^{-1}) \leqq l(\mathfrak{a}^{-1}) = d(\mathfrak{a}) - g + 1 + \delta(\mathfrak{a}) = g - 1 + \delta(\mathfrak{a})$, whence $\delta(\mathfrak{a}) > 0$. This means that there exists a differential $\omega \neq 0$ of R which is multiple of $\mathfrak{a}$. If $z \in \mathfrak{L}$, then $z\omega$ is of the first kind. Since $\dim \mathfrak{L} = g$, we see that every differential of the first kind of R is of the form $z\omega$, $z \in S$, from which it follows that R is hyperelliptic and that the field S' generated by the ratios of dfferentials of the first kind of R is contained in S. But we know from the first part of the proof that $[R:S'] = 2$; it follows that $S' = S$.

Consider now the case where $d(\mathfrak{q}_0) = 2$, $g \equiv 0 \pmod 2$. We then set $\mathfrak{a}_0 = \mathfrak{q}_0^{(g-2)/2}$, $\mathfrak{a} = \mathrm{Con}_{S/R}\mathfrak{a}_0$. By arguments similar to the ones which were used above, we see that $l(\mathfrak{a}_0^{-1}) = g - 1$, $\delta(\mathfrak{a}) > 0$. Let ω be a differential $\neq 0$ which is multiple of $\mathfrak{a}$; then the divisor $\mathfrak{d}(\omega)$ of ω is of the form $\mathfrak{a}\mathfrak{b}$, where $\mathfrak{b}$ is an integral divisor of degree 2. Since $d(\mathfrak{q}_0) = 2$, every place of S is of degree $\geqq 2$, and the same is true *a fortiori* of every place of R. Thus $\mathfrak{b}$ is a place $\mathfrak{p}$. If z is an element of S which is multiple of $\mathfrak{a}_0^{-1}$, then $z\omega$ is a multiple of $\mathfrak{p}$, whence $\delta(\mathfrak{p}) \geqq g - 1$. We have $l(\mathfrak{p}^{-1}) = d(\mathfrak{p}) - g + 1 + \delta(\mathfrak{p}) \geqq 2$. If w is a non constant element of R which is multiple of $\mathfrak{p}^{-1}$, then the divisor of poles of w is $\mathfrak{p}$, whence $[R:K\langle w \rangle] = 2$ by the corollary to Theorem 4, I, §8. Using instead of S the field $W = K\langle w \rangle$, which has a place of degree 1, we are in the case considered above, and we see that W is the field generated by the ratios of differentials of the first kind of R. We have $l(\mathfrak{q}_0^{-1}) = d(\mathfrak{q}_0) + 1 = 3$; let $\{1, u, v\}$ be a base for the space of elements of S which are multiples of $\mathfrak{q}_0^{-1}$. Then $\mathfrak{q}_0$ is the divisor of poles of u in S, whence $[S:K\langle u \rangle] = 2$. If $\mathfrak{q}_0'$ is the place of $K\langle u \rangle$ below $\mathfrak{q}_0$, then $l(\mathfrak{q}_0') = 2$, from which it follows immediately that $v \notin K\langle u \rangle$, whence $S = K\langle u, v \rangle$. If $g > 2$, then u and v are multiples of $\mathfrak{a}_0^{-1}$, which implies that $u\omega$ and $v\omega$ are of the first kind, and therefore that u and v are in W, whence $S \subset W$; but this is impossible, since S has no place of degree 1, while W has one. If $g = 2$, denote by $\mathfrak{p}_0$ the place of S below $\mathfrak{p}$, and by $\mathfrak{c}$ the divisor $\mathrm{Con}_{S/R}\mathfrak{p}_0$. Since $d(\mathfrak{p}) = 2$ and $d(\mathfrak{p}_0) \geqq 2$, we have $d(\mathfrak{p}_0) = 2$ and $d(\mathfrak{c}) = 4$ (Lemma 1, §7). Since $2g - 2 = 2$, we have $\delta(\mathfrak{c}) = 0$, whence $l(\mathfrak{c}^{-1}) = d(\mathfrak{c}) - 2 + 1 = 3$. On the other hand, we have $l(\mathfrak{p}_0^{-1}) = d(\mathfrak{p}_0) + 1 = 3$, and every element of S which is multiple of $\mathfrak{p}_0^{-1}$ is multiple of $\mathfrak{c}^{-1}$ in R. Since $l(\mathfrak{c}^{-1}) = l(\mathfrak{p}_0^{-1})$, we conclude that every element of R which is multiple of $\mathfrak{c}^{-1}$ is in S. This applies in particular to w, since $\mathfrak{p}$ divides $\mathfrak{c}$. Thus we see that S contains $K\langle w \rangle$. Since $[R:S] = [R:K\langle w \rangle] = 2$, we have $S = K\langle w \rangle$. But this is impossible for the same reason as above. Thus we see that the second case cannot occur. Theorem 9 is now completely proved.

The field S of Theorem 9, being of genus 0, has a place of degree $\leqq 2$, from which it follows that *a hyperelliptic field of genus $g > 1$ has a place of degree $\leqq 4$*. Moreover, the last part of the proof shows that, *if g is even, then R has a place of degree $\leqq 2$*. If R is separable over S (which certainly happens if R is of characteristic $\neq 2$), then it can be proved that at least one place of R is ramified with respect to S (cf. Corollary to Theorem 2, VI, §2) from which it follows that R has a place of degree $\leqq 2$ and has a place of degree 1 if g is even.

THEOREM 10. *Let R be a field of algebraic functions of one variable of genus $g > 1$. Assume that R has a place $\mathfrak{p}$ of degree 1 and is not hyperelliptic. Then, given any* *place* $\mathfrak{q} \neq \mathfrak{p}$ *of* R, *there exists a differential* ω *of the first kind of* R *such that* $\nu_{\mathfrak{p}}(\omega) = 1$, $\nu_{\mathfrak{q}}(\omega) = 0$.

We have $1 \leqq l(\mathfrak{p}^{-1}) = d(\mathfrak{p}) - g + 1 + \delta(\mathfrak{p}) = 2 - g + \delta(\mathfrak{p})$; since $g \geqq 2$, it follows that $\delta(\mathfrak{p}) > 0$; there exists at least one differential $\neq 0$ which is multiple of $\mathfrak{p}$. Let $\mathfrak{s}$ be the integral divisor of largest degree of which every differential which is a multiple of $\mathfrak{p}$ is a multiple. Then it follows from Lemma 2 that $d(\mathfrak{s}\mathfrak{p}^{-1}) \leqq d(\mathfrak{p}) = 1$, whence $d(\mathfrak{s}) \leqq 2$. If we had $d(\mathfrak{s}) = 2$, then it would follow from Lemma 2 that $l(\mathfrak{s}^{-1}) = l(\mathfrak{p}^{-1}) + 1 \geqq 2$, and there would exist a non constant $x \in R$ such that $x \equiv 0 \pmod{\mathfrak{s}^{-1}}$. The divisor of poles of x would be either $\mathfrak{p}$ or $\mathfrak{s}$, and we would have, by the corollary to Theorem 4, I, §8, $[R:K\langle x \rangle] =$ either 1 or 2; but then R would be either of genus 0 or 1 or hyperelliptic, which is impossible. Thus we have $\mathfrak{s} = \mathfrak{p}$. It follows that there exist differentials ω_1, ω_2 of R which are multiples of $\mathfrak{p}$ and such that $\nu_{\mathfrak{p}}(\omega_1) = 1$, $\nu_{\mathfrak{q}}(\omega_2) = 0$. If either $\nu_{\mathfrak{q}}(\omega_1) = 0$ or $\nu_{\mathfrak{p}}(\omega_2) = 1$, then either $\omega = \omega_1$ or $\omega = \omega_2$ satisfies our conditions; if $\nu_{\mathfrak{q}}(\omega_1) > 0$ and $\nu_{\mathfrak{p}}(\omega_2) > 1$, then we may take $\omega = \omega_1 + \omega_2$. Theorem 10 is thereby proved.

From the point of view of algebraic geometry, Theorem 10 allows one to construct the "normal model" of R, a curve in g-dimensional projective space which has no singularity, and whose field of rational functions is R.

CHAPTER V
EXTENSIONS OF THE FIELD OF CONSTANTS

§1. Separable transcendental extensions

We propose to generalize for arbitrary extensions the notion of separability, which is usually only defined for algebraic extensions. To do so, we shall make use of the notion of Kronecker product of commutative algebras, introduced in IV, §4.

Lemma 1. *Let K be a field, and let M be an overfield of K which is algebraic over K. If M is separable over K, then, if L is any overfield of K, the algebra $L \otimes M$ has no nilpotent element $\neq 0$. If M is not separable over K, then there exists an overfield L of K such that $L \otimes M$ contains a nilpotent element $\neq 0$; L may be taken to be of finite degree over K and such that $L^p \subset K$, where p is the characteristic of K.*

Assume first that M is separable over K, and let $x = \sum_{k=1}^{h} u_k v_k$ be a nilpotent element of $L \otimes M$ ($u_k \in L, v_k \in M, 1 \leqq k \leqq h$). Then the field $M_1 = K\langle v_1, \cdots, v_h\rangle$ is separable and of finite degree over K, and $x \in L \otimes M_1$. We may write $M_1 = K\langle v\rangle$, where v is some element of M_1; let $F(X)$ be an irreducible polynomial with coefficients in K such that $F(v) = 0$. If F is of degree n, then $1, v, \cdots, v^{n-1}$ form a base of M_1 with respect to K, and therefore also of $L \otimes M_1$ with respect to L. It follows immediately that $L \otimes M_1$ is isomorphic (as a ring) with the ring $L[X]/\mathfrak{f}$, where $\mathfrak{f}$ is the ideal generated in $L[X]$ by $F(X)$. We may write $x = G(v)$, where G is a polynomial with coefficients in L. If $x^a = 0$, then G^a must be divisible by F. Now, since v is separable over K, F is not divisible by the square of any irreducible polynomial with coefficients in L, and it follows that not only G^a but G itself is divisible by F, whence $x = 0$.

Now, assume that M is not separable over K. Then M contains an element v which is not separable over K. Let again $F(X)$ be an irreducible polynomial with coefficients in K such that $F(v) = 0$. Then we know that $F = F_1(X^p)$, where F_1 is a polynomial with coefficients in K. Let L be an overfield of K in which the coefficients of F are all p-th powers; we see that we may take for L an overfield of finite degree of K such that $L^p \subset K$. Then $F(X) = (G(X))^p$, where G is a polynomial with coefficients in L. Since G is of lower degree than F, it is clear that the element $G(v)$ of $L \otimes M$ is $\neq 0$; but we have $(G(v))^p = 0$. Lemma 1 is thereby proved.

If K is a field of characteristic $p > 0$, we shall say that an overfield L of K is a *radical extension* of K if $L^p \subset K$; if K is of characteristic 0, then we say that L is a radical extension of K if and only if $L = K$. A radical extension will be called *finite* if it is of finite degree. An overfield M of K will be called *separable* over K if the following condition is satisfied: L being any finite radical extension of K, the algebra $L \otimes M$ has no nilpotent element $\neq 0$. It follows immedately that every extension of a field of characteristic 0 is separable, and that our new notion of separability coincides with the classical one for algebraic extensions.

LEMMA 2. *Let L and M be overfields of a field K. If L is purely unseparable over K, then every zero divisor in $L \otimes M$ is nilpotent.*

Let $x = \sum_{k=1}^{h} u_k v_k$ ($u_k \in L$, $v_k \in M$) be a zero divisor in $L \otimes M$. Since L is purely unseparable over K, we can find an exponent $a > 0$ such that $u_k^{p^a} \in K$ ($1 \leqq k \leqq h$), whence $x^{p^a} = \sum_{k=1}^{h} u_k^{p^a} v_k^{p^a} \in M$. Were $x^{p^a} \neq 0$, it would have an inverse in M, which is obviously impossible since x is a zero divisor. Lemma 2 is thereby proved.

COROLLARY. *Let M be an overfield of a field K which is separable over K, and let L be a finite radical extension of K. Then $L \otimes M$ is a field and is a finite radical extension of M; we have $[L \otimes M:M] = [L:K]$.*

It follows from the definition and from Lemma 2 that $L \otimes M$ has no zero divisor $\neq 0$. If $L = K[u_1, \cdots, u_r]$, we have $L \otimes M = M[u_1, \cdots, u_r]$ and $u_k^p \in K \subset M$ ($1 \leqq k \leqq r$), which proves that $L \otimes M$ is a field and a finite radical extension of M. The last assertion of the corollary follows from the very definition of Kronecker products.

LEMMA 3. *Let M be an overfield of a field K, and let M_1 be an intermediary field between K and M. If M is separable over K, then the same is true of M_1. If M_1 is separable over K and M over M_1, then M is separable over K.*

Let L be a finite radical extension of K. The first assertion follows immediately from the fact that $L \otimes M_1 \subset L \otimes M$. Now, assume that M_1 is separable over K and M over M_1. Then we have $(L_M) = (L_{M_1})_M$. By the corollary to Lemma 2, L_{M_1} is a field and a finite radical extension of M_1. Since M is separable over M_1, L_M has no nilpotent element $\neq 0$.

LEMMA 4. *Let L and M be overfields of a field K. Assume that M is purely transcendental over K; then $L \otimes M$ has no zero divisor $\neq 0$, and the field of quotients U of this ring is purely transcendental over L. If $M = K\langle x_i \rangle_{i \in I}$, the x_i's being algebraically independent over K, then $U = L\langle x_i \rangle_{i \in I}$, and the elements x_i are algebraically independent over L.*

The elements $x_{i_1}^{e_1} \cdots x_{i_n}^{e_n}$ (where $i_1, \cdots, i_n$ are any distinct elements of I, and $e_1, \cdots, e_n$ exponents $\geqq 0$) are linearly independent over K in M, and therefore also over L in $L \otimes M$. It follows immediately that the ring $L[x_i]_{i \in I}$ has no zero divisor $\neq 0$. Let u, u' be elements of $L \otimes M$ such that $uu' = 0$; then we may write $u = \sum_{k=1}^{h} y_k z_k$, $u' = \sum_{k=1}^{h'} y_k' z_k'$, with y_k, y_k' in L and z_k, z_k' in M. We can find a $v \neq 0$ in $K[x_i]_{i \in I}$ such that the elements vz_k ($1 \leqq k \leqq h$), vz_k' ($1 \leqq k \leqq h'$) all lie in $K[x_i]_{i \in I}$. Then we have $vu \in L[x_i]_{i \in I}$, $vu' \in L[x_i]_{i \in I}$, and $(vu)(vu') = 0$. It follows that one at least of vu, vu' is 0. Since v has an inverse in M and therefore also in $L \otimes M$, it follows that one of u, u' is 0. The field of quotients of $L[x_i]_{i \in I}$, containing L and M, contains U, which proves Lemma 4.

We shall say that an overfield M of a field K is *separably generated* over K if there exists an intermediary field T between K and M which is purely transcendental over K and over which M is algebraic and separable. A field R of algebraic functions of one variable is said to be *separably generated* if it is separably generated over its field of constants K; if this is the case, any element x of R such that R is algebraic and separable over $K\langle x\rangle$ is called a *separating variable* in R. It follows immediately from Lemmas 3 and 4 that an overfield M of K which is separably generated over K is separable over K. The converse of this statement is not true in general. However, we have the following partial converse:

LEMMA 5. *Let M be an overfield of a field K which is separable over K and which can be obtained from K by adjunction of a finite number of elements. Then M is separably generated over K.*

Write $M = K\langle y_1, \cdots, y_r\rangle$, where $y_1, \cdots, y_r$ are elements of M. Denote by p the characteristic of K, which we may assume to be >0, and by P the field $K\langle M^p\rangle$ obtained by adjunction to K of the p-th powers of elements of M. Then we have $M = P\langle y_1, \cdots, y_r\rangle$. Let s be the smallest integer such that M can be obtained from P by adjunction of s of the elements y_i; assume that $M = P\langle y_1, \cdots, y_s\rangle$. We shall prove that $y_1, \cdots, y_s$ are algebraically independent over K. Assuming for a moment that this is not the case, then there exists an irreducible polynomial $F(Y_1, \cdots, Y_s)$ with coefficients in K such that

$$F(y_1, \cdots, y_s) = 0.$$

We may assume that F is a polynomial of the smallest possible degree with these properties; let then d be this degree. If $1 \leqq i \leqq s$, we can write

$$F = \sum_{k=0}^{p-1} G_k(X_1, \cdots, X_s)X_i^k,$$

where $G_k(X_1, \cdots, X_s)$ is a polynomial with coefficients in K in which X_i occurs only with exponents which are $\equiv 0 \pmod p$. Thus, $G_k(y_1, \cdots, y_s)$ belongs to the field P_i obtained by adjunction to P of the elements y_j for $j \neq i$ (because $y_i^p \in P$). Since $y_i^p \in P_i$ but $y_i \notin P_i$, y_i cannot satisfy any equation of degree $<p$ with coefficients in P_i; it follows that $G_k(y_1, \cdots, y_s) = 0$ $(0 \leqq k \leqq p - 1)$. The polynomial F having been selected of smallest possible degree, it follows that $F = G_0$, $G_k = 0$ for $k > 0$. This being true for every i $(1 \leqq i \leqq s)$, we see that we can write $F(Y_1, \cdots, Y_s) = H(Y_1^p, \cdots, Y_s^p)$, where H is a polynomial with coefficients in K. Let L be the field which is generated by adjunction to K of the p-th roots of the coefficients of H; then

$$F(Y_1, \cdots, Y_s) = (J(Y_1, \cdots, Y_s))^p,$$

where J is a polynomial with coefficients in L of degree $p^{-1}d$. In virtue of our choice of F as of smallest possible degree, the monomials in $y_1, \cdots, y_s$ of degrees $<d$ are linearly independent over K in M and therefore also over L in $L \otimes M$. It follows that $J(y_1, \cdots, y_s) \neq 0$, while $(J(y_1, \cdots, y_s))^p = 0$,

in contradiction to the assumption that M is separable over K. Our assertion that $y_1, \cdots, y_s$ are algebraically independent over K is thereby proved. Let T_0 be the field obtained by adjunction of $y_1, \cdots, y_s$ to K; then $T_0\langle M^p\rangle$ contains $P\langle y_1, \cdots, y_s\rangle = M$, whence $T_0\langle M^p\rangle = M$. If M is algebraic over T_0 (which, actually, necessarily happens, although we do not prove it here), then it is also separable over T_0. For, let N be the field of elements of M which are separable over T_0. Then we know that $N \supset M^{p^a}$ for some $a > 0$; on the other hand, since $M = T_0\langle M^p\rangle$, we see easily that $M = T_0\langle M^p\rangle = T_0\langle M^{p^2}\rangle = \cdots = T_0\langle M^{p^a}\rangle$, whence $M = N$. Were M not algebraic over T_0, we could extract from the set $\{y_{s+1}, \cdots, y_r\}$ a transcendence base of M over T_0. Adjoining the elements of this transcendence base to T_0, we would obtain a field T, purely transcendental over K and such that M is algebraic over T. Since $T_0 \subset T$, we would have $M = T\langle M^p\rangle$, and the same argument as above would show that M is separable over T.

COROLLARY 1. *A necessary and sufficient condition for a field R of algebraic functions of one variable to be separable over its field of constants is for R to be separably generated.*

COROLLARY 2. *Let L and M be overfields of a field K. If M is separable over K, the algebra $L \otimes M$ does not have any nilpotent element $\neq 0$.*

For, let $x = \sum_{k=1}^{h} u_k v_k$ ($u_k \in L$, $v_k \in M$, $1 \leqq k \leqq h$) be a nilpotent element of $L \otimes M$. Then x belongs to $L \otimes M'$, where $M' = K\langle v_1, \cdots, v_h\rangle$. Since M' is separable over K (Lemma 3), it is algebraic and separable over a field T containing K and purely transcendental over K (Lemma 5). We have $L_{M'} = (L_T)_{M'}$, and L_T is contained in a field U (Lemma 4), whence $L_{M'} \subset U_{M'}$ (where U is considered as an algebra over T). Since M' is algebraic and separable over T, $U_{M'}$ has no nilpotent element $\neq 0$ (Lemma 1), whence $x = 0$.

§2. RELATIVELY ALGEBRAICALLY CLOSED SUBFIELDS

A subfield K of a field L is called *relatively algebraically closed* in L if every element of L which is algebraic over K is already contained in K.

LEMMA 1. *Let K be a relatively algebraically closed subfield of a field L. If a polynomial $F(X_1, \cdots, X_n)$ in n letters with coefficients in K is irreducible in*

$$K[X_1, \cdots, X_n],$$

then it is also irreducible in $L[X_1, \cdots, X_n]$.

1) Let us first consider the case where $n = 1$. Let $G(X_1)$ be a factor of $F(X_1)$ in $L[X_1]$; assume that G is of degree >0 and has leading coefficient 1. In a suitable extension L' of L, G factors into polynomials of the first degree, say

$$G(X_1) = \prod_{i=1}^{r} (X_1 - x_i).$$

We have $F(x_i) = 0$ $(1 \leqq i \leqq r)$, which proves that the elements x_i are algebraic over K. It follows immediately that the coefficients of G are themselves algebraic over K. But these coefficients are in L; they are therefore in K. Since F is irreducible in $K[X_1]$, G can differ from F only by a constant factor, which proves Lemma 1 if $n = 1$.

2) In the general case, we shall make use of an artifice due to Kronecker. Let D be the degree of F; set $a_i = (D+1)^i \cdot i!$ $(1 \leqq i \leqq n)$. If $H(X_1, \cdots, X_n)$ is any polynomial in $X_1, \cdots, X_n$, denote by H^* the polynomial deduced from H by the substitution $X_i \to T^{a_i}$ $(1 \leqq i \leqq n)$, where T is a new letter. If $M_1 = X_1^{e_1} \cdots X_n^{e_n}$ and $M_2 = X_1^{f_1} \cdots X_n^{f_n}$ are monomials of degrees $\leqq D$, then the equality $M_1^* = M_2^*$ implies $M_1 = M_2$. For, assume that $\sum_{i=1}^n e_i a_i = \sum_{i=1}^n f_i a_i$ and that $e_i = f_i$ for $i > r$ (where $1 \leqq r \leqq n$). Then

$$a_r(f_r - e_r) = -\sum_{i<r} a_i(f_i - e_i).$$

But $|f_i - e_i| \leqq D$ for every i, whence

$$\left|\sum_{i<r} a_i(f_i - e_i)\right| \leqq rD(D+1)^{r-1}(r-1)! < (D+1)^r r! = a_r,$$

and therefore $e_r = f_r$. It follows that, if H is any polynomial of degree $\leqq D$, then the coefficients of H^* are those of H. Call leading coefficient of a polynomial H of degree $\leqq D$ the leading coefficient of H^*. We may then write

$$F = aG_1 \cdots G_h,$$

where $G_1, \cdots, G_h$ are irreducible polynomials with coefficients in L with leading coefficient 1; a is the leading coefficient of F. We have $F^* = aG_1^* \cdots G_h^*$, from which it follows that each G_i^* is a product of irreducible factors of F^* in $L[T]$ with leading coefficient 1. Now, it follows immediately from the first part of the proof that the irreducible factors of F^* in $L[T]$ with leading coefficient 1 are the irreducible factors of F^* in $K[T]$ with leading coefficient 1. Thus the coefficients of each G_i^* are in K, and the same is true of the coefficients of each G_i, since G_i is of degree $\leqq D$. This shows that F is irreducible in $L[X_1, \cdots, X_n]$.

COROLLARY. *Let K be a subfield of a field L and let $F(X_1, \cdots, X_n)$ be a polynomial in n letters with coefficients in K. We may then factor F in irreducible polynomials in $L[X_1, \cdots, X_n]$ whose coefficients are algebraic over K.*

The elements of L which are algebraic over K form a field K'; we may then take the irreducible factors of F in $K'[X_1, \cdots, X_n]$.

LEMMA 2. *Let K be a relatively algebraically closed subfield of a field L. Let $M = L\langle x_i\rangle_{i \epsilon I}$ be an overfield of L which can be obtained from L by adjunction of elements x_i which are algebraically independent over L. Then $K\langle x_i\rangle_{i\epsilon I}$ is relatively algebraically closed in M.*

Let y be an element of M which is algebraic over $K\langle x_i\rangle_{i\epsilon I}$. It is clear that we can find a finite number of indices $i_1, \cdots, i_m$ in I with the following properties: y is algebraic over the field $K\langle x_{i_1}, \cdots, x_{i_m}\rangle$ and is in $L\langle x_{i_1}, \cdots, x_{i_m}\rangle$. Let

$F(X_1, \cdots, X_m, Y)$ be an irreducible polynomial in $m + 1$ letters with coefficients in K such that $F(x_{i_1}, \cdots, x_{i_m}, y) = 0$. Then it follows from Lemma 1 that F is also irreducible in $L[X_1, \cdots, X_m, Y]$; it follows that

$$F(x_{i_1}, \cdots, x_{i_m}, Y)$$

is irreducible in $L\langle x_{i_1}, \cdots, x_{i_m}\rangle[Y]$. Since $y \in L\langle x_{i_1}, \cdots, x_{i_m}\rangle$, F is of degree 1 in Y, whence $y \in K\langle x_{i_1}, \cdots, x_{i_m}\rangle$.

LEMMA 3. *Let L and R be overfields of a field K. If every element of R which is algebraic over K is purely unseparable over K, then every zero divisor in $L \otimes R$ is nilpotent.*

Any zero divisor in $L \otimes R$ is already a zero divisor in an algebra of the form $L' \otimes R'$, where L' and R' are subfields of L and R respectively each of which can be obtained from K by adjunction of a finite number of elements. We may therefore assume without loss of generality that L and R can both be obtained from K by adjunction of a finite number of elements. Let R_1 be the field formed by the elements of R which are algebraic over K. Then, if K is of characteristic $p > 0$, there is an exponent e such that $R_1^{p^e} \subset K$. For, R is algebraic of finite degree over a field R_0 which is purely transcendental over K, from which it follows that there exists an $e > 0$ such that $R_0\langle R^{p^e}\rangle$ is separable over R_0. It follows from Lemmas 3 and 4, §1 that $R_0\langle R^{p^e}\rangle$ is separable over K. In particular, $K\langle R_1^{p^e}\rangle$ is separable over K; being also purely unseparable over K, it is identical with K. Let B be a transcendence base of L with respect to K, and set $T = K\langle B\rangle$. Then we know that $T \otimes R$ has no zero divisor $\neq 0$, and that, if S is the field of quotients of this ring, then $S = R\langle B\rangle$, the elements of B being algebraically independent over R in S (Lemma 4, §1). Let S_1 be the subfield $T\langle R_1\rangle$ of S; then it follows from Lemma 2 that S_1 is relatively algebraically closed in S. On the other hand, we have $S_1^{p^e} \subset T$. Thus we see that every element of S which is algebraic over T is purely unseparable over T. The algebra R_L is $(R_T)_L$ and is therefore contained in S_L (S and L being considered as algebras over T). Since L is of finite degree over T, we see that it will be sufficient to prove Lemma 3 in the case where L is of finite degree over K.

The field L is then purely unseparable and of finite degree over a field L_0 of the form $K\langle y\rangle$, where y is separable over K. Let $F(Y) = 0$ be an irreducible equation with coefficients in K which is satisfied by y. If n is the degree of F, then $1, y, \cdots, y^{n-1}$ form a base of L_0 over K and therefore also of $(L_0)_R$ over R. It follows that $(L_0)_R$ is isomorphic with $R[Y]/\mathfrak{f}$, where $\mathfrak{f}$ is the ideal generated by $F(Y)$ in the polynomial ring $R[Y]$. Let $F_1(Y)$ be an irreducible factor of $F(Y)$ in $R_1[Y]$. If K is of characteristic $p > 0$, then $F_1^{p^e}(Y) \in K[Y]$, and is therefore divisible by $F(Y)$, which proves that F is (except for a constant factor) a power of F_1. If K is of characteristic 0 we have $R_1 = K$, $F_1 = F$ (except for a constant factor). On the other hand, it follows from Lemma 1 that $F_1(Y)$ remains irreducible in $R[Y]$. Let $x = G(y)$ be any element of $(L_0)_R$, G being a polynomial with coefficients in R. If G is divisible by F_1, then x is nilpotent. If not, G is relatively prime to F_1, and therefore also to F, and there exist polynomials U and V with coefficients in R such that $GU + FV = 1$, whence $xU(y) = 1$. Thus we see that every element of $(L_0)_R$ either is nilpotent or has an inverse. This completes the proof of Lemma 3 in the case where the characteristic p of K is 0. Assume now that $p > 0$. If x' is a zero divisor in R_L, we may write $x' = \sum_{k=1}^{h} u_k v_k$, with $u_k \in L$, $v_k \in R$ $(1 \leqq k \leqq h)$. Since L is purely unseparable over R_0, there exists an exponent $c > 0$ such that $u_k^{p^c} \in L_0$ $(1 \leqq k \leqq h)$, whence $x'^{p^c} = \sum_{k=1}^{h} u_k^{p^c} v_k^{p^c} \in (L_0)_R$. Since x' is a zero divisor, x'^{p^c} cannot have an inverse, and is therefore nilpotent; Lemma 3 is thereby proved.

LEMMA 4. *Let R be an overfield of a field K which contains an element y which is algebraic but not purely unseparable over K. Then there exists an overfield L of K which is of finite degree and separable over K such that $L \otimes R$ contains a zero divisor which is not nilpotent.*

It is well known that some power z of y is separable over K but not contained in K; we take $L = K\langle z \rangle$. Since $L \subset R$, the identification of L with a subfield of $L \otimes R$ would create confusions in this case. Let us therefore denote by z' the element $z \otimes 1$ of $L \otimes R$. The element z is a root of an irreducible equation $F(Z) = 0$ with coefficients in K. If n is the degree of F, then $1, z', \cdots, z'^{n-1}$ form a base of L_R with respect to R. We have

$$0 = F(z) - F(z') = (z - z')G(z, z'),$$

where $G(Z, Z')$ is a polynomial with coefficients in K whose degree with respect to Z' is $n - 1$. It follows that $G(z, z') \neq 0$ and that $z - z'$ is a zero divisor. On the other hand, z is a simple root of the equation $F(Z) = 0$ which is of degree >1. It follows that no power of $Z - z$ is divisible by $F(Z)$ in $R[Z]$, which proves that $z' - z$ is not nilpotent.

§3. COMMUTATIVE ALGEBRAS

Let $\mathfrak{A}$ be a commutative algebra over a field K. Denote by $\mathfrak{n}$ the set of nilpotent elements of $\mathfrak{A}$. Then $\mathfrak{n}$ *is an ideal in* $\mathfrak{A}$. For, it is first clear that the conditions $x \in \mathfrak{n}$, $a \in K$, $z \in \mathfrak{A}$ imply $ax \in \mathfrak{n}$, $zx \in \mathfrak{n}$. On the other hand, let x and y be elements of $\mathfrak{n}$. Then there exist exponents a and b such that $x^a = 0$, $y^b = 0$; expanding $(x - y)^{a+b}$ by the binomial formula (which is permissible in a commutative ring), we see that each term of the expansion either involves x with an exponent $\geqq a$ or y with an exponent $\geqq b$, whence $(x - y)^{a+b} = 0$ and $x - y \in \mathfrak{n}$, which completes the proof of the fact that $\mathfrak{n}$ is an ideal. The ideal $\mathfrak{n}$ is called the *radical* of the algebra $\mathfrak{A}$.

Since the radical $\mathfrak{n}$ of the algebra $\mathfrak{A}$ is an ideal in $\mathfrak{A}$, we may construct the factor algebra $\mathfrak{A}/\mathfrak{n}$.

LEMMA 1. *Let $\mathfrak{n}$ be the radical of an algebra $\mathfrak{A}$. Then the radical of $\mathfrak{A}/\mathfrak{n}$ is $\{0\}$. If every zero divisor in $\mathfrak{A}$ is nilpotent, then $\mathfrak{A}/\mathfrak{n}$ has no zero divisor $\neq 0$.*

Let x^* and y^* be elements of $\mathfrak{A}/\mathfrak{n}$, and let x and y be elements of $\mathfrak{A}$ which belong to the residue classes x^* and y^* respectively modulo $\mathfrak{n}$. If x^* is nilpotent, let a be an exponent such that $(x^*)^a = 0$. Then $x^a \in \mathfrak{n}$, and there exists an exponent a' such that $(x^a)^{a'} = 0$; it follows that $x \in \mathfrak{n}$, whence $x^* = 0$. Now, assume that $x^*y^* = 0$ and that every zero divisor in $\mathfrak{A}$ is nilpotent. Then there exists an exponent m such that $x^m y^m = 0$. If $y^m = 0$, then y is nilpotent, whence $y^* = 0$. If $y^m \neq 0$, then either $x^m = 0$ or x^m is a zero divisor and therefore nilpotent, which implies that x is nilpotent, whence $x^* = 0$.

Lemma 2. *If the algebra $\mathfrak{A}$ is the direct sum of a certain number of its ideals, say of $\mathfrak{A}_1, \cdots, \mathfrak{A}_h$, then any product of an element of $\mathfrak{A}_i$ by an element of $\mathfrak{A}_j$ is 0 if $i \neq j$, and the radical of $\mathfrak{A}$ is the direct sum of the radicals of $\mathfrak{A}_1, \cdots \mathfrak{A}_h$.*

The product of an element of $\mathfrak{A}_i$ by an element of $\mathfrak{A}_j$ is in both $\mathfrak{A}_i$ and $\mathfrak{A}_j$, and is therefore 0 if $i \neq j$. It follows that, if x_i and y_i are elements of $\mathfrak{A}_i$ $(1 \leqq i \leqq h)$, then we have $(\sum_{i=1}^h x_i)(\sum_{i=1}^h y_i) = \sum_{i=1}^h x_i y_i$. In particular, we have, for any exponent $a > 0$, $(\sum_{i=1}^h x_i)^a = \sum_{i=1}^h x_i^a$, which proves that $\sum_{i=1}^h x_i$ is nilpotent if and only if each x_i is nilpotent.

Remark. The formula $(\sum_{i=1}^h x_i)(\sum_{i=1}^h y_i) = \sum_{i=1}^h x_i y_i$ means that, under the assumption of Lemma 2, $\mathfrak{A}$ is isomorphic with the (cartesian) product of the algebras $\mathfrak{A}_1, \cdots, \mathfrak{A}_h$.

We now turn to the consideration of commutative algebras which are of finite dimension over their basic field. An algebra of finite dimension is called *semi-simple* if its radical is $\{0\}$. If $\mathfrak{A}$ is any algebra of finite dimension, the factor algebra $\mathfrak{A}/\mathfrak{n}$ is clearly semi-simple.

An ideal $\mathfrak{a}$ of an algebra $\mathfrak{A}$ is called *minimal* if it is $\neq \{0\}$ but does not properly contain any other ideal of $\mathfrak{A}$ than $\{0\}$.

Lemma 3. *A commutative semi-simple algebra of finite dimension is the direct sum of its minimal ideals, and these minimal ideals are all fields.*

If $\mathfrak{A}$ is an algebra of finite dimension over a field K, then, among all ideals $\neq \{0\}$ of $\mathfrak{A}$, we can select one of smallest possible dimension (over K); such an ideal is clearly minimal, which shows that, if $\mathfrak{A} \neq \{0\}$, then $\mathfrak{A}$ has at least one minimal ideal. Assuming $\mathfrak{A}$ to be semi-simple, let $\mathfrak{a}$ be a minimal ideal in $\mathfrak{A}$. Let x be an element $\neq 0$ of $\mathfrak{a}$; then the set $\mathfrak{a}x$ of elements of the form yx, $y \in \mathfrak{a}$, is clearly an ideal contained in $\mathfrak{a}$, and is not $\{0\}$ because $x^2 \neq 0$ (since x would then be nilpotent). It follows that $\mathfrak{a}x = \mathfrak{a}$. This being true for every $x \neq 0$ in $\mathfrak{a}$, we see that $\mathfrak{a}$ is a field. Let $\mathfrak{a}_1, \cdots, \mathfrak{a}_h$ be distinct minimal ideals of $\mathfrak{A}$. Then the sum $\mathfrak{a}_1 + \cdots + \mathfrak{a}_h$ is direct. For, if $i \neq j$, then the ideal $\mathfrak{a}_i \cap \mathfrak{a}_j$, which is in $\mathfrak{a}_i$, can only be either $\{0\}$ or $\mathfrak{a}_i$; similarly, $\mathfrak{a}_i \cap \mathfrak{a}_j$ is either $\{0\}$ or $\mathfrak{a}_j$, whence $\mathfrak{a}_i \cap \mathfrak{a}_j = \{0\}$. This implies that $x_i x_j = 0$ if $x_i \in \mathfrak{a}_i$, $x_j \in \mathfrak{a}_j$. Assume then that $\sum_{i=1}^h x_i = 0$ with $x_i \in \mathfrak{a}_i$ $(1 \leqq i \leqq h)$; multiplying by x_i, we obtain $x_i^2 = 0$, whence $x_i = 0$, which proves that the sum $\mathfrak{a}_1 + \cdots + \mathfrak{a}_h$ is direct. The dimension of each $\mathfrak{a}_i$ being $\geqq 1$, we see that h is at most equal to the dimension of $\mathfrak{A}$; this proves that $\mathfrak{A}$ has only a finite number of minimal ideals. Assume now that $\mathfrak{a}_1, \cdots, \mathfrak{a}_h$ are all the minimal ideals of $\mathfrak{A}$. Each $\mathfrak{a}_i$, being a field, has a unit element e_i; set $e = \sum_{i=1}^h e_i$. Let $\mathfrak{a}'$ be the set of elements $x \in \mathfrak{A}$ such that $xe = 0$. Then $\mathfrak{a}'$ is clearly an ideal in $\mathfrak{A}$. Were $\mathfrak{a}' \neq \{0\}$, among all ideals of $\mathfrak{A}$ contained in $\mathfrak{a}'$, we could find one of smallest possible dimension. This ideal would be minimal and would therefore be one of $\mathfrak{a}_1, \cdots, \mathfrak{a}_h$. But this is impossible, because, since $e_ie_j = 0$ for $i \neq j$, we have $e_ie = e_i^2 = e_i \neq 0$. Thus we see that $\mathfrak{a}' = \{0\}$. If x is any element of $\mathfrak{A}$, we have $x = xe + (x - xe)$, and, since $e^2 = (\sum_{i=1}^h e_i)e = e$, $(x - xe)e = 0$, whence $x - xe = 0$. Since $xe = \sum_{i=1}^h xe_i \in \mathfrak{a}_1 + \cdots + \mathfrak{a}_h$, we have $\mathfrak{A} = \mathfrak{a}_1 + \cdots + \mathfrak{a}_h$. Lemma 3 is thereby proved.

LEMMA 4. *Let $\mathfrak{A}$ be a commutative algebra of finite dimension over a field K. Denote by $\mathfrak{n}$ the radical of $\mathfrak{A}$, and represent $\mathfrak{A}/\mathfrak{n}$ as the direct sum of fields $Z_1, \cdots, Z_h$ which are ideals in $\mathfrak{A}/\mathfrak{n}$. Then there exists for each i $(1 \leqq i \leqq h)$ a homomorphism φ_i of $\mathfrak{A}$ onto Z_i. Conversely, if φ is a homomorphism of $\mathfrak{A}$ into any field Z (containing K as a subfield) and if $\varphi(\mathfrak{A}) \neq \{0\}$, then there exists a uniquely determined index i such that φ can be written in the form $\psi\varphi_i$, where ψ is an isomorphism of Z_i into Z.*

REMARK. The homomorphisms which are considered in this statement are not only ring homomorphisms, but homomorphisms of algebras; i.e., if θ is any one of them, we have $\theta(ax) = a\theta(x)$ whenever $a \in K$.

Let x be any element of $\mathfrak{A}$. Denote by x^* the residue class of x modulo $\mathfrak{n}$; we may then write $x^* = \sum_{i=1}^h z_i$, with $z_i \in Z_i$ $(1 \leqq i \leqq h)$. If we assign to x the element z_i, we clearly obtain a homomorphism φ_i of $\mathfrak{A}$ onto Z_i. Let conversely φ be a homomorphism of $\mathfrak{A}$ into a field Z containing K. If $u \in \mathfrak{n}$, we have $(\varphi(u))^m = 0$ for some exponent $m > 0$, whence $\varphi(u) = 0$, since Z is a field. Thus, φ maps $\mathfrak{n}$ upon $\{0\}$ and defines in a natural way a homomorphism φ^* of $\mathfrak{A}/\mathfrak{n}$ into Z. If $\varphi(\mathfrak{A}) \neq \{0\}$, there exists an index i and an element $z_i \in Z_i$ such that $\varphi^*(z_i) \neq 0$. If $z_j \in Z_j$, $j \neq i$, then $z_iz_j = 0$, whence $\varphi^*(z_j) = 0$ since Z is a field. On the other hand, the restriction of φ^* to Z_i is a homomorphism of Z_i into Z, and therefore an isomorphism since Z_i is a field. Lemma 4 is thereby proved.

LEMMA 5. *Let L and M be overfields of a field K, M being of finite degree over K. Considering M as an algebra over K, construct the algebra M_L over L, deduced from M by extending the basic field from K to L. Assume that the factor algebra of M_L by its radical $\mathfrak{n}$ is the sum of h minimal ideals. Then there exists a subfield L' of L, which is of finite degree and separable over K which has the property that $M_{L'}$ is already the sum of h minimal ideals.*

Let L_0 be the set of elements of L which are algebraic and separable over K. Then L_0 is a field and is algebraic and separable over K. It follows that M_{L_0} is semi-simple (Lemma 1, §1). Let $T_1, \cdots, T_{h_0}$ be the distinct minimal ideals of M_{L_0}. Then M_{L_0} is the direct sum of $T_1, \cdots, T_{h_0}$, from which it follows immediately that $M_L = (M_{L_0})_L$ is the direct sum of the algebras $(T_1)_L, \cdots, (T_{h_0})_L$ which are ideals in it. Every element of L which is algebraic over L_0 is clearly purely unseparable over L_0; it follows that every zero divisor in $(T_i)_L$ is nilpotent

(Lemma 3, §2) and therefore that the factor algebra of $(T_i)_L$ by its radical $\mathfrak{n}_i$ has no zero divisor (Lemma 1). It follows immediately from Lemma 3 that this factor algebra is a field Z_i. Making use of Lemma 2, we see that $\mathfrak{n} = \mathfrak{n}_1 + \cdots + \mathfrak{n}_{h_0}$, and therefore that $(M_L)/\mathfrak{n}$ is the direct sum of $Z_1, \cdots, Z_{h_0}$ which are clearly ideals in it. It follows that $h = h_0$. Let e_i be the unit element of T_i. Then it is clear that there exists a subfield L' of L_0, containing K and obtainable from K by adjunction of a finite number of elements which is such that the h elements e_i $(1 \leqq i \leqq h)$ already belong to $M \otimes L'$. The principal ideal generated by e_i in $M_{L'}$ is contained in T_i and has therefore no zero divisor. It follows immediately from Lemma 3 that this ideal is a field T'_i. Since $e_ie_j = 0$ for $i \neq j$, the fields $T'_1, \cdots, T'_h$ are distinct. Since $\sum_{i=1}^h e_i$ is the unit element of $M_{L'}$, we have, for every $x \in M_{L'}$, $x = \sum_{i=1}^h xe_i \in T'_1 + \cdots + T'_h$, which proves that $M_{L'}$ is the sum of h minimal ideals.

§4. Definition of the extended field

Let R be a field of algebraic functions of one variable, and let K be the field of constants of R. Let L be an overfield of K; we propose to associate to the pair (R, L) a field S of algebraic functions of one variable over L. In order to do this, we first construct the algebra R_L deduced from R (regarded as an algebra over K) by extending the basic field from K to L. Denote by $\mathfrak{S}$ the factor ring of R_L by its radical $\mathfrak{n}$. Since K is relatively algebraically closed in R, it follows from Lemma 3, §2 and Lemma 1, §3 that $\mathfrak{S}$ has no zero divisor $\neq 0$. Let S be the field of quotients of $\mathfrak{S}$. The field R may be considered as a subalgebra of $R \otimes L$; being a field, it is mapped isomorphically onto a subfield R^* of $\mathfrak{S}$ under the natural homomorphism of R_L onto $\mathfrak{S} = R_L/\mathfrak{n}$. It is clear that every element of $\mathfrak{S}$ can be written as a linear combination of elements of R^* with coefficients in L; it follows that $S = R^* \langle L \rangle$. Let x be an element of R not in K, and let x^* be its residue class modulo $\mathfrak{n}$. Then R^* is of finite degree over $K\langle x^* \rangle$, from which it follows that S is of finite degree over $L\langle x^* \rangle$. We shall see that x^* is transcendental over L. Were this not the case, there would exist a polynomial $F \neq 0$ with coefficients in L such that $F(x^*) = 0$; there would therefore exist an exponent $m > 0$ such that $(F(x))^m = F^m(x) = 0$. But this is impossible, because the elements $1, x, \cdots, x^k, \cdots$, being linearly independent over K in R, are linearly independent over L in R_L. Thus we have proved that S is a field of algebraic functions over L.

From now on, we shall identify R with the subfield R^* of S, and we shall therefore write $S = L\langle R \rangle = R\langle L \rangle$; we shall say that *$S$ is the field deduced from R by adjoining the elements of L to its field of constants*. It should be observed that the notation $R\langle L \rangle$ may be misleading in the case where R and L are given *a priori* as subfields of some common field Ω, since in that case, $R\langle L \rangle$ already represents the subfield S_1 of Ω generated by R and L. However, we shall prove that, if S_1 is not algebraic over L, then there exists an isomorphism of S with S_1 which maps upon itself every element of R and every element of L. More generally, we shall prove the following result:

LEMMA 1. *Let R be a field of algebraic functions of one variable, and let L be an overfield of the field of constants K of R. Let Ω be a field containing K as a subfield. Assume that we are given an isomorphism ρ of R into Ω and an isomorphism λ of L into Ω, and that ρ and λ coincide with the identity on K. Then, if the subfield S_1 of Ω generated by $\rho(R)$ and $\lambda(L)$ is not algebraic over $\lambda(L)$, there exists an isomorphism σ of $R\langle L\rangle$ with S_1 which coincides with ρ on R and with λ on L.*

It follows from Lemma 1, IV, §4 that there exists a homomorphism φ of $R \otimes L$ into S_1 which coincides with ρ on R and with λ on L. Since S_1 has no zero divisor $\neq 0$, φ maps the radical $\mathfrak{n}$ of $R \otimes L$ upon $\{0\}$ and defines in a natural way a homomorphism φ^* of $\mathfrak{S} = (R \otimes L)/\mathfrak{n}$ into S_1; in view of our identification of R with a subring of $R_L/\mathfrak{n}$, we can say that φ^* coincides with ρ on R and with λ on L. We shall see that φ^* is actually an isomorphism. Set $R_1 = \rho(R)$ and $L_1 = \lambda(L)$. Let x be an element of R not in K; then R_1 is algebraic over $K\langle x_1\rangle$, where $x_1 = \rho(x)$, and S_1 is algebraic over $L_1\langle x_1\rangle$. Since S_1 is not algebraic over L_1, x_1 is transcendental over L_1. Let y be an element $\neq 0$ of $\mathfrak{S}$; then y satisfies an equation of the form $\sum_{i=0}^{n} F_i(x)y^{n-i} = 0$, where each F_i is a polynomial with coefficients in L and $F_n \neq 0$. Denote by G_i the polynomial deduced from F_i by submitting its coefficients to the operation λ. Then, if $y_1 = \varphi^*(y)$, we have $\sum_{i=0}^{n} G_i(x_1)y_1^{n-i} = 0$. Since $G_n \neq 0$ and x_1 is transcendental over L_1, we have $G_n(x_1) \neq 0$, whence $y_1 \neq 0$. This proves that φ^* is an isomorphism. We may therefore extend φ^* to an isomorphism σ of the field of quotients $R\langle L\rangle$ of $\mathfrak{S}$ into S_1; the image of $R\langle L\rangle$, containing R_1 and L_1 and being a field, is the whole of S_1. Lemma 1 is thereby proved.

COROLLARY 1. *Let R be a field of algebraic functions of one variable, and let L be an overfield of the field of constants K of R. Then every automorphism of L over K can be extended in a unique way to an automorphism of $R\langle L\rangle$ over R.*

This follows immediately from Lemma 1.

COROLLARY 2. *Let R be a field of algebraic functions of one variable, and let $L = K\langle B\rangle$ be a purely transcendental extension of K, the elements of B being algebraically independent over K. Then we have $R\langle L\rangle = R\langle B\rangle$ and the elements of B are algebraically independent over R.*

We may construct an overfield Ω of R of the form $R\langle B_1\rangle$, the elements of B_1 being in a one-to-one correspondence with those of B and being algebraically independent over R. Since Ω is not algebraic over $K\langle B_1\rangle$, it follows from Lemma 1 that $R\langle L\rangle$ is isomorphic with Ω under an isomorphism which maps B onto B_1.

LEMMA 2. *Let R be a separably generated field of algebraic functions of one variable, and let L be an overfield of the field of constants K of R. Then $R\langle L\rangle$ is separably generated.*

We can find an $x \in R$ such that R is algebraic and separable over $K\langle x\rangle$, whence $R = K\langle x, y\rangle$, with y separable over $K\langle x\rangle$. We have $R\langle L\rangle = L\langle x, y\rangle$, and y is obviously algebraic and separable over $L\langle x\rangle$.

THEOREM 1. *Let R be a field of algebraic functions of one variable and let L be an overfield of the field of constants K of R. Assume that one at least of the fields L or R is separable over K. Then any set of elements of R which are linearly independent over K remain linearly independent over L in $R\langle L\rangle$; any set of elements of L which are linearly independent over K remain linearly independent over R in $R\langle L\rangle$.*

It follows from our assumption that $R \otimes L$ has no nilpotent element $\neq 0$ (Corollary 2 to Lemma 5, §1). Therefore $R\langle L\rangle$ is the field of quotients of $R \otimes L$, and Theorem 1 follows immediately from the definition of a Kronecker product.

COROLLARY 1. *The assumptions being as in Theorem 1, if L is algebraic over K, then any base of L over K is also a base of $R\langle L\rangle$ over R. If L is of finite degree over K, then $[R\langle L\rangle : R] = [L:K]$, and we have, for any $a \in L$, $\mathrm{Sp}_{R\langle L\rangle/R}a = \mathrm{Sp}_{L/K}a$, $\mathrm{N}_{R\langle L\rangle/R}a = \mathrm{N}_{L/K}a$.*

The elements of a base of L over K are linearly independent over R in virtue of Theorem 1. Any element y of $R\langle L\rangle$ belongs to a field of the form $R\langle L'\rangle$, where L' can be obtained from K by adjunction of a finite number of elements of L. Since L is algebraic over K, L' is of finite degree over K, and it follows from Lemma 2, IV, §1 that y can be written as a linear combination of elements of L' (and therefore also of L) with coefficients in K. It follows immediately that any base of L over K is also a base of $R\langle L\rangle$ over R. The other assertions of Corollary 1 follow immediately.

COROLLARY 2. *The assumptions being as in Theorem 1, let also S be a subfield of R containing K and over which R is algebraic. Then any base of R over S is also a base of $R\langle L\rangle$ over $S\langle L\rangle$. We have $[R\langle L\rangle : S\langle L\rangle] = [R:S]$, and, if x is any element of R, $\mathrm{Sp}_{R\langle L\rangle/S\langle L\rangle}x = \mathrm{Sp}_{R/S}x$, $\mathrm{N}_{R\langle L\rangle/S\langle L\rangle}x = \mathrm{N}_{R/S}x$.*

If u is an element of S not in K, then R is of finite degree over $K\langle u\rangle$ and *a fortiori*, over S. Let $\{y_1, \cdots, y_n\}$ be a base of R over S. Since, obviously, $R\langle L\rangle = (S\langle L\rangle)\langle R\rangle$ and R is of finite degree over S, it follows from Lemma 2, IV, §1 that every element of $R\langle L\rangle$ can be written as a linear combination of $y_1, \cdots, y_n$ with coefficients in $S\langle L\rangle$. We shall prove that $y_1, \cdots, y_n$ are linearly independent over $S\langle L\rangle$. Assume that $\sum_{i=1}^n u_i y_i = 0$ with $u_i \in S\langle L\rangle$ $(1 \leqq i \leqq n)$. Since S is a subfield of R, one at least of the fields S or L is separable over K (Lemma 3, §1) and $S\langle L\rangle$ is therefore the field of quotients of $S \otimes L$ (cf. proof of Theorem 1). It follows that we can find an element $v \neq 0$ in $S\langle L\rangle$ such that $vu_i \in S \otimes L$ $(1 \leqq i \leqq n)$. Let $\{s_j\}_{j\in J}$ be a base of S with respect to K (J being some infinite set of indices). Then the elements $y_i s_j$ $(1 \leqq i \leqq n, j \in J)$ are linearly independent over K in R; for if $\sum_{i=1}^n \sum_{j\in J} a_{ij} y_i s_j = 0$, $a_{ij} \in K$, then we have $\sum_{i=1}^n (\sum_{j\in J} a_{ij} s_j) y_i = 0$, whence $\sum_{j\in J} a_{ij} s_j = 0$ for every i (because $y_1, \cdots, y_n$ are linearly independent over S) and therefore $a_{ij} = 0$ for every i and j. It follows that the elements $y_i s_j$ are linearly independent over L in $R \otimes L$. On the other hand, we may write $u_i v = \sum_{j\in J} b_{ij} s_j$ with $b_{ij} \in L$ $(1 \leqq i \leqq n, j \in J)$, whence $\sum_{i=1}^n \sum_{j\in J} b_{ij} y_i s_j = 0$ and therefore $b_{ij} = 0$ for every i and j, and $vu_i = 0$,

$u_i = 0$ $(1 \leqq i \leqq n)$. This shows that any base of R over S is also a base of $R\langle L\rangle$ over $S\langle L\rangle$. The other assertions follow immediately.

Let R be a field of algebraic functions of one variable, and let L be an overfield of the field of constants K of R. Then, although $S = R\langle L\rangle$ is a field of algebraic functions over L, it is not always true that L is the field of constants of S. Assume for instance that $K = P\langle a, b\rangle$, where P is any field of characteristic $p > 0$, while a and b are algebraically independent over P. Let R be of the form $K\langle x, y\rangle$, where x is transcendental over K, while y satisfies the equation $y^p = ax^p + b$. Then it is not difficult to see that K is the field of constants of R. Now, set $L = K\langle a^{1/p}\rangle$; then b is a p-th power in $R\langle L\rangle$, because $b = (y - a^{1/p}x)^p$, but not in $L = P\langle a^{1/p}, b\rangle$; it follows that L is not the whole field of constants of $R\langle L\rangle$. However, we have the following result:

THEOREM 2. *Let R be a field of algebraic functions of one variable, and let L be an overfield of the field of constants K of R. Then the field of constants of $R\langle L\rangle$ is purely unseparable over L. If either one of the fields R or L is separable over K, then L is the field of constants of $R\langle L\rangle$.*

If c is a constant of $R\langle L\rangle$, then there exists a subfield L' of L which may be obtained from K by adjunction of a finite number of elements and which is such that c belongs to $R\langle L'\rangle$ and is algebraic over L'. Furthermore, if L is separable over K, then so is L' (Lemma 3, §1). It will therefore be sufficient to prove Theorem 2 in the case where L can be obtained from K by adjunction of a finite number of elements. Let then B be a transcendence base of L over K such that, if L is separable over K, then it is also separable over $K\langle B\rangle$ (Lemma 5, §1). Construct the field $R_1 = R\langle K\langle B\rangle\rangle$ deduced from R by extending the field of constants from K to $K\langle B\rangle$. Making use of the Corollary 2 to Lemma 1 and of Lemma 2, §2, we see that the field of constants of R_1 is $K\langle B\rangle$. Moreover, it follows from Lemma 2 that, if R is separable over K, then R_1 is separable over $K\langle B\rangle$. It is clear that $R\langle L\rangle$ is the field $R_1\langle L\rangle$ deduced from R_1 by adjoining the elements of L to its field of constants. Thus we see that it will be sufficient to consider the case where L is algebraic of finite degree over K. Then L is purely unseparable over a field L_0 of the form $K\langle a\rangle$, with a separable over K. If n is the degree of a over K, then $1, a, \cdots, a^{n-1}$ form a base of L_0 over K, and therefore also of $R\langle L_0\rangle$ over R (Corollary 1 to Theorem 1). Let $c = \sum_{i=0}^{n-1} r_i a^i$ be a constant of $R\langle L_0\rangle$. Denote by $a_1 = a, \cdots, a_n$ the conjugates of a with respect to K (in a suitable normal extension L' of K containing L_0). For each j $(1 \leqq j \leqq n)$ there exists an automorphism σ_j of L' over K which maps a upon a_j. It follows from Corollary 1 to Lemma 1 that σ_j may be extended to an automorphism (also denoted by σ_j) of $R\langle L'\rangle$ over R. Set $c_j = \sigma_j(c)$; then we have

$$(1) \qquad c_j = \sum_{i=0}^{n-1} r_i a_j^i .$$

Since c is algebraic over L_0, it is algebraic over K, and the same is true of each c_j. On the other hand, since a is separable over K, the determinant of the system

(1) (considered as a system of equations in $r_0, \cdots, r_{n-1}$) is $\neq 0$. The coefficients and the left sides of the equations (1) being algebraic over K, it follows that $r_0, \cdots, r_{n-1}$ are algebraic over K, and therefore contained in K, whence $c \in L_0$. Thus we see that L_0 is the field of constants of $R\langle L_0\rangle$. Since L is purely unseparable over L_0, $R\langle L\rangle$ is purely unseparable over $R\langle L_0\rangle$. If c' is a constant of $R\langle L\rangle$, then there exists an exponent $e > 0$ such that $c'^{p^e} \in R\langle L_0\rangle$ (where p is the characteristic of K, which we may assume here to be $\neq 0$). Since c'^{p^e} is algebraic over L_0, it is in L_0, which proves that c' is purely unseparable over L. If R is separable over K, then $R\langle L\rangle$ is separable over L (Lemma 2), and so is its field of constants (Lemma 3, §1). This field of constants, being at the same time separable and purely unseparable over L, coincides with L. Theorem 2 is thereby proved.

COROLLARY. *Let R be a field of algebraic functions of one variable, and let L be an overfield of the field of constants K of R. Assume that one at least of the fields R or L is separable over K. Then, if $\mathfrak{a}$ is any divisor of R, we have $d(\mathrm{Con}_{R/R\langle L\rangle}\mathfrak{a}) = d(\mathfrak{a})$.*

It is sufficient to prove it in the case where $\mathfrak{a}$ is a place $\mathfrak{p}$. Let then x be an element of R which has $\mathfrak{p}$ as its only zero (Lemma 1, IV, §1). If $\mathfrak{p}^h$ is the divisor of zeros of x in R, then we have $hd(\mathfrak{p}) = [R:K\langle x\rangle]$ (Theorem 4, I, §8). The divisor of zeros of x in $R\langle L\rangle$ is $(\mathrm{Con}_{R/R\langle L\rangle}\mathfrak{p})^h$, whence, since L is the field of constants of $R\langle L\rangle$, $hd(\mathrm{Con}_{R/R\langle L\rangle}\mathfrak{a}) = [R\langle L\rangle:L\langle x\rangle]$. Making use of Corollary 2 to Theorem 1, this number is seen to be equal to $[R:K\langle x\rangle]$, which proves our assertion.

§ 5. THE EFFECT ON A PLACE

Let R be a field of algebraic functions of one variable, and let L be an overfield of its field of constants. We propose to study the places of the field $R\langle L\rangle$ which lie above a given place of R.

THEOREM 3. *Let $\mathfrak{p}$ be a place of a field R of algebraic functions of one variable and let L be an overfield of the field of constants K of R. Denote by Σ the residue field of $\mathfrak{p}$, and form the algebra Σ_L deduced from Σ (regarded as an algebra over K) by extending the basic field from K to L. Let $\mathfrak{n}$ be the radical of Σ_L, and let $Z_1, \cdots, Z_h$ be the distinct minimal ideals of $\Sigma_L/\mathfrak{n}$. Then there is a one-to-one correspondence between the fields Z_i and the places $\mathfrak{P}_i$ of $R\langle L\rangle$ above $\mathfrak{p}$. If Z_i corresponds to the place $\mathfrak{P}_i$, then there exists an isomorphism φ_i of Z_i with a subfield of the residue field Σ_i of $\mathfrak{P}_i$, and Σ_i is purely unseparable over $\varphi_i(Z_i)$. If $\xi \in \Sigma$, let ξ^* be the residue class of ξ modulo $\mathfrak{n}$, and set $\xi^* = \sum_{i=1}^h \zeta_i$, $\zeta_i \in Z_i$ $(1 \leqq i \leqq h)$; then $\varphi_i(\zeta_i)$ is the image of ξ under the natural isomorphism of Σ into Σ_i. If Σ_L is semi-simple, then $\varphi_i(Z_i) = \Sigma_i$ $(1 \leqq i \leqq h)$, and none of the places $\mathfrak{P}_i$ is ramified with respect to R. Moreover, in this case, $R \otimes L$ has no zero divisor $\neq 0$.*

Let $\mathfrak{o}$ be the ring of $\mathfrak{p}$. We denote by $\mathfrak{O}$ the subring $\mathfrak{o} \otimes L$ of $R \otimes L$. To every element $\sum_{i=1}^r x_i a_i$ $(x_i \in \mathfrak{o}, a_i \in L)$ of $\mathfrak{O}$ let us assign the element $\sum_{i=1}^r \xi_i a_i$ of Σ_L, where ξ_i is the residue class of x_i modulo $\mathfrak{p}$. Then we clearly obtain a homomorphism ψ of $\mathfrak{O}$ onto Σ_L. For each i $(1 \leqq i \leqq h)$, there is a homomorphism

θ_i of Σ_L onto Z_i (Lemma 4, §3); then $\theta_i\psi$ is a homomorphism of $\mathfrak{O}$ onto Z_i. Let $\mathfrak{N}$ be the radical of $R \otimes L$; then we know that $R\langle L\rangle$ is the field of quotients of $(R \otimes L)/\mathfrak{N}$. The image of $\mathfrak{O}$ under the natural homomorphism of $R \otimes L$ onto $(R \otimes L)/\mathfrak{N}$ is the subring $\mathfrak{O}'$ of $R\langle L\rangle$ generated by $\mathfrak{o}$ and L, and $\mathfrak{O}'$ is the factor ring $\mathfrak{O}/(\mathfrak{O} \cap \mathfrak{N})$. Every element of $\mathfrak{O} \cap \mathfrak{N}$, being nilpotent, is mapped upon 0 by $\theta_i\psi$ (because Z_i is a field). It follows that $\theta_i\psi$ defines in a natural way a homomorphism χ_i of $\mathfrak{O}'$ onto Z_i. Let $\mathfrak{Q}_i$ be the kernel of this homomorphism; since Z_i is a field, $\mathfrak{Q}_i$ is a prime ideal. Making use of Theorem 1, I, §4, we see that there is a place $\mathfrak{P}_i$ of $R\langle L\rangle$ such that $\mathfrak{P}_i \cap \mathfrak{O}' \supset \mathfrak{Q}_i$. It is clear that χ_i maps upon 0 every linear combination of elements of $\mathfrak{p}$ with coefficients in L; it follows that $\mathfrak{p} \subset \mathfrak{Q}_i \subset \mathfrak{P}_i$ and that $\mathfrak{P}_i$ lies above $\mathfrak{p}$. Since Z_i is a field, $\mathfrak{Q}_i$ is a maximal prime ideal of $\mathfrak{O}'$, whence $\mathfrak{Q}_i = \mathfrak{O}' \cap \mathfrak{P}_i$. It follows that $\mathfrak{O}'/\mathfrak{Q}_i$ is isomorphic in a natural way with a subring of the residue field Σ_i of $\mathfrak{P}_i$; since $\mathfrak{O}'/\mathfrak{Q}_i$ is isomorphic with Z_i, we obtain an isomorphism φ_i of Z_i with a subfield of Σ_i. Using the notation of the statement of Theorem 3, denote by u an element of R which takes the value ξ at $\mathfrak{p}$. Then $\theta_i\psi$ maps the element $u\cdot 1$ of $\mathfrak{o} \otimes L$ upon ζ_i, and it follows that $\varphi_i(\zeta_i)$ is the residue class of u modulo $\mathfrak{P}_i$; this is also the image of ξ under the natural isomorphism of Σ into Σ_i. We have not yet proved however that the place $\mathfrak{P}_i$ is uniquely determined when Z_i is given, nor that Σ_i is purely unseparable over $\varphi_i(Z_i)$. Before we establish these points, let us consider first the case where Σ_L is semi-simple.

Let x be an element of R which admits $\mathfrak{p}$ as its only zero (Lemma 1, IV, §1). Then we have $[R:K\langle x\rangle] = \nu_{\mathfrak{p}}(x)[\Sigma:K]$ (Theorem 4, I, §8). Since Σ_L is semi-simple, we have $[\Sigma:K] = [\Sigma_L:L] = \sum_{i=1}^{h} [Z_i:L]$ (Lemma 3, §3). On the other hand, since R is of finite degree over $K\langle x\rangle$, it follows from Lemma 2, IV, §1 that $[R:K\langle x\rangle] \geqq [R\langle L\rangle:L\langle x\rangle]$. This we write in the form $[R\langle L\rangle:M\langle x\rangle][M\langle x\rangle:L\langle x\rangle]$, where M stands for the field of constants of $R\langle L\rangle$. We have $[M\langle x\rangle:L\langle x\rangle] = [M:L]$ (Lemma 3, IV, §1). On the other hand, if $\mathfrak{P}_1, \cdots, \mathfrak{P}_{h'}$ are all the places of $R\langle L\rangle$ above $\mathfrak{p}$ (with $h' \geqq h$), and if e_j is the ramification index of $\mathfrak{P}_j$ with respect to R, then we have, by Theorem 4, I, §8, $[R\langle L\rangle:M\langle x\rangle] = \nu_{\mathfrak{p}}(x)\cdot\sum_{j=1}^{h'} e_j[\Sigma_j:M]$, and therefore $\sum_{i=1}^{h} [Z_i:L] \geqq \sum_{j=1}^{h'} e_j[\Sigma_j:L]$. If $j \leqq h$, we have $[\Sigma_j:L] = [\Sigma_j:\varphi_j(Z_j)]\,[Z_j:L]$. Therefore, we may conclude from our inequality that $h' = h$, that $e_j = 1$ $(1 \leqq j \leqq h)$ and that $\Sigma_j = \varphi_j(Z_j)$ $(1 \leqq j \leqq h)$. Let furthermore $\sum_{k=1}^{r} v_k a_k$ be a nilpotent element of $R \otimes L$ $(v_k \in R,\ a_k \in L)$; we may assume the elements a_k to be linearly independent over K. Were the elements v_k not all 0, then we could find an element w of R such that the elements wv_k should all be in $\mathfrak{o}$, but not all in $\mathfrak{p}$. Then $\sum_{k=1}^{r} wv_ka_k$ would be in $\mathfrak{O}$, and its image under $\theta_i\psi$ would be a nilpotent element of Σ_L, and therefore 0. On the other hand, it is clear that the kernel of $\theta_i\psi$ consists of all linear combinations of elements of $\mathfrak{p}$ with coefficients in L. Including the elements a_k in a base of L over K, we see that $\sum_{k=1}^{r} wv_ka_k$ could be written in the form $\sum_{k=1}^{r'} z_ka_k$, with $z_k \in \mathfrak{p}$, $a_k \in L$ $(1 \leqq k \leqq r')$ and $a_1, \cdots, a_{r'}$ linearly independent over K. This would imply $wv_k = z_k$ $(1 \leqq k \leqq r)$, which is impossible since one at least of the elements wv_k is not in $\mathfrak{p}$. This proves that, if Σ_L is semi-simple, then $R \otimes L$ has no nilpotent element $\neq 0$. Making use of Lemma 3, §2, we conclude that $R \otimes L$ has no zero divisor $\neq 0$.

Let us now return to the general case. The field L is algebraic over a field T which contains K and which is purely transcendental over K. Let L' be the field of elements of L which are separable over T. Then L' is separable over T and T over K (Lemma 4, §1), from which it follows that L' is separable over K (Lemma 3, §1). On the other hand, L is purely unseparable over L' and therefore $R\langle L\rangle$ is purely unseparable over $R\langle L'\rangle$. Now we prove

LEMMA 1. *Let R and S be fields of algebraic functions of one variable. Assume that R is a subfield of S and that S is purely unseparable over R. Then above any place $\mathfrak{p}$ of R there lies only one place $\mathfrak{P}$ of S, and the residue field of $\mathfrak{P}$ is purely unseparable over that of $\mathfrak{p}$.*

If u is any element of S, then there is an exponent a such that $u^{p^a} \in R$, where p is the characteristic of R, which we may assume to be >0. A necessary and sufficient condition for u to be in some place of S above $\mathfrak{p}$ is for u^{p^a} to be in $\mathfrak{p}$, which proves that there lies only one place $\mathfrak{P}$ of S above $\mathfrak{p}$. On the other hand, the p^a-th power of the value taken by u at $\mathfrak{P}$ (assuming that u is in the ring of $\mathfrak{P}$) is in the residue field of $\mathfrak{p}$, which proves that the residue field of $\mathfrak{P}$ is purely unseparable over that of $\mathfrak{p}$.

This being said, we see that the number of places of $R\langle L\rangle$ above $\mathfrak{p}$ is equal to the number of places of $R\langle L'\rangle$ above $\mathfrak{p}$. But, L' being separable over K, $\Sigma_{L'}$ is semi-simple (Corollary 2 to Lemma 5, §1), and our number is therefore equal to the number of minimal prime ideals in $\Sigma_{L'}$. Now, we know (Lemma 5, §3) that there is a subfield L'' of L which is algebraic and separable over K with the property that $\Sigma_{L''}$ is already the sum of h minimal prime ideals. Since L is purely unseparable over L', we have $L'\langle L''\rangle = L'$, $L'' \subset L'$, and we see that the numbers of minimal prime ideals in $\Sigma_{L'}$, $\Sigma_{L''}$, and $\Sigma_L/\mathfrak{n}$ are all equal to h, which proves that there are h places of $R\langle L\rangle$ above $\mathfrak{p}$ and that the correspondence $Z_i \rightarrow \mathfrak{P}_i$ is a one-to-one correspondence between the fields Z_i and all the places of $R\langle L\rangle$ above $\mathfrak{p}$. On the other hand, if $\mathfrak{P}_i'$ is the place of $R\langle L'\rangle$ below $\mathfrak{P}_i$, then, since $\Sigma_{L'}$ is semi-simple, the residue field Σ_i' of $\mathfrak{P}_i'$ is generated by adjunction of the elements of L to the image of Σ under the natural isomorphism of Σ into Σ_i'. The residue field of $\mathfrak{P}_i$ being purely unseparable over Σ_i' (Lemma 1) is also purely unseparable over $\varphi_i(Z_i)$. Theorem 3 is thereby proved.

COROLLARY 1. *The notation being as in Theorem 3, assume furthermore that every element of L which is algebraic over K is purely unseparable over K. Then above any place of R there lies only one place of $R\langle L\rangle$.*

For, in that case, every zero divisor in Σ_L is nilpotent (Lemma 3, §2) and $\Sigma_L/\mathfrak{n}$ has no zero divisor $\neq 0$ (Lemma 1, §3), from which it follows that $\Sigma_L/\mathfrak{n}$ is a field.

COROLLARY 2. *The notation being as in Theorem* 3, *assume furthermore that* L *is purely transcendental over* K. *Then there is only one place* $\mathfrak{P}$ *of* $R\langle L\rangle$ *above* $\mathfrak{p}$; $\mathfrak{P}$ *is unramified with respect to* R; *its residue field is obtained by adjunction of the elements of* L *to the residue field of* $\mathfrak{p}$, *and* $d(\mathfrak{P}) = d(\mathfrak{p})$.

We know that L is then separable over K (Lemma 4, §1). The assertions of Corollary 2 then follow from Theorem 3, from Corollary 1, and from the Corollary to Theorem 2, §4.

COROLLARY 3. *The notation being as in Theorem* 3, *denote also by* Σ_s *the field of elements of* Σ *which are separable over* K. *Then the subalgebra* $(\Sigma_s)_L$ *of* Σ_L *is the direct sum of* h *fields* Z'_i $(1 \leqq i \leqq h)$ *such that, for each* i, $Z''_i = (Z'_i + \mathfrak{n})/\mathfrak{n}$ *is the field of elements of* Z_i *which are separable over* L. *The field* $\varphi_i(Z''_i)$ *is then the field of elements of* Σ_i *which are separable over* L.

We may assume K to be of characteristic $p > 0$. Since Σ_s is separable over K, $(\Sigma_s)_L$ is semi-simple (Lemma 1, §1) and has therefore only 0 in common with $\mathfrak{n}$. It follows that the natural homomorphism of Σ_L onto $\Sigma_L/\mathfrak{n}$ maps $(\Sigma_s)_L$ isomorphically onto a subalgebra $\mathfrak{A}$ of $\Sigma_L/\mathfrak{n}$. There exists an integer $m > 0$ such that the p^m-th power of any element of Σ lies in Σ_s. If $\eta = \sum_{j=1}^{r} a_j\xi_j \in \Sigma_L$ $(a_j \in L, \xi_j \in \Sigma, 1 \leqq j \leqq r)$, then $\eta^{p^m} = \sum_{j=1}^{r} a_j^{p^m}\xi_j^{p^m}$ is in $(\Sigma_s)_L$, and it follows that the p^m-th power of every element of $\Sigma_L/\mathfrak{n}$ is in $\mathfrak{A}$. Let ε_i be the unit element of Z_i $(1 \leqq i \leqq h)$; since $\varepsilon_i^{p^m} = \varepsilon_i$, we have $\varepsilon_i \in \mathfrak{A}$. If $\zeta = \sum_{i=1}^{h} \zeta_i$ is in $\mathfrak{A}$, with $\zeta_i \in Z_i$ $(1 \leqq i \leqq h)$, we have $\zeta_i = \zeta\varepsilon_i \in \mathfrak{A}$, from which it follows immediately that $\mathfrak{A}$ is the direct sum of the algebras $\mathfrak{A}\varepsilon_i$ $(1 \leqq i \leqq h)$. Since $\mathfrak{A}\varepsilon_i$ is a subalgebra of the field Z_i and Z_i is of finite degree over L, $\mathfrak{A}\varepsilon_i$ is itself a field Z''_i, and it is clear that the p^m-th power of any element of Z_i is in Z''_i, which proves that Z_i is purely unseparable over Z''_i. Every element of Z''_i can be written as a linear combination with coefficients in L of products by ε_i of elements of the field $(\Sigma_s + \mathfrak{n})/\mathfrak{n}$, which is separable over the subfield K of L; it follows that Z''_i is separable over L, and is therefore the field of elements of Z_i which are separable over L. The last assertion of Corollary 3 follows immediately from the others and from Theorem 3.

COROLLARY 4. *The notation being as in Theorem* 3, *assume furthermore that* L *is algebraically closed. Then the natural isomorphisms of* Σ *into the residue fields of the places of* $R\langle L\rangle$ *above* $\mathfrak{p}$ *are all the distinct isomorphisms of* Σ *into* L *which coincide with the identity on* K. *The number of places of* $R\langle L\rangle$ *above* $\mathfrak{p}$ *is equal to* $[\Sigma_s:K]$, *where* Σ_s *is the field of elements of* Σ *which are separable over* K.

Since L is algebraically closed, the fields Z_i and the fields Σ_i (which are of finite degree over L) are of degree 1 over L, and the first assertion of Corollary 4 follows from Lemma 4, §3. The second assertion follows from the first and from a well-known theorem in Galois theory.

REMARK. In the case where Σ_L is not semi-simple, two phenomena may *a priori* take place: either some places of $R\langle L\rangle$ lying above $\mathfrak{p}$ are ramified with respect to R, or we may have $\Sigma_i \neq \varphi_i(Z_i)$ for some of the places $\mathfrak{P}_i$. We shall show by examples that these two possibilities can actually be realized even when R is separable over K. Let K be a field of characteristic $p > 0$ which contains two elements a and b such that $[K\langle a^{1/p}, b^{1/p}\rangle : K] = p^2$. Set $R_1 = K\langle x\rangle$, where x is transcendental over K, and let $\mathfrak{p}$ be the unique zero of $x^p - a$ in R_1. Since $x^p - a$ is irreducible in $K[x]$, we have $\nu_{\mathfrak{p}}(x^p - a) = 1$. Set $L = K\langle a^{1/p}\rangle$; then $x^p - a$ is in $R_1\langle L\rangle$ the p-th power of $x - a^{1/p}$. If $\mathfrak{P}$ is the place of $R_1\langle L\rangle$ above $\mathfrak{p}$, $\nu_{\mathfrak{P}}(x^p - a)$ is divisible by p, which shows that $\mathfrak{P}$ is ramified with respect to $K\langle x\rangle$. Now, set $R_2 = R_1\langle y\rangle$, where y satisfies the equation $y^p = (b + y)(x^p - a)$. Then y is separable over R_1, which proves that R_2 is separable over K. Let $\mathfrak{q}$ be a place of R_2 above $\mathfrak{p}$; then it is easily seen that $\mathfrak{q}$ is a zero of y, whence $\nu_{\mathfrak{q}}(x^p - a) \equiv 0 \pmod{p}$. This means that the ramification index of $\mathfrak{q}$ with respect to R_1 is $\geqq p$; since R_2 is of degree $\leqq p$ over R_1, we see that the ramification index of $\mathfrak{q}$ with respect to R_1 is p, whence $\nu_{\mathfrak{q}}(y) = 1$, and that the residue field of $\mathfrak{q}$ is the same as that of $\mathfrak{p}$; this residue field is therefore isomorphic to the field L introduced above. If $\mathfrak{Q}$ is the place of $R_2\langle L\rangle$ above $\mathfrak{q}$, then the formula $(y(x - a^{1/p})^{-1})^p = b + y$ shows that b is a p-th power in the residue field of $\mathfrak{Q}$. We conclude that the residue field of $\mathfrak{Q}$ cannot be obtained from that of $\mathfrak{q}$ by adjunction of the elements of L. We observe also that, if we adjoin a p-th root of b to $R_2\langle L\rangle$, the place of the field we obtain which lies above $\mathfrak{Q}$ is ramified with respect to $R_2\langle L\rangle$. We shall see later that this is a general phenomenon: if it happens (in the notation of Theorem 3) that some field Σ_i is different from $\varphi_i(Z_i)$, then there exist extensions M of the field L with the property that some place of $R\langle M\rangle$ above $\mathfrak{P}_i$ is ramified with respect to $R\langle L\rangle$.

§6. The effect on the genus

THEOREM 4. *Let R be a field of algebraic functions of one variable. Denote by K the field of constants of R and by L an overfield of K which we assume to be separable over K. Let $\mathfrak{a}$ be any divisor of R. If y is an element of $R\langle L\rangle$ which is* $\equiv 0$ $(\text{mod Con}_{R/R\langle L\rangle}\mathfrak{a})$, *then y can be written as a linear combination with coefficients in L of elements of R which are* $\equiv 0 \pmod{\mathfrak{a}}$.

The element y belongs to some field of the form $R\langle L_1\rangle$, where L_1 is a subfield of L which results from the adjunction of a finite number of elements to K. The field L_1 is separable over K (Lemma 3, §1). On the other hand, we have $\text{Con}_{R/R\langle L\rangle}\mathfrak{a} = \text{Con}_{R\langle L_1\rangle/R\langle L\rangle}(\text{Con}_{R/R\langle L_1\rangle}\mathfrak{a})$ (Lemma 3, IV, §7). Set $\mathfrak{a}_1 = \text{Con}_{R/R\langle L_1\rangle}\mathfrak{a}$; since y is in $R\langle L_1\rangle$ and is $\equiv 0$ $(\text{mod Con}_{R\langle L_1\rangle/R\langle L\rangle}\mathfrak{a}_1)$ when it is considered as an element of $R\langle L\rangle$, we see immediately that y is $\equiv 0 \pmod{\mathfrak{a}_1}$ in $R\langle L_1\rangle$. This shows that it will be sufficient to prove Theorem 5 in the case where L can be obtained from K by adjunction of a finite number of elements. We shall assume from now on that this is the case.

There exists an intermediary field M between K and L which is purely transcendental over K and over which L is of finite degree and separable (Lemma 5, §1). Let $\{b_1, \cdots, b_r\}$ be a base of L over M. Then $b_1, \cdots, b_r$ are linearly independent over $R\langle M\rangle$ (cf. Corollary 1 to Theorem 1, §4), and they form a base of $R\langle L\rangle$ over $R\langle M\rangle$. We write $y = \sum_{i=1}^{r} b_i z_i$, with $z_i \in R\langle M\rangle$ $(1 \leqq i \leqq r)$, and we shall prove that $z_i \equiv 0 \pmod{\mathfrak{b}}$, where $\mathfrak{b}$ is the divisor $\mathrm{Con}_{R/R\langle M\rangle}\mathfrak{a}$. Let $\mathfrak{q}$ be any place of $R\langle M\rangle$, and let q be the exponent with which $\mathfrak{q}$ enters in $\mathfrak{b}$; let u be a uniformizing variable at $\mathfrak{q}$. Since $y \equiv 0$ $(\mathrm{mod}\ \mathrm{Con}_{R\langle M\rangle/R\langle L\rangle}\mathfrak{b})$, it follows immediately that $u^{-q}y$ is integral at every place of $R\langle L\rangle$ which lies above $\mathfrak{q}$. On the other hand, the discriminant of the base $\{b_1, \cdots, b_r\}$ is an element $\neq 0$ of M, from which it follows that this base is integral at $\mathfrak{q}$ (Lemma 2, IV, §8). It follows that $\nu_{\mathfrak{q}}(u^{-q}z_i) \geqq 0$ $(1 \leqq i \leqq r)$. This being true for every place $\mathfrak{q}$ of $R\langle M\rangle$, we conclude that the elements z_i are $\equiv 0 \pmod{\mathfrak{b}}$. Theorem 5 will be proved if we can show that each z_i is a linear combination with coefficients in M of elements of R which are $\equiv 0 \pmod{\mathfrak{a}}$. In other words, it will be sufficient to prove Theorem 5 in the case where $L = K\langle t_1, \cdots, t_n\rangle$, with $t_1, \cdots, t_n$ algebraically independent over K. In this case, we proceed by induction on n. There is nothing to prove if $n = 0$. Assume that $n > 0$ and that the theorem is true for $n - 1$. We set $N = K\langle t_1, \cdots, t_{n-1}\rangle$, $S = R\langle N\rangle$, and we write $y = P(t_n)/Q(t_n)$, where P and Q are polynomials with coefficients in S, relatively prime to each other, and where Q has leading coefficient 1. We shall see that all coefficients of Q are in N. Assume for a moment that this is not the case. If we decompose Q into linear factors in a suitable algebraic extension of $R\langle L\rangle$, we see that one at least, say s, of the zeros of Q in this extension is transcendental over N (since, otherwise, the coefficients of Q would all be algebraic over N, and therefore contained in N in virtue of Lemma 2, §2). We set $U = (R\langle L\rangle)\langle s\rangle$. The element s is algebraic over S but not over N; since the degree of transcendency of S over N is 1, it follows that $S\langle s\rangle$ is algebraic over $N\langle s\rangle$ and that U is algebraic over $L\langle s\rangle$. Considering U as a field of algebraic functions of one variable over L, we see that $s - t_n$ has at least one zero $\mathfrak{Q}$ in U. The value t_n taken by s at $\mathfrak{Q}$ is transcendental over N, which is the field of constants of S; it follows that the residue field of $\mathfrak{Q}$ is transcendental over N, and therefore that $\mathfrak{Q}$ is a variable place with respect to the subfield $S\langle s\rangle$ of U. In particular, no element of S can have a pole at $\mathfrak{Q}$. Since P and Q have their coefficients in S and $\nu_{\mathfrak{Q}}(t_n - s) > 0$, we see that $\nu_{\mathfrak{Q}}(Q(t_n)) > 0$ and that $\nu_{\mathfrak{Q}}(P(t_n) - P(s)) > 0$. Since P and Q are relatively prime to each other, we have $P(s) \neq 0$; but $P(s)$ belongs to $S\langle s\rangle$, with respect to which $\mathfrak{Q}$ is variable. Therefore we have $\nu_{\mathfrak{Q}}(P(s)) = 0$, whence $\nu_{\mathfrak{Q}}(P(t_n)) = 0$. This shows that $\mathfrak{Q}$ is a pole of y. Let $\mathfrak{q}$ be the place of $R\langle L\rangle$ which lies below $\mathfrak{Q}$; then $\mathfrak{q}$ is a pole of y in $R\langle L\rangle$, and is obviously a variable place with respect to $R\langle N\rangle = S$ and therefore also with respect to R. It follows that $\mathfrak{q}$ does not occur in the divisor $\mathrm{Con}_{R/R\langle L\rangle}\mathfrak{a}$, and we obtain a contradiction with the assumption that y is congruent to 0 modulo this divisor. Thus we have proved that $Q(t_n) \in L$. Write $P(t_n) = \sum_{k=0}^{d} t_n^k y_k$, $y_k \in R\langle N\rangle$, whence $y = \sum_{k=0}^{d} (t_n^k/Q(t_n))y_k$. We shall prove that $y_k \equiv 0$ $(\mathrm{mod}\ \mathrm{Con}_{R/R\langle N\rangle}\mathfrak{a})$ $(0 \leqq k \leqq d)$. Let $\mathfrak{p}$ be any place of R, and let v be a uniformizing variable at $\mathfrak{p}$. Then we know that there exists exactly one place $\mathfrak{P}$ of $R\langle L\rangle$ above $\mathfrak{p}$, and that $\nu_{\mathfrak{P}}(v) = 1$ (Corollary 2 to Theorem 3, §5).

Let a be an integer such that $\nu_{\mathfrak{P}}(v^{-a}y_k) \geqq 0$ for all k, but $\nu_{\mathfrak{P}}(v^{-a}y_k) = 0$ for at least one k. If we denote by η_k the value taken at $\mathfrak{P}$ by $v^{-a}y_k$, then the value taken at $\mathfrak{P}$ by $v^{-a}y$ is $\sum_{k=0}^{d} (t_n^k/Q(t_n))\eta_k$. Now, each η_k belongs to the residue field of the place of $R\langle N\rangle$ which lies above $\mathfrak{p}$, and t_n is clearly transcendental over this residue field. It follows that $\nu_{\mathfrak{P}}(v^{-a}y) = 0$, whence $\nu_{\mathfrak{P}}(y) = a$. Since $\nu_{\mathfrak{P}}(y)$ is at least equal to the exponent a' with which $\mathfrak{p}$ enters in $\mathfrak{a}$, we see that $a \geqq a'$, and this proves that each y_k is $\equiv 0 \pmod{\mathrm{Con}_{R/R\langle N\rangle}\mathfrak{a}}$. In virtue of our induction assumption, we may conclude that each y_k is representable as a linear combination with coefficients in N of elements of R which are $\equiv 0 \pmod{\mathfrak{a}}$. Since $t_n^k/Q(t_n) \in L$, we see that y is representable as a linear combination with coefficients in L of elements of R which are $\equiv 0 \pmod{\mathfrak{a}}$. Theorem 5 is thereby proved.

COROLLARY 1. *The assumptions being as in Theorem 4, we have*

$$l(\mathrm{Con}_{R/R\langle L\rangle}\mathfrak{a}) = l(\mathfrak{a}).$$

Set $l(\mathfrak{a}) = l$, and let $x_1, \cdots, x_l$ be l elements of R which are linearly independent over K and which are $\equiv 0 \pmod{\mathfrak{a}}$. Considered as elements of $R\langle L\rangle$, $x_1, \cdots, x_l$ are $\equiv 0 \pmod{\mathrm{Con}_{R/R\langle L\rangle}\mathfrak{a}}$ and are linearly independent over L (Theorem 1, §4). It follows from Theorem 5 that every element of $R\langle L\rangle$ which is $\equiv 0 \pmod{\mathrm{Con}_{R/R\langle L\rangle}\mathfrak{a}}$ is a linear combination of $x_1, \cdots, x_l$ with coefficients in L. Since L is the field of constants of $R\langle L\rangle$ (Theorem 2, §4), the corollary is proved.

COROLLARY 2. *The assumptions being as in Theorem 4, a necessary and sufficient condition for an element $y \in R\langle L\rangle$ to be expressible as a linear combination of elements of R with coefficients in L is for y not to have any variable pole with respect to R.*

The condition is clearly necessary. Conversely, assume that it is satisfied. Let $\mathfrak{P}_1, \cdots, \mathfrak{P}_h$ be the poles of y. Then each $\mathfrak{P}_i$ lies above some place $\mathfrak{p}_i$ of R, and it is clear that, for n large enough, we have $y \equiv 0 \pmod{\mathrm{Con}_{R/R\langle L\rangle}(\mathfrak{p}_1 \cdots \mathfrak{p}_h)^{-n}}$. It then follows from Theorem 5 that y is expressible as a linear combination of elements of R with coefficients in L.

REMARK. *The conclusion of Corollary 2 remains true if, without making any assumption on L, we suppose that the field R is separably generated.*

In order to prove this, we proceed as follows. Let x be a separating variable in R, and let $(a_i)_{i\in I}$ be a base of L with respect to K, I being some set of indices. If z is any element of R, we construct the element $u_z = \mathrm{Sp}_{R\langle L\rangle/L\langle x\rangle} yz$ of $L\langle x\rangle$. If $\mathfrak{P}_0$ is a place of $L\langle x\rangle$ which is variable with respect to $K\langle x\rangle$, then any place $\mathfrak{P}$ of $R\langle L\rangle$ above $\mathfrak{P}_0$ is variable with respect to R; for, were this not the case, then $\mathfrak{P} \cap R$ would be a place of R and therefore $\mathfrak{P} \cap K\langle x\rangle = \mathfrak{P}_0 \cap K\langle x\rangle$ would be a place of $K\langle x\rangle$, which is not the case. Thus we see that no place of $R\langle L\rangle$ above $\mathfrak{P}_0$ is a pole of yz, from which it follows that $\mathfrak{P}_0$ is not a pole of u_z; in other words, u_z has no variable pole with respect to $K\langle x\rangle$. The divisor of poles of u_z consequently divides a divisor of the form Con $_{K\langle x\rangle/L\langle x\rangle}\mathfrak{u}_z$, where $\mathfrak{u}_z$ is an integral divisor of $K\langle x\rangle$. We can find a polynomial $V_z \neq 0$ with coefficients in K whose divisor is of the form $\mathfrak{u}_z\mathfrak{q}_0^{-m_z}$, where $\mathfrak{q}_0$ is the pole of x in $K\langle x\rangle$ and m_z some integer $\geqq 0$. Then the only pole of $V_z(x)u_z$ in $L\langle x\rangle$ is the pole of x, and we may therefore write $V_z(x)u_z = U_z(x)$, where U_z is a polynomial with coefficients in L. Since V_z has its coefficients in K, we have an expression of the form $u_z = \sum_{i\in I} a_i\varphi_i(z)$, with $\varphi_i(z) \in K\langle x\rangle$. It is clear that $\varphi_i(vz) = v\varphi_i(z)$ for every $i \in I$ and $v \in K\langle x\rangle$. If $\{z_1, \cdots, z_n\}$ is a base of R with respect to $K\langle x\rangle$, we may express every $z \in R$ in the form $\sum_{j=1}^n v_jz_j$, whence $\varphi_i(z) = \sum_{j=1}^n v_j\varphi_i(z_j)$. For each j there are only a finite number of indices i for which $\varphi_i(z_j) \neq 0$; it follows that there are only a finite number of indices i for which the function φ_i is not identically 0. Making use of Lemma 1, III, §5, we see that there exists for each $i \in I$ an element $y_i \in R$ such that $\varphi_i(z) = \mathrm{Sp}_{R/K\langle x\rangle}y_iz$ for every $z \in R$, and this is equal (in virtue of the Corollary 2 to Theorem 1, §4) to $\mathrm{Sp}_{R\langle L\rangle/L\langle x\rangle}y_iz$. There are only a finite number of indices i for which $y_i \neq 0$; set $y' = \sum_{i\in I} a_iy_i$. Then we have $\mathrm{Sp}_{R\langle L\rangle/L\langle x\rangle}(y - y')z = 0$ for every $z \in R$. This formula applies in particular to $z_1, \cdots, z_n$, which form a base of $R\langle L\rangle$ with respect to $L\langle x\rangle$ (by the result quoted above), from which it follows that our formula is still valid for any $z \in R\langle L\rangle$. Since $R\langle L\rangle$ is separable over $L\langle x\rangle$, it follows from Lemma 1, III, §5 that $y = y'$, which proves our assertion.

THEOREM 5. *Let R be a field of algebraic functions of one variable, and let L be an overfield of the field of constants K of R. Then the genus g' of $R\langle L\rangle$ is at most equal to the genus g of R, and the equality $g = g'$ is true whenever L is separable over K.*

We first consider the case where L is separable over K. Let $\mathfrak{a}$ be a divisor of R such that $d(\mathfrak{a}^{-1}) > 2g - 2$ and $d(\mathrm{Con}_{R/R\langle L\rangle}\mathfrak{a}^{-1}) > 2g' - 2$. Then we have, by Riemann-Roch's theorem, $l(\mathfrak{a}) = -d(\mathfrak{a}) - g + 1$, $l(\mathrm{Con}_{R/R\langle L\rangle}\mathfrak{a}) = -d(\mathrm{Con}_{R/R\langle L\rangle}\mathfrak{a}^{-1}) - g' + 1$. The left sides of these two equalities are equal to each other in virtue of the Corollary 1 to Theorem 5. On the other hand, we have $d(\mathfrak{a}) = d(\mathrm{Con}_{R/R\langle L\rangle}\mathfrak{a})$ in virtue of the corollary to Theorem 2, §4. It follows that $g = g'$.

Let us now consider the general case. The field L is algebraic over a field M_0 which is purely transcendental over K. Let M be the field of elements of L which are separable over M_0. Then M is separable over M_0 and M_0 over K (Lemma 4, §1), from which it follows that M is separable over K (Lemma 3, §1). On the other hand, L is purely unseparable over M. The genus of $R\langle M\rangle$ is equal to g, as was already proved. This shows that it will be sufficient to prove Theorem 6 under the supplementary assumption that L is algebraic and purely unseparable over K. Let then L' be the field of constants of $R\langle L\rangle$, and let $\{a_i\}_{i\in I}$ be a base of L' over K, I being a set of indices which may be finite or infinite. Let $\Omega_1, \cdots, \Omega_{g'}$ be g' linearly independent differentials of the first kind of $R\langle L\rangle$. If $\mathfrak{x}$ is any repartition in R, then $\mathrm{Cosp}_{R/R\langle L\rangle}\mathfrak{x}$ is a repartition in $R\langle L\rangle$, and we may write $\Omega_k(\mathrm{Cosp}_{R/R\langle L\rangle}\mathfrak{x}) = \sum_{i\in I}\omega_{ik}(\mathfrak{x})a_i$, with $\omega_{ik}(\mathfrak{x}) \in K$. It is clear that each ω_{ik} is a linear function on the vector space of repartitions in R. Let $\mathfrak{e}$ be the unit divisor of R; if $\mathfrak{x} \equiv 0 \pmod{\mathfrak{e}}$, then $\mathrm{Cosp}_{R/R\langle L\rangle}\mathfrak{x}$ is congruent to 0 modulo the unit divisor of $R\langle L\rangle$, whence $\omega_{ik}(\mathfrak{x}) = 0$ for all i and k. On the other hand, if $\mathfrak{x}$ is the repartition x_R attached to an element x of R, then $\mathrm{Cosp}_{R/R\langle L\rangle}\mathfrak{x}$ is the repartition $x_{R\langle L\rangle}$ attached to x in $R\langle L\rangle$, whence again $\omega_{ik}(\mathfrak{x}) = 0$ for all i and k. This shows that each ω_{ik} is a differential of the first kind in R. We shall see that only a finite number of these differentials are $\neq 0$. Let $\mathfrak{b}$ be an integral divisor of R of degree $d > 2g - 2$; then it follows from Theorem 2, II, §5 that every repartition $\mathfrak{x}$ of R is congruent modulo $\mathfrak{b}^{-1}$ to a repartition of the form x_R, $x \in R$. Making use of Lemma 1, II, §4, we see that we can find d repartitions $\mathfrak{x}_1, \cdots, \mathfrak{x}_d$ which are $\equiv 0 \pmod{\mathfrak{b}^{-1}}$ such that every repartition $\mathfrak{x}'$ which is $\equiv 0 \pmod{\mathfrak{b}^{-1}}$ can be written in the form $\sum_{j=1}^{d} c_j\mathfrak{x}_j + \mathfrak{x}''$, with $\mathfrak{x}'' \equiv 0 \pmod{\mathfrak{e}}$ and $c_j \in K$. It follows that every repartition $\mathfrak{x}$ of R can be written in the form $x_R + \sum_{j=1}^{d} c_j\mathfrak{x}_j + \mathfrak{x}''$, with $\mathfrak{x}'' \equiv 0 \pmod{\mathfrak{e}}$, whence $\Omega_k(\mathrm{Cosp}_{R/R\langle L\rangle}\mathfrak{x}) = \sum_{j=1}^{d} c_j\Omega_k(\mathrm{Cosp}_{R/R\langle L\rangle}\mathfrak{x}_j)$. We can find a finite subset I_0 of I such that the quantities $\Omega_k(\mathrm{Cosp}_{R/R\langle L\rangle}\mathfrak{x}_j)$ can all be expressed as linear combinations of the elements a_i for $i \in I_0$ with coefficients in K; it is then clear that $\omega_{ik} = 0$ if $i \notin I_0$. This being said, if we had $g' > g$, then we could find a differential Ω of the first kind of $R\langle L\rangle$ such that $\Omega \neq 0$, $\Omega(\mathrm{Cosp}_{R/R\langle L\rangle}\mathfrak{x}) = 0$ for every repartition $\mathfrak{x}$ of R. We shall derive a contradiction from the existence of such a differential. Since $\Omega \neq 0$, there exists a repartition $\mathfrak{y}$ of $R\langle L\rangle$ such that $\Omega(\mathfrak{y}) \neq 0$, and we may even assume that $\mathfrak{y}(\mathfrak{P}) \in R\langle L\rangle$ for all places $\mathfrak{P}$ of $R\langle L\rangle$ (i.e., we take a repartition $\mathfrak{y}$ in the sense of the first definition of this concept, as given in II, §4). For each place $\mathfrak{P}$ of $R\langle L\rangle$, denote by $\mathfrak{y}_{\mathfrak{P}}$ the repartition which assigns $\mathfrak{y}(\mathfrak{P})$ to $\mathfrak{P}$ and 0 to every other place. Then only a finite number of the elements $\Omega(\mathfrak{y}_{\mathfrak{P}})$ are $\neq 0$, and $\Omega(\mathfrak{y}) = \sum_{\mathfrak{P}} \Omega(\mathfrak{y}_{\mathfrak{P}})$. Let $\mathfrak{P}$ be a place such that $\Omega(\mathfrak{y}_{\mathfrak{P}}) \neq 0$. Since L is algebraic over K, it follows easily from Lemma 2, IV, §1 that $\mathfrak{y}(\mathfrak{P})$ can be written in the form $\sum_{i \in I} a_i x_i$, with $x_i \in R$. Let $\mathfrak{p}$ be the place of R below $\mathfrak{P}$, and let $\mathfrak{x}_i$ be the repartition of R which assigns x_i to $\mathfrak{p}$ and 0 to every other place. Since $\mathfrak{P}$ is the only place of $R\langle L\rangle$ above $\mathfrak{p}$ (Corollary 1 to Theorem 3, §4), we have $\mathfrak{y}_{\mathfrak{P}} = \sum_{i \in I} a_i \,\mathrm{Cosp}_{R/R\langle L\rangle}\mathfrak{x}_i$, whence $\Omega(\mathfrak{y}_{\mathfrak{P}}) = 0$ which brings the desired contradiction. Theorem 6 is thereby proved.

It will be observed that the strict inequality $g' < g$ may be true even in the case where R is separably generated over K.

CHAPTER VI

EXACT DIFFERENTIALS

§1. The differential dx in $K\langle x\rangle$

Let R be a field of algebraic functions of one variable. Our purpose in this chapter is to associate to every element x of R a differential dx of the field R and to show that the differentials obtained in this manner have the formal properties of the differentials one is used to handling in the calculus.

We shall first consider the case where the field R is of the form $K\langle x\rangle$, K being any field and x transcendental over K; in that case, we shall define in this section the differential of the particular element x. We shall pass to the general case later by making use of the notion of the cotrace of a differential.

In II, §5, we have determined the differentials of $K\langle x\rangle$ when K is the field of complex numbers. They were the expressions udx, where u was an arbitrary element of $K\langle x\rangle$. Taking $u = 1$, we obtain what we shall call the differential of x. For any u, udx is obviously the product of u by the differential dx. We have seen that the order of udx at a place $\mathfrak{p}$ which is not the pole of x is $\nu_{\mathfrak{p}}(u)$, while the order of udx at the pole $\mathfrak{p}(\infty)$ of x is $\nu_{\mathfrak{p}(\infty)}(u) - 2$. Moreover, if $\mathfrak{x}$ is the repartition which assigns x^{-1} to $\mathfrak{p}(\infty)$ and 0 to every other place, then it is easily seen that $(dx)(\mathfrak{x}) = -1$. It follows immediately from this that the divisor of dx is $(\mathfrak{p}(\infty))^{-2}$.

Now, let K be an arbitrary field, and set $R = K\langle x\rangle$, where x is transcendental over K. Then x has a single pole $\mathfrak{p}(\infty)$ in R, and $\mathfrak{p}(\infty)$ is of degree 1. Applying the theorem of Riemann-Roch and remembering that $K\langle x\rangle$ is of genus 0, we obtain $l((\mathfrak{p}(\infty))^2) = d((\mathfrak{p}(\infty))^{-2}) + 1 + \delta((\mathfrak{p}(\infty))^{-2}) = \delta((\mathfrak{p}(\infty))^{-2}) - 1$, where $\delta((\mathfrak{p}(\infty))^{-2})$ is the dimension of the space of differentials which are multiples of $(\mathfrak{p}(\infty))^{-2}$. Thus we see that there exists a differential $\omega_1 \neq 0$ of R which is multiple of $(\mathfrak{p}(\infty))^{-2}$. The degree of $\mathfrak{d}(\omega_1)$ being $-2 = d((\mathfrak{p}(\infty))^{-2})$, we have $\mathfrak{d}(\omega_1) = (\mathfrak{p}(\infty))^{-2}$. Let $\mathfrak{x}$ be the repartition which assigns x^{-1} to $\mathfrak{p}(\infty)$ and 0 to every other place; if $\mathfrak{a}$ is any divisor, let $\mathfrak{X}(\mathfrak{a})$ be the space of repartitions which are multiples of $\mathfrak{a}$. Then it follows from Lemma 1, II, §4 that $\mathfrak{X}(\mathfrak{p}(\infty))/\mathfrak{X}((\mathfrak{p}(\infty))^2)$ is of dimension 1 over K. The repartition $\mathfrak{x}$ is in $\mathfrak{X}(\mathfrak{p}(\infty))$ but not in $\mathfrak{X}((\mathfrak{p}(\infty))^2)$; it follows that every $\mathfrak{y}$ in $\mathfrak{X}(\mathfrak{p}(\infty))$ may be written in one and only one way in the form $a\mathfrak{x} + \mathfrak{y}'$, with $a \in K$ and $\mathfrak{y}' \in \mathfrak{X}((\mathfrak{p}(\infty))^2)$. We have $\omega_1(\mathfrak{y}') = 0$; were $\omega_1(\mathfrak{x}) = 0$, then $\omega_1(\mathfrak{y})$ would be 0 for every multiple $\mathfrak{y}$ of $\mathfrak{p}(\infty)$, and ω_1 would be a multiple of $(\mathfrak{p}(\infty))^{-1}$, which is not the case. If $h = \omega_1(\mathfrak{x})$, we set $\omega = -h^{-1}\omega_1$; ω is then a differential of R which is a multiple of $(\mathfrak{p}(\infty))^{-2}$ and for which $\omega(\mathfrak{x}) = -1$. The differential ω is uniquely determined by these two conditions. For, if a differential ω' satisfies the same conditions, then $\omega' - \omega$ is a multiple of $(\mathfrak{p}(\infty))^{-2}$ and $(\omega' - \omega)(\mathfrak{x}) = 0$. It follows immediately that $(\omega' - \omega)(\mathfrak{y}) = 0$ for every repartition $\mathfrak{y} \in \mathfrak{X}(\mathfrak{p}(\infty))$, i.e., $\mathfrak{d}(\omega' - \omega)$ is a multiple of $(\mathfrak{p}(\infty))^{-1}$. The degree of the divisor of a differential $\neq 0$ of R being -2, which is $< d((\mathfrak{p}(\infty))^{-1})$, we see that $\omega' - \omega = 0$.

We shall denote by dx the differential ω whose existence and uniqueness we have just established. We shall determine more explicitly dx by computing its $\mathfrak{p}$-components $(dx)^{\mathfrak{p}}$ for the various places $\mathfrak{p}$ of R (cf. II, §7). It will be sufficient to compute $(dx)^{\mathfrak{p}}(u)$ for every $u \in R$.

Let first $\mathfrak{p}$ be a place $\neq \mathfrak{p}(\infty)$. Then there is an irreducible polynomial $f \in K[x]$ which admits $\mathfrak{p}$ as a zero of order 1; we may assume that the leading coefficient of f is 1, and f is then uniquely determined. It is well known that every $u \in R$ may be represented in the form

$$u = \frac{g_r}{f^r} + \cdots + \frac{g_1}{f} + v \tag{1}$$

where v is a rational fraction in which the denominator is not divisible by f, whence $\nu_{\mathfrak{p}}(v) \geqq 0$; $g_1, \cdots, g_r$ are in $K[x]$, and each is either 0 or of degree $<d$, if d is the degree of f. Since $\mathfrak{p}$ does not occur in $\mathfrak{d}(dx)$, we have $(dx)^{\mathfrak{p}}(v) = 0$ (Lemma 2, II, §7). Similarly, if $\mathfrak{q}$ is a place distinct from $\mathfrak{p}$ and $\mathfrak{p}(\infty)$, then $\nu_{\mathfrak{q}}(g_i f^{-i}) \geqq 0$, and therefore $(dx)^{\mathfrak{q}}(g_i f^{-i}) = 0$. We have (by Lemma 1, II, §7) $0 = (dx)\ (g_i f^{-i}) = \sum_{\mathfrak{r}} (dx)^{\mathfrak{r}}(g_i f^{-i})$, the summation being extended over all places $\mathfrak{r}$. Therefore, $(dx)^{\mathfrak{p}}(g_i f^{-i}) = -(dx)^{\mathfrak{p}(\infty)}(g_i f^{-i})$. Let γ_i be the degree of g_i (if $g_i \neq 0$); then $\nu_{\mathfrak{p}(\infty)}(g_i f^{-i}) = id - \gamma_i$. Since $\gamma_i < d$, this number is $\geqq 2$ if $i > 1$, whence then $(dx)^{\mathfrak{p}}(g_i f^{-i}) = 0$. Set $g_1 = \sum_{k=0}^{d-1} a_k x^k$; then $xg_1 - a_{d-1} f$ is either 0 or of degree $<d - 1$, whence $\nu_{\mathfrak{p}(\infty)}(g_1 f^{-1} - a_{d-1} x^{-1}) \geqq 2$. It follows that $(dx)^{\mathfrak{p}}(g_1 f^{-1}) = -(dx)^{\mathfrak{p}(\infty)}(a_{d-1} x^{-1}) = a_{d-1}$. Thus we have obtained the following result:

LEMMA 1. *Let f be an irreducible element of $K[x]$, of degree d with leading coefficient 1, and let $\mathfrak{p}$ be the zero of f in $K\langle x\rangle$. Let u be an element of $K\langle x\rangle$, represented in the form* (1) *above, where $\nu_{\mathfrak{p}}(v) \geqq 0$ and $g_1, \cdots, g_r$ are elements of $K[x]$ of degrees $<d$. Then $(dx)^{\mathfrak{p}}(u)$ is the coefficient of x^{d-1} in g_1.*

The simplest case is the one in which $d = 1$, i.e., $f = x - a$, $a \in K$. In that case, the $\mathfrak{p}$-adic completion of $K\langle x\rangle$ may be identified with the field of formal power series in $x - a$ with coefficients in K(III, §3). Our result can then be formulated as follows.

LEMMA 2. *Let a be an element of K, and let $\mathfrak{p}$ be the zero of $x - a$ in $K\langle x\rangle$. If $y = \sum_{k=r}^{\infty} c_k(x - a)^k$ ($c_k \in K, r \leqq k < \infty$) is any element of the $\mathfrak{p}$-adic completion of $K\langle x\rangle$, then $(dx)^{\mathfrak{p}}(y) = c_{-1}$.*

Let us also consider the case where $\mathfrak{p}$, without necessarily being of degree 1, nevertheless has a residue field Σ which is separable over K. In this case, we shall consider a repartition $\mathfrak{x}$ for which $\nu_{\mathfrak{p}}(\mathfrak{x}) \geqq -1$, and we shall give another expression for the value of $(dx)^{\mathfrak{p}}(\mathfrak{x})$. If ξ is the value taken by x at $\mathfrak{p}$, we have $\Sigma = K\langle \xi\rangle$, and ξ is a root of the equation $f(X) = 0$. Since ξ is separable over K, we have $f'(\xi) \neq 0$ (where f' stands for the derivative of the polynomial f). On the other hand, since $\nu_{\mathfrak{p}}(\mathfrak{x}) \geqq -1$, we have $\nu_{\mathfrak{p}}(f\mathfrak{x}(\mathfrak{p})) \geqq 0$; we shall denote by φ the residue class of $f\mathfrak{x}(\mathfrak{p})$ modulo $\mathfrak{p}$. With this notation, we shall prove that

$$(dx)^{\mathfrak{p}}(\mathfrak{x}) = \mathrm{Sp}_{\Sigma/K} \frac{\varphi}{f'(\xi)}. \tag{2}$$

Write $\mathfrak{x}(\mathfrak{p}) = g(x)/f(x) + v$, where g is a polynomial of degree $<d$, while $\nu_{\mathfrak{p}}(v) \geqq 0$. Then φ is equal to $g(\xi)$. On the other hand, we know by Lemma 1 that $(dx)^{\mathfrak{p}}(\mathfrak{x})$ is the coefficient of x^{d-1} in $g(x)$. Formula (2) will therefore follow immediately from

LEMMA 3. *Let Σ be a field which is obtained from a field K by adjunction of an element $\xi \neq 0$ which is algebraic and separable over K. Denote by $f(X)$ the characteristic polynomial of ξ with respect to K and by d the degree of f. Then we have*

$$\mathrm{Sp}_{\Sigma/K} \frac{\xi^k}{f'(\xi)} = 0 \text{ for } 0 \leqq k < d-1; \qquad \mathrm{Sp}_{\Sigma/K} \frac{\xi^{d-1}}{f'(\xi)} = 1.$$

In a suitable extension field of Σ, the equation $f(X) = 0$ has d distinct roots $\xi_1 = \xi, \cdots, \xi_d$. Set

$$g_i(X) = \frac{f(X)}{(X - \xi_i)f'(\xi_i)} \qquad (1 \leqq i \leqq d).$$

Then g_i is a polynomial of degree $d - 1$, and we have $g_i(\xi_i) = 1$, $g_i(\xi_j) = 0$ for $j \neq i$. If $0 \leqq k \leqq d - 1$, the polynomial $h_k(X) = \sum_{i=1}^{d} \xi_i^k \, g_i(X)$ is of degree $\leqq d - 1$, and we have $h_k(\xi_i) = \xi_i^k$ for $1 \leqq i \leqq d$, from which it follows immediately that $h_k(X) = X^k$. Setting $X = 0$, we obtain

$$f(0) \, \mathrm{Sp}_{\Sigma/K} \frac{\xi^{k-1}}{f'(\xi)} = \begin{cases} -1 & \text{if } k = 0, \\ 0 & \text{if } 1 \leqq k \leqq d-1. \end{cases}$$

If $f(X) = X^d + \sum_{k=0}^{d-1} a_k X^{d-k}$, we have $f(0) = a_d$ and $\xi^{d-1} = -\sum_{k=0}^{d-1} a_k \xi^{d-k-1}$ whence $\mathrm{Sp}_{\Sigma/K}\xi^{d-1}/f'(\xi) = 1$. Lemma 1, and therefore also Formula (2), are thereby proved.

In order to make our investigation of dx complete, we still have to consider the $\mathfrak{p}(\infty)$-component $(dx)^{\mathfrak{p}(\infty)}$ of dx. We may write any element u of $K\langle x\rangle$ in the form

$$(3) \qquad u = \sum_{k=0}^{r} a_k x^k + a_{-1} x^{-1} + v \qquad (\text{with } a_{-1}, a_0, \cdots, a_r \text{ in } K)$$

where v is such that $\nu_{\mathfrak{p}(\infty)}(v) \geqq 2$. If $k \geqq 0$, then we shall see that $(dx)^{\mathfrak{p}(\infty)}(x^k) = 0$. We have $(dx)^{\mathfrak{p}(\infty)}(x^k) = -\sum_{\mathfrak{p}} (dx)^{\mathfrak{p}}(x^k)$, the summation being extended to all places $\mathfrak{p} \neq \mathfrak{p}(\infty)$. None of these places being a pole of either x or dx, we have, for each of them, $(dx)^{\mathfrak{p}}(x^k) = 0$, which proves our assertion. Remembering that $(dx)^{\mathfrak{p}(\infty)}(x^{-1}) = -1$ and that $\mathfrak{d}(dx) = (\mathfrak{p}(\infty))^{-2}$, we obtain

LEMMA 4. *An element u of $K\langle x\rangle$ being represented in the form (3), with $\nu_{\mathfrak{p}(\infty)}(v) \geqq 2$, we have $(dx)^{\mathfrak{p}(\infty)}(u) = -a_{-1}$.*

§2. TRACE AND COTRACE OF DIFFERENTIALS

Let R be a field of algebraic functions of one variable, and let S be an overfield of finite degree of R; then we may also consider S as a field of algebraic functions of one variable. We propose to associate to every differential Ω of S a differential ω of R. Denote by K and L the fields of constants of R and S respectively. If $\mathfrak{x}$ is any repartition in R, set $\omega(\mathfrak{x}) = \mathrm{Sp}_{L/K}\Omega(\mathrm{Cosp}_{R/S}\mathfrak{x})$; it is clear that the function

ω defined in this way is linear. If $\mathfrak{x} = x_R$, with $x \in R$, then $\mathrm{Cosp}_{R/S}\mathfrak{x} = x_S$, whence $\omega(\mathfrak{x}) = 0$ (cf. IV, §7; the reader is reminded that x_R and x_S are the repartitions attached to x in R and S respectively). If $\Omega = 0$, then $\omega = 0$ is a differential. If not, let $\mathfrak{A}$ be the divisor of Ω, and set $\mathfrak{A} = \prod_{\mathfrak{P}} \mathfrak{P}^{a(\mathfrak{P})}$. Denote by $\mathfrak{p}$ a place of S, by $\mathfrak{P}_1, \cdots, \mathfrak{P}_h$ the distinct places of S above $\mathfrak{p}$, and by e_i the ramification index of $\mathfrak{P}_i$ with respect to R $(1 \leqq i \leqq h)$. Let $a'(\mathfrak{p})$ be the largest integer such that $e_i a'(\mathfrak{p}) \leqq a(\mathfrak{P}_i)$ $(1 \leqq i \leqq h)$. Then $a'(\mathfrak{p}) = 0$ except for a finite number of places $\mathfrak{p}$, and $\mathfrak{a} = \prod_{\mathfrak{p}} \mathfrak{p}^{a'(\mathfrak{p})}$ is a divisor of R. Let $\mathfrak{x}$ be any repartition of R which is $\equiv 0 \pmod{\mathfrak{a}^{-1}}$. Then, for every place $\mathfrak{p}$ of R, we have $\nu_{\mathfrak{p}}(\mathfrak{x}(\mathfrak{p})) \geqq -a'(\mathfrak{p})$, whence, if $\mathfrak{P}$ is a place of S lying above $\mathfrak{p}$, $\nu_{\mathfrak{P}}((\mathrm{Cosp}_{R/S}\mathfrak{x})(\mathfrak{P})) \geqq -ea'(\mathfrak{p}) \geqq -a(\mathfrak{P})$, where e is the ramification index of $\mathfrak{P}$ with respect to R. It follows that $\Omega(\mathrm{Cosp}_{R/S}\mathfrak{x}) = 0$, whence $\omega(\mathfrak{x}) = 0$. This proves that ω is a differential of R. The differential ω will be called the *trace* (from S to R) of Ω, and will be denoted by $\mathrm{Sp}_{S/R}\Omega$.

It is clear that, if Ω and Ω' are differentials of S, then

$$\mathrm{Sp}_{S/R}(\Omega + \Omega') = \mathrm{Sp}_{S/R}\Omega + \mathrm{Sp}_{S/R}\Omega'. \tag{1}$$

On the other hand, if $x \in R$, we have

$$\mathrm{Sp}_{S/R}x\Omega = x\,\mathrm{Sp}_{S/R}\Omega. \tag{2}$$

For, let $\mathfrak{x}$ be any repartition in R. Then, $(\mathrm{Sp}_{S/R}x\Omega)(\mathfrak{x}) = \mathrm{Sp}_{L/K}((x\Omega)(\mathrm{Cosp}_{R/S}\mathfrak{x})) = \mathrm{Sp}_{L/K}(\Omega(\mathrm{Cosp}_{R/S}x\mathfrak{x})) = (\mathrm{Sp}_{S/R}\Omega)(x\mathfrak{x})$, which proves our assertion.

If T is an intermediary field between R and S, we have

$$\mathrm{Sp}_{S/R}\Omega = \mathrm{Sp}_{T/R}(\mathrm{Sp}_{S/T}\Omega). \tag{3}$$

For, let M be the field of constants of T, and let $\mathfrak{x}$ be any repartition in R. Making use of Lemma 3, IV, §7, we have

$$\begin{aligned}(\mathrm{Sp}_{T/R}(\mathrm{Sp}_{S/T}\Omega))(\mathfrak{x}) &= \mathrm{Sp}_{M/K}((\mathrm{Sp}_{S/T}\Omega)(\mathrm{Cosp}_{R/T}\mathfrak{x}))\\ &= \mathrm{Sp}_{M/K}(\Omega(\mathrm{Cosp}_{T/S}(\mathrm{Cosp}_{R/T}\mathfrak{x})))\\ &= \mathrm{Sp}_{M/K}(\Omega(\mathrm{Cosp}_{R/S}\mathfrak{x})) = (\mathrm{Sp}_{S/R}\Omega)(\mathfrak{x}),\end{aligned}$$

and this proves our formula.

THEOREM 1. *Let R and S be fields of algebraic functions of one variable. Assume that S is an overfield of finite degree of R. Denote by Ω a differential of S, by $\mathfrak{p}$ a place of R, and by $\mathfrak{P}_1, \cdots, \mathfrak{P}_h$ the distinct places of S above $\mathfrak{p}$, by K and L the fields of constants of R and S respectively, by $\Sigma_s(\mathfrak{p})$ the field of elements of the residue field of $\mathfrak{p}$ which are separable over K, by $\Sigma_s(\mathfrak{P}_i)$ the field of elements of the residue field of $\mathfrak{P}_i$ which are separable over L $(1 \leqq i \leqq h)$. Then we have*

$$\mathrm{res}_{\mathfrak{p}}\,\mathrm{Sp}_{S/R}\Omega = \sum_{i=1}^{h} \mathrm{Sp}_{\Sigma_s(\mathfrak{P}_i)/\Sigma_s(\mathfrak{p})}\,\mathrm{res}_{\mathfrak{P}_i}\Omega.$$

REMARK. In this theorem we assume as usual (cf. IV, §1) that L contains K; it follows immediately that $\Sigma_s(\mathfrak{p})$ is a subfield of $\Sigma_s(\mathfrak{P}_i)$.

Let γ be any element of $\Sigma_s(\mathfrak{p})$, and let $\bar{\gamma}$ be the element of the $\mathfrak{p}$-adic completion of R which is algebraic over K and which belongs to the residue class γ modulo $\mathfrak{p}$ (cf. Theorem 1, III, §3). Denote also by α and β the two sides of the formula to be proved. We have $\mathrm{Sp}_{\Sigma_s(\mathfrak{p})/K}\alpha\gamma = (\mathrm{Sp}_{S/R}\Omega)(\mathfrak{c})$, where $\mathfrak{c}$ is the repartition of R which assigns $\bar{\gamma}$ to $\mathfrak{p}$ and 0 to every other place. By definition of the trace, this is equal to $\mathrm{Sp}_{L/K}(\Omega(\mathrm{Cosp}_{R/S}\mathfrak{c})) = \sum_{i=1}^{h} \mathrm{Sp}_{L/K}(\Omega^{\mathfrak{P}_i}(\bar{\gamma}))$. Since $\bar{\gamma}$ is algebraic over L and belongs to the residue class γ modulo $\mathfrak{P}_i$, we have $\Omega^{\mathfrak{P}_i}(\bar{\gamma}) = \mathrm{Sp}_{\Sigma_s(\mathfrak{P}_i)/L}\gamma \operatorname{res}_{\mathfrak{P}_i}\Omega$, whence

$$\mathrm{Sp}_{L/K}\Omega^{\mathfrak{P}_i}(\bar{\gamma}) = \mathrm{Sp}_{\Sigma_s(\mathfrak{P}_i)/K}\gamma \operatorname{res}_{\mathfrak{P}_i}\Omega = \mathrm{Sp}_{\Sigma_s(\mathfrak{p})/K}(\gamma\, \mathrm{Sp}_{\Sigma_s(\mathfrak{P}_i)/\Sigma_s(\mathfrak{p})} \operatorname{res}_{\mathfrak{P}_i}\Omega).$$

Thus we see that $\mathrm{Sp}_{\Sigma_s(\mathfrak{p})/K}\alpha\gamma = \mathrm{Sp}_{\Sigma_s(\mathfrak{p})/K}\beta\gamma$. This being true for every $\gamma \in \Sigma_s(\mathfrak{p})$, it follows from Lemma 1, III, §5 that we have $\alpha = \beta$. Theorem 1 is thereby proved.

To the operation of cotrace for repartitions we have just associated an operation of trace for differentials. Conversely, to the operation of trace for repartitions, we shall associate an operation of cotrace for differentials. However, in so doing, we shall limit ourselves to the case of pairs of fields of algebraic functions of one variable which have the same field of constants. It is only later (after we have defined the exact differentials in arbitrary fields of algebraic functions of one variable) that we shall be able to define the operation of cotrace in the case where the fields of constants are distinct.

Let then R and S be fields of algebraic functions of one variable which satisfy the following conditions: S is an overfield of R, R and S have the same field of constants K. The field S is then of finite degree over R. Let ω be any differential in R. If we set

$$\Omega(\mathfrak{y}) = \omega(\mathrm{Sp}_{S/R}\mathfrak{y}),$$

we clearly obtain a linear function Ω on the space of repartitions $\mathfrak{y}$ of S. We shall prove that Ω is a differential. If $\mathfrak{y}$ is the repartition y_S attached to an element y of S, then $\mathrm{Sp}_{S/R}\mathfrak{y} = (\mathrm{Sp}_{S/R}y)_R$, whence $\Omega(y_S) = 0$. If S is not separable over R, then we have seen in IV, §7 that $\mathrm{Sp}_{S/R}\mathfrak{y} = 0$ for every repartition $\mathfrak{y}$ in S, whence $\Omega = 0$. Now, assume that S is separable over R and that $\omega \neq 0$; we denote as usual by $\mathfrak{d}(\omega)$ the divisor of ω. Let $\mathfrak{P}$ be any place of S; we denote by $\mathfrak{p}$ the place of R below $\mathfrak{P}$, by $a(\mathfrak{p})$ the exponent of $\mathfrak{p}$ in $\mathfrak{d}(\omega)$, by $e(\mathfrak{P})$ the ramification index of $\mathfrak{P}$ with respect to R and by $m(\mathfrak{P})$ its differential exponent. If u is any element of the $\mathfrak{P}$-adic completion $\bar{S}$ of S, we have $\Omega^{\mathfrak{P}}(u) = \omega^{\mathfrak{p}}(\mathrm{Sp}_{\bar{S}/\bar{R}}u)$, where $\bar{R}$ represents the $\mathfrak{p}$-adic completion of R. Let t be a uniformizing variable at $\mathfrak{p}$ in R. If $\nu_{\mathfrak{P}}(u) \geqq -e(\mathfrak{P})a(\mathfrak{p}) - m(\mathfrak{P})$, then $\nu_{\mathfrak{P}}(t^{a(\mathfrak{p})}u) \geqq -m(\mathfrak{P})$, whence $\nu_{\mathfrak{p}}(\mathrm{Sp}_{\bar{S}/\bar{R}}t^{a(\mathfrak{p})}u) \geqq 0$, and therefore $\nu_{\mathfrak{p}}(\mathrm{Sp}_{\bar{S}/\bar{R}}u) \geqq -a(\mathfrak{p})$, and $\Omega^{\mathfrak{P}}(u) = 0$. On the other hand, there exists an element z of $\bar{S}$ such that $\nu_{\mathfrak{P}}(z) = -m(\mathfrak{P}) - 1$, $\nu_{\mathfrak{p}}(\mathrm{Sp}_{\bar{S}/\bar{R}}z) < 0$. Since $\nu_{\mathfrak{P}}(tz) \geqq -m(\mathfrak{P})$, we have $\nu_{\mathfrak{p}}(t\ \mathrm{Sp}_{\bar{S}/\bar{R}}z) \geqq 0$, whence $\nu_{\mathfrak{p}}(\mathrm{Sp}_{\bar{S}/\bar{R}}z) = -1$. The exponent of $\mathfrak{p}$ in $\mathfrak{d}(\omega)$ being $a(\mathfrak{p})$, we can find an element w of R such that $\nu_{\mathfrak{p}}(w) = -a(\mathfrak{p}) - 1$, $\omega^{\mathfrak{p}}(w) \neq 0$. We then see easily that $\nu_{\mathfrak{P}}(wz(\mathrm{Sp}_{\bar{S}/\bar{R}}z)^{-1}) = -e(\mathfrak{P})a(\mathfrak{p}) - m(\mathfrak{P}) - 1$ and that $\Omega^{\mathfrak{P}}(wz(\mathrm{Sp}_{\bar{S}/\bar{R}}z)^{-1}) =$

$\omega^{\mathfrak{p}}(w) \neq 0$. Let $\mathfrak{D}$ be the different of S with respect to R; then the number $e(\mathfrak{P})a(\mathfrak{p}) + m(\mathfrak{P})$ is the exponent of $\mathfrak{P}$ in $\mathfrak{D}\ \mathrm{Con}_{R/S}\mathfrak{d}(\omega)$. Thus we see that, whether S is separable over R or not, Ω is a differential of S, and that, if S is separable over R, the divisor of Ω is $\mathfrak{D}\ \mathrm{Con}_{R/S}\mathfrak{d}(\omega)$.

The differential Ω is called the *cotrace of* ω, *from* R *to* S, and is denoted by $\mathrm{Cosp}_{R/S}\ \omega$. We have proved the following result:

THEOREM 2. *Let R and S be fields of algebraic functions of one variable; assume that S is an overfield of R and has the same field of constants as R. Let ω be a differential $\neq 0$ in R. If S is not separable over R, then* $\mathrm{Cosp}_{R/S}\omega = 0$. *If S is separable over R, then $\mathrm{Cosp}_{R/S}\omega$ is a differential $\neq 0$ of S, whose divisor is $\mathfrak{D}\ \mathrm{Con}_{R/S}\mathfrak{d}(\omega)$, where $\mathfrak{D}$ represents the different of S with respect to R.*

COROLLARY 1. *The notation being as in Theorem 2, the cotrace from R to S of any differential of the first kind of R is a differential of the first kind of S.*

COROLLARY 2. *Let the notation be as in Theorem 2, and assume that S is separable over R. Denote by g and G the genera of R and S respectively. Then we have* $G - 1 = [S:R](g - 1) + (1/2)\ d(\mathfrak{D})$.

We have (Theorem 6, II, §6)

$$2G - 2 = d(\mathfrak{d}(\mathrm{Cosp}_{R/S}\omega)) = d(\mathrm{Con}_{R/S}\mathfrak{d}(\omega)) + d(\mathfrak{D}),$$

and this is equal, in virtue of Lemma 1, IV, §7 to $[S:R](2g - 2) + d(\mathfrak{D})$.

COROLLARY 3 (THEOREM OF LÜROTH). *Let R be a field of the form $K\langle x\rangle$, where K is a field and x transcendental over K, and let T be an intermediary field between K and R, distinct from K itself. Then T contains an element t such that $T = K\langle t\rangle$.*

It is clear that R is algebraic over T. Let U be the field of elements of R which are separable over T. If $U \neq R$, then R is of characteristic $p \neq 0$, and the degree of x over U is a power p^n of p. Since x is purely unseparable over U, we have $x^{p^n} \in U$. On the other hand, the divisor of poles of x^{p^n} in R is of degree p^n, whence, by the corollary to Theorem 4, I, §8, $[R:K\langle x^{p^n}\rangle] = p^n$. Since $[R:U] \geqq p^n$ and $K\langle x^{p^n}\rangle \subset U$, we have $U = K\langle x^{p^n}\rangle$, and U is of genus 0. Applying Corollary 2 to T and U (instead of R and S) and observing that the different of U with respect to T is integral, we have $-1 \geqq [U:T](g - 1)$, where g is the genus of T. It follows that $g = 0$. Since R has a place of degree 1, the same is true of T and T is a purely transcendental extension of K (cf. II, §2).

Let R and S be fields of algebraic functions of one variable such that S is an overfield of R and has the same field of constants as R. It is clear that, if ω and ω' are differentials of R, then we have

$$\mathrm{Cosp}_{R/S}\ (\omega + \omega') = \mathrm{Cosp}_{R/S}\omega + \mathrm{Cosp}_{R/S}\omega'. \tag{4}$$

If x is any element of R, we have

$$\mathrm{Cosp}_{R/S}x\omega = x\ \mathrm{Cosp}_{R/S}\omega. \tag{5}$$

For, let $\mathfrak{y}$ be any repartition in S. Then $(\mathrm{Cosp}_{R/S}x\omega)(\mathfrak{y}) = (x\omega)(\mathrm{Sp}_{S/R}\mathfrak{y}) = \omega(x\,\mathrm{Sp}_{S/R}\mathfrak{y}) = \omega(\mathrm{Sp}_{S/R}x\mathfrak{y}) = (\mathrm{Cosp}_{R/S}\omega)(x\mathfrak{y}) = (x\,\mathrm{Cosp}_{R/S}\omega)(\mathfrak{y})$, which proves our assertion. On the other hand, if T is an intermediary field between R and S, then we have

$$\mathrm{Cosp}_{R/S}\omega = \mathrm{Cosp}_{T/S}(\mathrm{Cosp}_{R/T}\omega). \tag{6}$$

For if $\mathfrak{y}$ is any repartition in S, we have $(\mathrm{Cosp}_{R/S}\omega)(\mathfrak{y}) = \omega(\mathrm{Sp}_{S/R}\mathfrak{y}) = \omega(\mathrm{Sp}_{T/R}(\mathrm{Sp}_{S/T}\mathfrak{y})) = (\mathrm{Cosp}_{R/T}\omega)\,(\mathrm{Sp}_{S/T}\mathfrak{y}) = (\mathrm{Cosp}_{T/S}(\mathrm{Cosp}_{R/T}\omega))(\mathfrak{y})$, which proves our assertion.

If ω is a differential of R and $z \in S$, then we have

$$\mathrm{Sp}_{S/R}z\,\mathrm{Cosp}_{R/S}\omega = (\mathrm{Sp}_{S/R}z)\omega. \tag{7}$$

For, let $\mathfrak{x}$ be a repartition of R. Then $(\mathrm{Sp}_{S/R}z\,\mathrm{Cosp}_{R/S}\omega)(\mathfrak{x}) = (z\,\mathrm{Cosp}_{R/S}\omega)(\mathrm{Cosp}_{R/S}\mathfrak{x}) = (\mathrm{Cosp}_{R/S}\omega)(z\,\mathrm{Cosp}_{R/S}\mathfrak{x}) = \omega(\mathrm{Sp}_{S/R}(z\,\mathrm{Cosp}_{R/S}\mathfrak{x}))$, and this is equal (by Formula (1), IV, §7), to $\omega((\mathrm{Sp}_{S/R}z)\mathfrak{x}) = ((\mathrm{Sp}_{S/R}z)\omega)(\mathfrak{x})$, which proves our formula.

THEOREM 3. *Let R and S be fields of algebraic functions of one variable; assume that R is a subfield of S and that R and S have the same field of constants. Denote by ω a differential of R, by $\mathfrak{p}$ a place of R, and by $\mathfrak{P}$ a place of S above $\mathfrak{p}$. Let e be the ramification index of $\mathfrak{P}$ with respect to R. If the residue field of $\mathfrak{P}$ is not separable over that of $\mathfrak{p}$, then* $\mathrm{res}_{\mathfrak{P}}\,\mathrm{Cosp}_{R/S}\omega = 0$. *If the residue field of $\mathfrak{P}$ is separable over that of $\mathfrak{p}$, then we have* $\mathrm{res}_{\mathfrak{P}}\,\mathrm{Cosp}_{R/S}\omega = e\,\mathrm{res}_{\mathfrak{p}}\,\omega$.

Denote by $\Sigma(\mathfrak{p})$ and $\Sigma(\mathfrak{P})$ the residue fields of $\mathfrak{p}$ and $\mathfrak{P}$ respectively, by $\Sigma_s(\mathfrak{p})$ and $\Sigma_s(\mathfrak{P})$ the largest subfields of $\Sigma(\mathfrak{p})$ and $\Sigma(\mathfrak{P})$ respectively which are separable over the common field of constants K of R and S. We form the $\mathfrak{p}$-adic completion $\bar{R}$ of R and the $\mathfrak{P}$-adic completion $\bar{S}$ of S; we may consider $\bar{R}$ as a subfield of $\bar{S}$. Let $\bar{\Sigma}_s(\mathfrak{p})$ and $\bar{\Sigma}_s(\mathfrak{P})$ be the subfields of $\bar{R}$ and $\bar{S}$ respectively which contain K and which are systems of representatives for $\Sigma_s(\mathfrak{p})$ and $\Sigma_s(\mathfrak{P})$ (cf. Theorem 1, III, §3). Denote by Ω the differential $\mathrm{Cosp}_{R/S}\,\omega$, and let u be any element of $\Sigma_s(\mathfrak{P})$. If $\bar{u}$ is the representative of u in $\bar{\Sigma}_s(\mathfrak{P})$, we have $\Omega^{\mathfrak{P}}(\bar{u}) = \mathrm{Sp}_{\Sigma_s(\mathfrak{P})/K}\,u\,\mathrm{res}_{\mathfrak{P}}\Omega$. By definition of the cotrace of a differential, this is also equal to $\omega^{\mathfrak{p}}(\mathrm{Sp}_{\bar{S}/\bar{R}}\bar{u})$. Now, $\bar{u}$ is algebraic over K, and therefore also over $\bar{R}$. The conjugates of $\bar{u}$ with respect to $\bar{R}$ (in a suitable normal extension of $\bar{R}$) are also conjugates of $\bar{u}$ with respect to K, and are therefore algebraic and separable over K. It follows that $\mathrm{Sp}_{\bar{S}/\bar{R}}\bar{u}$ is algebraic and separable over K, and therefore contained in $\bar{\Sigma}_s(\mathfrak{p})$. We know that the residue class of $\mathrm{Sp}_{\bar{S}/\bar{R}}\bar{u}$ is $e\,\mathrm{Sp}_{\Sigma(\mathfrak{P})/\Sigma(\mathfrak{p})}u$ (cf. Formula (1), IV, §8). This residue class is 0 if $\Sigma(\mathfrak{P})$ is not separable over $\Sigma(\mathfrak{p})$, whence, in that case, $\mathrm{Sp}_{\bar{S}/\bar{R}}\bar{u} = 0$, $\Omega^{\mathfrak{P}}(\bar{u}) = 0$, and, this being true for every $u \in \Sigma_s(\mathfrak{P})$, $\mathrm{res}_{\mathfrak{P}}\,\Omega = 0$. Assume now that $\Sigma(\mathfrak{P})$ is separable over $\Sigma(\mathfrak{p})$. Let f be the degree of $\Sigma(\mathfrak{P})$ over $\Sigma(\mathfrak{p})$, and let $\bar{K}$ be an algebraically closed overfield of K containing $\Sigma(\mathfrak{P})$. Then there are f distinct isomorphisms of $\Sigma(\mathfrak{P})$ into $\bar{K}$ which coincide with the identity on $\Sigma(\mathfrak{p})$. Since $\Sigma(\mathfrak{P})$ is purely unseparable over $\Sigma_s(\mathfrak{P})$, these isomorphisms induce f distinct isomorphisms of $\Sigma_s(\mathfrak{P})$

into $\bar{K}$ which coincide with the identity on $\Sigma_s(\mathfrak{p})$. Conversely, any isomorphism of $\Sigma_s(\mathfrak{P})$ into $\bar{K}$ which coincides with the identity on $\Sigma_s(\mathfrak{p})$ can be extended to an isomorphism of $\Sigma(\mathfrak{P})$, and, since $\Sigma(\mathfrak{p})$ is purely unseparable over $\Sigma_s(\mathfrak{p})$, the extended isomorphism coincides with the identity on $\Sigma(\mathfrak{p})$. It follows that, in the case under consideration, we have $[\Sigma(\mathfrak{P}):\Sigma(\mathfrak{p})] = [\Sigma_s(\mathfrak{P}):\Sigma_s(\mathfrak{p})]$ and $\mathrm{Sp}_{\Sigma(\mathfrak{P})/\Sigma(\mathfrak{p})}u = \mathrm{Sp}_{\Sigma_s(\mathfrak{P})/\Sigma_s(\mathfrak{p})}u$. We conclude that $\mathrm{Sp}_{\Sigma_s(\mathfrak{P})/K}u \ \mathrm{res}_{\mathfrak{P}}\,\Omega = e\omega^{\mathfrak{p}}(\bar{v})$, where $\bar{v}$ is the representative of $\mathrm{Sp}_{\Sigma_s(\mathfrak{P})/\Sigma_s(\mathfrak{p})}u$ in $\bar{\Sigma}_s(\mathfrak{p})$. But $\omega^{\mathfrak{p}}(\bar{v}) = \mathrm{Sp}_{\Sigma_s(\mathfrak{p})/K}((\mathrm{Sp}_{\Sigma_s(\mathfrak{P})/\Sigma_s(\mathfrak{p})}u)\ \mathrm{res}_{\mathfrak{p}}\omega) = \mathrm{Sp}_{\Sigma_s(\mathfrak{P})/K}u\ \mathrm{res}_{\mathfrak{p}}\,\omega$. Thus we see that $\Omega^{\mathfrak{p}}(\bar{u}) = \mathrm{Sp}_{\Sigma_s(\mathfrak{P})/K}u(e\ \mathrm{res}_{\mathfrak{p}}\,\omega)$, which proves that $\mathrm{res}_{\mathfrak{P}}\,\Omega = e\ \mathrm{res}_{\mathfrak{p}}\omega$. Theorem 3 is thereby proved.

§3. The differential dx in an arbitrary field

We are now able to accomplish our purpose of associating to every element x of a field R of algebraic functions of one variable a differential of the field R. This differential will be denoted by $(dx)_R$, or, if no confusion is possible, by dx.

Let K be the field of constants of R. If x is not in K, it is transcendental over K, and the differential $(dx)_{K\langle x\rangle}$ has been defined in §1. We set in this case

$$(dx)_R = \mathrm{Cosp}_{K\langle x\rangle/R}\,(dx)_{K\langle x\rangle}\,.$$

This definition clearly does not conflict with the previous one in the case where $R = K\langle x\rangle$. If x is an element of K, we define $(dx)_R$ to be the zero differential. Any differential of R which is of the form $(dx)_R$, for some $x \in R$, is called an *exact differential.*

Theorem 4. *Let R be a field of algebraic functions of one variable, and let K be the field of constants of R. If R is not separably generated, then every exact differential in R is zero. If R is separably generated, and $x \in R$, a necessary and sufficient condition for $(dx)_R$ to be $\neq 0$ is for x to be a separating variable in R.*

Let x be any element of R. If R is not algebraic over $K\langle x\rangle$, then x is in K and $(dx)_R = 0$. If R is algebraic over $K\langle x\rangle$, it follows immediately from Theorem 2 that $(dx)_R$ is $\neq 0$ if and only if R is separable over $K\langle x\rangle$.

Theorem 5. *An exact differential of a field R of algebraic functions of one variable has no residue $\neq 0$. If R is of characteristic $p > 0$ and $x \in R$, then, for any integer $m > 0$, the residues of $x^{pm-1}(dx)_R$ are all equal to 0. Let K be the field of constants of R, and let $\mathfrak{p}$ be a place of R. Then, if $x \neq 0$, $\mathrm{res}_{\mathfrak{p}}\, x^{-1}(dx)_R$ is equal to 0 if the residue field of $\mathfrak{p}$ is not separable over K and is equal to $\nu_{\mathfrak{p}}(x)$ if the residue field of $\mathfrak{p}$ is separable over K.*

We may assume that x is not in K. Making use of Formula (5), §2, we have $x^h(dx)_R = \mathrm{Cosp}_{K\langle x\rangle/R}x^h(dx)_{K\langle x\rangle}$. Let $\mathfrak{p}(0)$ and $\mathfrak{p}(\infty)$ be respectively the zero and the pole of x in $K\langle x\rangle$; then $x^h(dx)_{K\langle x\rangle}$ has no pole outside $\mathfrak{p}(0)$ and $\mathfrak{p}(\infty)$. On the other hand, $\mathfrak{p}(0)$ and $\mathfrak{p}(\infty)$ are of degree 1. If $c \in K$, then $(x^h(dx)_{K\langle x\rangle})^{\mathfrak{p}(\infty)}(c) = ((dx)_{K\langle x\rangle})^{\mathfrak{p}(\infty)}(cx^h)$; making use of Lemma 4, §1, we see that this is equal to 0 if $h \neq -1$ and to $-c$ if $h = -1$. It follows that $\mathrm{res}_{\mathfrak{p}(\infty)}\, x^h(dx)_{K\langle x\rangle}$ is 0 if $h \neq -1$

and -1 if $h = -1$. It follows from Theorem 3, III, §5 that $\operatorname{res}_{\mathfrak{p}(0)} x^h (dx)_{K\langle x\rangle}$ is 0 if $h \neq -1$ and is 1 if $h = -1$. Theorem 5 follows immediately from this and from Theorem 3, §2.

LEMMA 1. *Let R be a field of algebraic functions of one variable; denote by K the field of constants of R and by x an element of R not in K. If R is separable over $K\langle x\rangle$, then there are infinitely many places of R whose residue fields are separable over K and which are not ramified with respect to $K\langle x\rangle$.*

First we observe that there are infinitely many essentially distinct irreducible polynomials $f(X)$ with coefficients in K such that $f'(X) \neq 0$. For, if K is infinite, we may take the polynomials $f(X) = X - a$, $a \in K$, while, if K is finite, it is well known that there exists for each integer $n > 0$ an irreducible polynomial of degree n with coefficients in K, and the derivative of this polynomial is $\neq 0$ because K is perfect. This proves that there are infinitely many places of $K\langle x\rangle$ whose residue fields are separable over K. Only for a finite number of these places $\mathfrak{p}_0$ can there exist a place $\mathfrak{p}$ of R above $\mathfrak{p}_0$ which either is ramified with respect to $K\langle x\rangle$ or has a residue field which is not separable over that of $\mathfrak{p}_0$ (Lemma 3, IV, §8). Lemma 1 is thereby proved.

This being said, let $\mathfrak{p}$ be a place of R which is not ramified with respect to $K\langle x\rangle$, whose residue field is separable over K, and which is not a pole of x. Let $f(x)$ be the irreducible element of $K[x]$ with leading coefficient 1 which admits $\mathfrak{p}$ as a zero. Since $\mathfrak{p}$ is not ramified with respect to $K\langle x\rangle$, we have $\nu_{\mathfrak{p}}(f) = 1$. Let $\mathfrak{x}$ be a repartition of R such that $\nu_{\mathfrak{p}}(\mathfrak{x}(\mathfrak{p})) \geqq -1$. Then $\nu_{\mathfrak{p}}(f\mathfrak{x}(\mathfrak{p})) \geqq 0$; denote by φ the residue class of $f\mathfrak{x}(\mathfrak{p})$ modulo $\mathfrak{p}$. We shall prove the formula

$$(1) \qquad (dx)_R^{\mathfrak{p}}(\mathfrak{x}) = \operatorname{Sp}_{\Sigma/K} \frac{\varphi}{f'(\xi)}$$

where Σ is the residue field of $\mathfrak{p}$ and ξ the residue class of x modulo $\mathfrak{p}$.

Let $\mathfrak{x}^{\mathfrak{p}}$ be the repartition which assigns $\mathfrak{x}(\mathfrak{p})$ to $\mathfrak{p}$ and 0 to every other place. Then $(dx)_R^{\mathfrak{p}}(\mathfrak{x}) = (dx)_R(\mathfrak{x}^{\mathfrak{p}}) = (dx)_{K\langle x\rangle}(\operatorname{Sp}_{R/K\langle x\rangle}\mathfrak{x}^{\mathfrak{p}})$. Let $\mathfrak{p}_0$ be the place of $K\langle x\rangle$ below $\mathfrak{p}$; then it is clear that the $\mathfrak{p}_0$-component of $\operatorname{Sp}_{R/K\langle x\rangle}\mathfrak{x}^{\mathfrak{p}}$ is $\operatorname{Sp}^{\mathfrak{p}}\mathfrak{x}(\mathfrak{p})$, where $\operatorname{Sp}^{\mathfrak{p}}$ denotes the trace from the $\mathfrak{p}$-adic completion of R to the $\mathfrak{p}_0$-adic completion of $K\langle x\rangle$. Let Σ_0 be the residue field of $\mathfrak{p}_0$, and let φ_0 be the residue class of $f\operatorname{Sp}^{\mathfrak{p}}\mathfrak{x}(\mathfrak{p})$ modulo $\mathfrak{p}_0$. Then it follows from Formula (2), §1 that $(dx)_{K\langle x\rangle}(\operatorname{Sp}_{R/K\langle x\rangle}\mathfrak{x}^{\mathfrak{p}}) = \operatorname{Sp}_{\Sigma_0/K}\varphi_0/f'(\xi)$. On the other hand, since $\mathfrak{p}$ is not ramified with respect to $K\langle x\rangle$ and $f\operatorname{Sp}^{\mathfrak{p}}\mathfrak{x}(\mathfrak{p}) = \operatorname{Sp}^{\mathfrak{p}}f\mathfrak{x}(\mathfrak{p})$, it follows immediately from Formula (1), IV, §8 that $\varphi_0 = \operatorname{Sp}_{\Sigma/\Sigma_0}\varphi$. Since $f'(\xi) \in \Sigma_0$, we have

$$(dx)_R^{\mathfrak{p}}(x) = \operatorname{Sp}_{\Sigma_0/K}\left(\operatorname{Sp}_{\Sigma/\Sigma_0}\frac{\varphi}{f'(\xi)}\right) = \operatorname{Sp}_{\Sigma/K}\frac{\varphi}{f'(\xi)},$$

which proves Formula (1).

Let us consider more specifically the case where $\mathfrak{p}$ is of degree one. Then $f = x - a$, with $a \in K$, and the $\mathfrak{p}$-adic completion $\bar{R}$ of R may be identified with the field of formal power series in $x - a$ with coefficients in K; on the other hand, $\bar{R}$ coincides with the $\mathfrak{p}_0$-adic completion of $K\langle x\rangle$, whence $\mathrm{Sp}^{\mathfrak{p}} y = y$ for every $y \in \bar{R}$. Making use of Lemma 2, §1, we obtain

THEOREM 6. *Let R be a field of algebraic functions of one variable and let K be the field of constants of R. Assume that R is algebraic and separable over $K\langle x\rangle$, where x is some element of R. Let $\mathfrak{p}$ be a place of degree 1 of R which is not ramified with respect to $K\langle x\rangle$ and which is not a pole of x. Let $u = \sum_{k=r}^{\infty} c_k(x - a)^k$ ($c_k \in K$, $r \leqq k < \infty$; $r < 0$; a is the value taken by x at $\mathfrak{p}$) be an element of the $\mathfrak{p}$-adic completion of R. Then we have $(dx)^{\mathfrak{p}}(u) = c_{-1}$.*

COROLLARY. *Let the notation be as in Theorem 6, and let y be any element of R. Let $y = \sum_{k=s}^{\infty} b_k(x - a)^k$ ($c_k \in K$, $s \leqq k < \infty$; $s < 0$) be the expansion of y in the $\mathfrak{p}$-adic completion of R. Then we have $\mathrm{res}_{\mathfrak{p}}\, y\, dx = b_{-1}$.*

Let c be any element of K. Then we have $(y\, dx)^{\mathfrak{p}}(c) = (dx)^{\mathfrak{p}}(cy) = cb_{-1}$, which proves our assertion.

Still assuming R to be separable over $K\langle x\rangle$, the divisor of the differential $(dx)_R$ can easily be determined. Let $\mathfrak{r}$ be the pole of x in $K\langle x\rangle$; then the divisor of $(dx)_{K\langle x\rangle}$ is $\mathfrak{r}^{-2}$. Making use of Theorem 2, §2, we see that $\mathfrak{d}((dx)_R) = \mathfrak{d}_x\mathfrak{n}^{-2}$, where $\mathfrak{d}_x$ is the different of R with respect to $K\langle x\rangle$, while $\mathfrak{n} = \mathrm{Con}_{K\langle x\rangle/R}\mathfrak{r}$. It is clear that $\mathfrak{n}$ is the divisor of poles of x. Thus we obtain

THEOREM 7. *Let R be a field of algebraic functions of one variable and let K be the field of constants of R. Assume that R is algebraic and separable over $K\langle x\rangle$, where x is some element of R. Then the divisor of the differential dx in R is $\mathfrak{d}_x\mathfrak{n}^{-2}$, where $\mathfrak{d}_x$ is the different of R with respect to $K\langle x\rangle$, while $\mathfrak{n}$ is the divisor of poles of x.*

COROLLARY. *The notation being as in Theorem 7, the poles of dx are to be found among the poles of x.*

The assumptions being the same as in the previous theorem, the fact that $dx \neq 0$ allows us to represent every differential ω of R in the form $u\, dx$, with $u \in R$. On the other hand, since R is separable over $K\langle x\rangle$, we may write $R = K\langle x, y_1\rangle$, with some $y_1 \in R$. The characteristic polynomial of y_1 with respect to $K\langle x\rangle$ is of the form $Y^n + \sum_{i=1}^{n} a_i(x)(a_0(x))^{-1}Y^{n-i}$ where $a_0, \cdots, a_n$ are polynomials with coefficients in K. It follows easily that the characteristic polynomial $F(Y)$ of $y = a_0(x)y_1$ has its coefficients in $K[x]$, i.e., that y is integral over $K[x]$. Now, assume furthermore that the only poles of ω (if any) occur among the poles of x in R. Then we shall see that u is of the form $u = G(x, y)/F'(y)$, where G is a polynomial in two letters with coefficients in K, while F' is the derivative dF/dY of F. It is clear that y is integral over the ring of any place $\mathfrak{p}_0$ of $K\langle x\rangle$ which is not a pole of x; it follows that the only poles of y in R occur among the poles of x. If k is any exponent $\geqq 0$, the poles of $y^k\omega$ are among the poles of x, from which it follows immediately that the only pole (if any) of $\mathrm{Sp}_{R/K\langle x\rangle}\, y^k\omega$ is the pole $\mathfrak{p}_\infty$ of x in $K\langle x\rangle$. Now we have $(dx)_R = \mathrm{Cosp}_{K\langle x\rangle/R}\,(dx)_{K\langle x\rangle}$, and it follows from Formula (7), §2 that $\mathrm{Sp}_{R/K\langle x\rangle}\, y^k\omega = (\mathrm{Sp}_{R/K\langle x\rangle}\, y^k u)\,(dx)_{K\langle x\rangle}$. Since $(dx)_{K\langle x\rangle}$ has no zero, we conclude that $\mathrm{Sp}_{R/K\langle x\rangle}\, y^k u$

has no pole outside $\mathfrak{p}_\infty$ and therefore lies in $K[x]$. Our result will therefore follow from

LEMMA 2. *Let S be a field, $\mathfrak{o}$ a subring of S containing* 1, *and R an overfield of S which is obtained from S by adjunction of an element y which is algebraic and separable over S and whose characteristic polynomial $F(Y)$ with respect to S has its coefficients in $\mathfrak{o}$. Then, if $u \in R$, a necessary and sufficient condition for the elements $\mathrm{Sp}_{R/S}\, y^k u$ $(0 \leqq k < \infty)$ to all be in $\mathfrak{o}$ is for u to be expressible in the form $v/F'(y)$, with $v \in \mathfrak{o}[y]$.*

If n is the degree of F, the elements $1, y, \cdots, y^{n-1}$ form a base of R with respect to S, and we may write $uF'(y) = \sum_{i=0}^{n-1} a_i y^i$ with $a_i \in S$ $(0 \leqq i \leqq n-1)$. If $0 \leqq k \leqq n-1$, we have, by Lemma 3, §1,

$$(2) \qquad a_k = \mathrm{Sp}_{R/S}\, y^{n-1-k} u - \sum_{k<i\leqq n-1} a_i \,\mathrm{Sp}_{R/S} \frac{y^{i+n-1-k}}{F'(y)}.$$

Now, we prove by induction on l that $\mathrm{Sp}_{R/S}\, y^l/F'(y) \in \mathfrak{o}$ for all $l \geqq 0$. This is true for $l \leqq n-1$ in virtue of the result quoted above. Assume that it is true for some $l \geqq n-1$. If $F(Y) = Y^n + \sum_{j=1}^n b_j Y^{n-j}$, we have $y^{l+1} = -\sum_{j=1}^n b_j y^{l+1-j}$, from which it follows immediately (since $b_j \in \mathfrak{o}$) that our assertion is true for $l+1$. From this, we conclude immediately that the conditions $a_i \in \mathfrak{o}$ $(0 \leqq i \leqq n-1)$ imply $\mathrm{Sp}_{R/S}\, y^k u \in \mathfrak{o}$ $(0 \leqq k < \infty)$. Conversely, assume these last conditions to be satisfied. Then $a_{n-1} = \mathrm{Sp}_{R/S}\, u \in \mathfrak{o}$, and formula (2) shows that the conditions $a_k \in \mathfrak{o}$ for $i < k \leqq n-1$ imply $a_i \in \mathfrak{o}$, which shows that $a_0, \cdots, a_{n-1}$ all belong to $\mathfrak{o}$. Lemma 2 is thereby proved.

We have established

THEOREM 8. *Let R be a field of algebraic functions of one variable of the form $K\langle x, y\rangle$, where K is the field of constants of R, x is transcendental over K, and y algebraic and separable over $K\langle x\rangle$. Assume furthermore that the characteristic polynomial $F(Y)$ of y with respect to $K\langle x\rangle$ has its coefficients in $K[x]$. Let ω be a differential of R which has no pole outside the set of poles of x. Then ω can be expressed in the form*

$$\omega = \frac{G(x, y)}{F'(y)}\, dx$$

where G is a polynomial with coefficients in K.

§4. DERIVATIONS OF FIELDS

Let R be a separably generated field of algebraic functions of one variable, and let x be a separating variable in R. Since $dx \neq 0$, every differential in R can be expressed in the form $u\, dx$, with $u \in R$. This applies in particular to the differential dy of an element $y \in R$; we propose to find out how it is possible to compute u in terms of x and y. In order to do this, we have first to study the notion of derivation of a field.

Let R be a field and S an overfield of R. By a *derivation* of R into S is meant a mapping D of R into S which satisfies the following conditions:

$$\begin{aligned} &(1) \qquad D(y_1 + y_2) = Dy_1 + Dy_2 \\ &(2) \qquad D(y_1y_2) = (Dy_1)y_2 + y_1(Dy_2) \end{aligned} \qquad \text{for } y_1, y_2 \in R.$$

The first condition means that D is a homomorphism of the additive group of R into that of S. It follows immediately that we have

$$(3) \qquad D(y_1 - y_2) = Dy_1 - Dy_2 \; (y_1 , y_2 \in R); \qquad D(0) = 0.$$

We have $D(1) = D(1 \cdot 1) = 2D(1)$, whence $D(1) = 0$. By induction on k, we deduce easily from (2) that $D(y^k) = ky^{k-1}Dy$ for all $k \geqq 0$. Assuming that y_1 , y_2 are in R and $y_2 \neq 0$, we shall see that

$$(4) \qquad D\left(\frac{y_1}{y_2}\right) = \frac{(Dy_1)y_2 - y_1(Dy_2)}{y_2^2}.$$

For, if $z = y_1y_2^{-1}$, we have $y_1 = zy_2$, whence $Dy_1 = (Dz)y_2 + z(Dy_2)$; solving for Dz and replacing z by its value, we obtain the desired formula.

If D is a derivation of R into S, and $u \in S$, then we see immediately that the mapping uD of R into S defined by $(uD)(y) = u(Dy)$ is again a derivation. From this we conclude that the derivations of R into S form a vector space over S.

A derivation of a field R into itself is called simply a *derivation of* R. Let D_1 and D_2 be derivations of R; then the operation $D = [D_1 , D_2]$ defined by $Dy = D_2(D_1y) - D_1(D_2y)$ is again a derivation of R. It is obvious that D satisfies condition (1) above. On the other hand, we have

$$\begin{aligned} D(y_1y_2) &= D_2((D_1y_1)y_2 + y_1(D_1y_2)) - D_1((D_2y_1)y_2 + y_1(D_2y_2)) \\ &= (D_2(D_1y_1))y_2 + y_1(D_2(D_1y_2)) - (D_1(D_2y_1))y_2 - y_1(D_1(D_2y_2)) \\ &= (Dy_1)y_2 + y_1(Dy_2), \end{aligned}$$

which proves our assertion.

LEMMA 1. *Let D and D' be derivations of a field R into an overfield S. Assume that D and D' coincide on some subset E of R. Then they coincide on the subfield F of R generated by the elements of E.*

For it follows immediately from Formulas (1), (2), (3), (4) that the set of elements $y \in R$ such that $Dy = D'y$ is a field.

LEMMA 2. *Let D be a derivation of a field R into an overfield S, and let z be an element of S which is algebraic and separable over R. Then it is possible in one and only way to extend D to a derivation of $R\langle z\rangle$ into S.*

Introduce an indeterminate Z. Let $G(Z) = \sum_{k=0}^{m} c_kZ^k$ be any polynomial with coefficients in R; then we set $(DG)(Z) = \sum_{k=0}^{m} (Dc_k)Z^k$. It is easy to see that, if G and G' are polynomials with coefficients in R, then $D(G + G') = DG + DG'$, $D(GG') = (DG)G' + G(DG')$. Assume that it is possible to extend D to a derivation D' of $R\langle z\rangle$ into S. Then we have, for $c \in R$, $D'(cz^k) = (Dc)z^k + kcz^{k-1}D'z$, whence $D'(G(z)) = (DG)(z) + G'(z)D'z$, where G' is the derivative of G. Apply this to the characteristic polynomial $F(Z)$ of z with respect to R; since $F(z) = 0$, we find $0 = (DF)(z) + F'(z)D'z$. Since z is separable over R, $F'(z)$ is $\neq 0$, and the preceding formula determines complete $D'z$. Making use of Lemma 1, we see that, if the extension D' exists, it is unique.

To prove the existence of D', we observe that any $u \in R\langle z\rangle$ may be written in one and only one way in the form $u = G(z)$ where G is a polynomial with coefficients in R of degree smaller than the degree n of F. We then define $D'u$ by the formula

$$D'u = (DG)(z) - \frac{(DF)(z)}{F'(z)} G'(z)$$

and we have to prove that D' is a derivation which extends D. First, if $u \in R$, then $G = u$ is of degree 0, and $D'u = Du$, which shows that the mapping D' coincides with D on R. Now, let u_1 and u_2 be elements of $R\langle z\rangle$, with $u_1 = G_1(z)$, $u_2 = G_2(z)$, G_1 and G_2 being polynomials of degrees $<n$ with coefficients in R. Then $u_1 + u_2 = (G_1 + G_2)(z)$, and $G_1 + G_2$ is of degree $< n$. Since

$$D(G_1 + G_2) = DG_1 + DG_2, \qquad (G_1 + G_2)' = G_1' + G_2',$$

we see that $D'(u_1 + u_2) = D'u_1 + D'u_2$. Now, if we consider the product u_1u_2, it is equal to $(G_1G_2)(z)$, but G_1G_2 may be of degree $\geqq n$. To find the correct representation of u_1u_2, we divide G_1G_2 by $F: G_1G_2 = FQ + G_3$, where G_3 is of degree $<n$, and we write $u_1u_2 = G_3(z)$. Since $F(z) = 0$, we have

$$(DG_3)(z) = (DG_1)(z)\cdot u_2 + u_1\cdot(DG_2)(z) - ((DF)(z))Q(z),$$

$$G_3'(z) = G_1'(z)u_2 + u_1G_2'(Z) - F'(z)Q(z)$$

and it is easy to verify that

$$D'(u_1u_2) = (DG_3)(z) - \frac{(DF)(z)}{F'(z)} G_3'(z)$$

is equal to $(D'u_1)u_2 + u_1(D'u_2)$. Lemma 2 is thereby proved.

LEMMA 3. *Assume that a derivation D of a field S maps a subfield R of S into itself, and that S is of finite degree over R. Then we have, for any $y \in S$,*

$$\mathrm{Sp}_{S/R}Dy = D(\mathrm{Sp}_{S/R}y).$$

We may assume that S is separable over R, since the formula is trivial if this is not the case. We can then find an overfield S^* of S which is normal, separable, and of finite degree over R. Let j_k $(1 \leqq k \leqq n)$ be the distinct isomorphisms of S with subfields of S^* which coincide with the identity on R. Then we have $\mathrm{Sp}_{S/R}y = \sum_{k=1}^n j_k(y)$, $\mathrm{Sp}_{S/R}Dy = \sum_{k=1}^n j_k(Dy)$. We can extend D to a derivation (which we also denote by D) of S^*. From the fact that j_k is an isomorphism, it follows immediately that the mappings j_kD and Dj_k are derivations of S into S^*. Since j_k coincides with the identity on R, j_kD and Dj_k coincide with each other on R, and therefore also on S by Lemma 2. Thus

$$\mathrm{Sp}_{S/R}Dy = \sum_{k=1}^{n} D(j_ky) = D(\mathrm{Sp}_{S/R}y).$$

LEMMA 4. *Let R be a field of algebraic functions of one variable. Let $\mathfrak{p}$ be a place of R and let D be a derivation of R which maps into itself the field of constants K of R. Then there exists an integer s such that $\nu_{\mathfrak{p}}(Dy) \geqq \nu_{\mathfrak{p}}(y) - s$ for all $y \in R$.*

Let x be a uniformizing variable at $\mathfrak{p}$. Set $s_0 = \max\{0, -\nu_{\mathfrak{p}}(Dx)\}$ (with the convention that $s_0 = 0$ if $Dx = 0$). If $f(x) = \sum_{k=0}^{m} a_kx^k$ is an element of $K[x]$, then $D(f(x)) = \sum_{k=0}^{m} (Da_k)x^k + (\sum_{k=1}^{m} ka_kx^{k-1})Dx$, whence

$$\nu_{\mathfrak{p}}(D(f(x))) \geqq -s_0 .$$

If y is any element $\neq 0$ of $K\langle x\rangle$, we may write $y = x^{\nu_{\mathfrak{p}}(y)}f(x)/g(x)$, where f and g are polynomials with coefficients in K, and $g(0) \neq 0$. Observing that the formula $Dx^{\nu} = \nu x^{\nu-1}Dx$ is true even if $\nu < 0$, we have

$$Dy = \nu_{\mathfrak{p}}(y)x^{\nu_{\mathfrak{p}}(y)-1}\frac{f(x)}{g(x)}Dx + x^{\nu_{\mathfrak{p}}(y)}\frac{D(f(x))g(x) - f(x)D(g(x))}{g^2(x)}$$

whence $\nu_{\mathfrak{p}}(Dy) \geqq \nu_{\mathfrak{p}}(y) - s_0 - 1$ if $y \in K\langle x\rangle$. Now, let $\mathfrak{p}_0$ be the zero of x in $K\langle x\rangle$, and let $\{z_1, \cdots, z_n\}$ be a base of R with respect to $K\langle x\rangle$ which is integral at $\mathfrak{p}_0$. Denote by $\mathfrak{p}_1, \cdots, \mathfrak{p}_h$ the places of R different from $\mathfrak{p}$ which lie above $\mathfrak{p}_0$. Then we can find an element u of R such that $\nu_{\mathfrak{p}}(u) = 0$, $\nu_{\mathfrak{p}_i}(u) > 0$ $(1 \leqq i \leqq h)$ (Corollary to Theorem 3, I, §6). Any element $y \neq 0$ of R may be written in the form $y = x^{\nu_{\mathfrak{p}}(y)}u^ry'$, where r is some integer and

$$\nu_{\mathfrak{p}}(y') = 0, \qquad \nu_{\mathfrak{p}_i}(y') \geqq 0 \qquad (1 \leqq i \leqq h).$$

We have

$$\frac{Dy}{y} = \nu_{\mathfrak{p}}(y)\frac{Dx}{x} + r\frac{Du}{u} + \frac{Dy'}{y'}$$

whence $\nu_{\mathfrak{p}}(Dy) \geqq \nu_{\mathfrak{p}}(y) + \min\{\nu_{\mathfrak{p}}(Dx/x), \nu_{\mathfrak{p}}(Du/u), \nu_{\mathfrak{p}}(Dy')\}$. We may write $y' = \sum_{i=1}^{n} y_iz_i$, with $y_1, \cdots, y_n$ in $K\langle x\rangle$; then we have

$$\nu_{\mathfrak{p}_0}(y_i) \geqq 0 \qquad (1 \leqq i \leqq n).$$

Thus $Dy' = \sum_{i=1}^{n} (Dy_i)z_i + \sum_{i=1}^{n} y_i(Dz_i)$. We have $\nu_{\mathfrak{p}}(Dy_i) \geqq -s_0 - 1$, whence

$$\nu_{\mathfrak{p}}(Dy') \geqq \min_{1\leqq i\leqq n}\{-s_0 - 1 + \nu_{\mathfrak{p}}(z_i), \nu_{\mathfrak{p}}(Dz_i)\}.$$

Lemma 4 is thereby proved.

LEMMA 5. *The assumptions being as in Lemma 4, denote also by $\bar{R}$ the $\mathfrak{p}$-adic completion of R. Then D can be extended to a derivation $\bar{D}$ of $\bar{R}$ which is continuous* (i.e., *if* $z \in \bar{R}$ *is the limit of a sequence* (z_n) *of elements of* $\bar{R}$, *then* $\bar{D}z = \lim_{n\to\infty}\bar{D}z_n$). *This extension is unique.*

If a sequence (x_n) of elements of R converges to an element $z \in \bar{R}$, then the sequence (Dx_n) is also convergent. In fact, $\nu_{\mathfrak{p}}(x_{n+1} - x_n)$ increases indefinitely with n, and the same is therefore true, in virtue of Lemma 4, of $\nu_{\mathfrak{p}}(Dx_{n+1} - Dx_n)$. Moreover, the limit $\lim_{n\to\infty} Dx_n$ depends only on z, not on the sequence (x_n), for, if $z = \lim_{n\to\infty} x_n = \lim_{n\to\infty} x'_n$, then $\nu_{\mathfrak{p}}(x'_n - x_n)$ increases indefinitely with n, and the same is true by Lemma 4 of $\nu_{\mathfrak{p}}(Dx'_n - Dx_n)$, whence $\lim_{n\to\infty} Dx'_n = \lim_{n\to\infty} Dx_n$. If we set $\bar{D}z = \lim_{n\to\infty} Dx_n$, then $\bar{D}$ is a mapping of $\bar{R}$ into itself. If (x_n) and (y_n) are convergent sequences of elements of R, then we have

$$\lim_{n\to\infty} (x_n + y_n) = \lim_{n\to\infty} x_n + \lim_{n\to\infty} y_n ,$$

and a similar formula for the product; it follows immediately from this that $\bar{D}$ is a derivation. Using the same notation as above, we have (by Lemma 4), $\nu_{\mathfrak{p}}(\bar{D}z) = \lim_{n\to\infty} \nu_{\mathfrak{p}}(Dx_n) \geqq \lim_{n\to\infty} \nu_{\mathfrak{p}}(x_n) - s = \nu_{\mathfrak{p}}(z) - s$ where s is an integer which does not depend on z. If z is the limit of a sequence (z_n) of elements of $\bar{R}$, we have $\lim_{n\to\infty} \nu_{\mathfrak{p}}(z - z_n) = \infty$, whence $\lim_{n\to\infty} \nu_{\mathfrak{p}}(\bar{D}z - \bar{D}z_n) = \infty$, and $\bar{D}z = \lim_{n\to\infty} \bar{D}z_n$. It is clear that $\bar{D}$ is the only continuous mapping of $\bar{R}$ into itself which extends D.

Let us consider in particular the case where the residue field Σ of $\mathfrak{p}$ is separable over K. Let then x be a uniformizing variable at $\mathfrak{p}$. We know (III, §3) that $\bar{R}$ contains a subfield Σ_1 isomorphic with Σ (under an isomorphism which maps upon themselves the elements of K) and that $\bar{R}$ may be identified with the field of formal power series in x with coefficients in Σ_1. Since Σ is separable over K, the restriction of D to K may be extended to a derivation D_1 of Σ_1, and it follows from the uniqueness assertion in Lemma 3 that D_1 is the restriction of $\bar{D}$ to Σ_1, which shows that $\bar{D}$ maps Σ_1 into itself. From this and from the continuity of $\bar{D}$, we deduce immediately the formula

$$\bar{D}\left(\sum_{k=r}^{\infty} c_k x^k\right) = \sum_{k=r}^{\infty} (D_1 c_k) x^k + \left(\sum_{k=r}^{\infty} k c_k x^{k-1}\right) Dx. \tag{5}$$

Conversely, let there be given any field Σ_1 and a derivation D_1 of this field. Construct the field $\bar{R}$ of formal power series in a symbol x with coefficients in Σ_1, and let u be any given element of $\bar{R}$. Then the formula

$$\bar{D}\left(\sum_{k=r}^{\infty} c_k x^k\right) = \sum_{k=r}^{\infty} (D_1 c_k) x^k + \left(\sum_{k=r}^{\infty} k c_k x^{k-1}\right) u$$

(where the c_k's are arbitrary elements of Σ_1) defines a derivation $\bar{D}$ of $\bar{R}$. For, let $y = \sum_{k=r}^{\infty} c_k x^k$ and $y' = \sum_{k=r'}^{\infty} c'_k x^k$ be elements of $\bar{R}$. It is first obvious that

$$\bar{D}(y + y') = \bar{D}y + \bar{D}y'.$$

On the other hand, we have $yy' = \sum_{l=r+r'}^{\infty} e_l x^l$, where e_l is given by

$$e_l = \sum_{k+k'=l} c_k c'_{k'} .$$

We have $D_1 e_l = \sum_{k+k'=l} (D_1 c_k) c'_{k'} + \sum_{k+k'=l} c_k (D_1 c'_{k'})$ and

$$l e_l = \sum_{k+k'=l} (k c_k) c'_{k'} + \sum_{k+k'=l} c_k (k' c'_{k'}),$$

from which it follows that $\bar{D}(yy') = (\bar{D}y)y' + y(\bar{D}y')$. It is clear that the derivation $\bar{D}$ extends D_1 and maps x upon u.

LEMMA 6. *Let R be a separably generated field of algebraic functions of one variable. If x is a separating variable in R, there exists a unique derivation of R which maps every constant upon* 0 *and x upon* 1.

Let K be the field of constants of R. We may consider the field $K\langle x \rangle$ as a subfield of the field $\bar{S}$ of formal power series in x with coefficients in K. We have just seen that there exists a derivation $\bar{D}$ of $\bar{S}$ which maps every element of K upon 0 and x upon 1. Making use of Formula (4) above, we see that $\bar{D}$ maps $K\langle x \rangle$ into itself. Since R is separable over $K\langle x \rangle$, the restriction of $\bar{D}$ to $K\langle x \rangle$ may be extended to a derivation D of R with the required properties. The uniqueness of D follows immediately from Lemmas 1 and 2.

The derivation whose existence and uniqueness are asserted in Lemma 6 is called the *derivation with respect to x* in R.

LEMMA 7. *Let R be a separably generated field of algebraic functions of one variable, and let D be a derivation of the field of constants K of R. If x is a separating variable in R, then D can be extended to a derivation of R which maps x upon* 0, *and this extension is unique.*

We use the notation of the proof of Lemma 6. From what we saw above, it follows that D may be extended to a derivation $\bar{D}$ of $\bar{S}$ which maps x upon 0. As in the proof of Lemma 6, we see that $\bar{D}$ maps $K\langle x \rangle$ into itself and that the restriction of $\bar{D}$ to $K\langle x \rangle$ may be extended to a derivation of R. Here again, the extension is unique in virtue of Lemmas 1 and 2.

§5. DERIVATIONS AND DIFFERENTIALS

We are now able to solve the problem which was proposed at the beginning of §4. The solution will be provided by

THEOREM 9. *Let R be a separably generated field of algebraic functions of one variable, and let x be a separating variable in R. Then, if y is any element of R, we have $dy = (D_x y)\, dx$, where D_x is the derivation with respect to x in R.*

If y is not a separating variable in R, then we know that $dy = 0$. Let in that case $F(X, Y)$ be an irreducible polynomial with coefficients in K such that $F(x, y) = 0$. Then $F(x, Y)$ is an irreducible polynomial with coefficients in $K\langle x \rangle$. Since y is separable over $K\langle x \rangle$, we have $(\partial F/\partial Y)(x, y) \neq 0$. On the other hand, x is not separable over $K\langle y \rangle$, for, were this the case, R, which is separable over $K\langle x \rangle$, and, *a fortiori*, over $K\langle x, y \rangle$, would be separable over $K\langle y \rangle$; it follows that $(\partial F/\partial X)(x, y) = 0$. Now, it is easily seen that $0 = D_x F(x, y) = (\partial F/\partial X)(x, y) D_x x + (\partial F/\partial Y)(x, y) D_x y$; we therefore have $D_x y = 0$, whence $dy = (D_x y)\, dx$.

Let us assume from now on that x and y are separating variables. Making use of Lemma 3, IV, §8 and Lemma 1, §3, we see that there exist infinitely many places $\mathfrak{p}$ of R which satisfy the following conditions:

1) the residue field of $\mathfrak{p}$ is separable over K;
2) $\mathfrak{p}$ is not ramified with respect to either one of the fields $K\langle x\rangle$ or $K\langle y\rangle$;
3) $\mathfrak{p}$ is not a pole of either x or y.

We shall prove that, if $\mathfrak{p}$ is any one of these places and $\mathfrak{x}$ a repartition such that $\nu_\mathfrak{p}(\mathfrak{x}) \geqq -1$, then $(dy - (D_x y)\, dx)^\mathfrak{p}(\mathfrak{x}) = 0$. When this will be done, Theorem 15 will follow immediately. For, were the differential $dy - (D_x y)\, dx \neq 0$, then it would follow from Lemma 2, II, §7 that each place $\mathfrak{p}$ which satisfies the conditions 1), 2), 3) would occur with an exponent >0 in the divisor of this differential, which is clearly impossible.

Let then $\mathfrak{p}$ be a place which satisfies conditions 1), 2), 3). Denote by Σ the residue field of $\mathfrak{p}$, by ξ and η the residue classes of x and y respectively modulo $\mathfrak{p}$, and by f and g the irreducible polynomials with coefficients in the field of constants K of R and leading coefficient 1 such that $f(\xi) = 0$, $g(\eta) = 0$. Since $\mathfrak{p}$ is not ramified with respect to either $K\langle x\rangle$ or $K\langle y\rangle$, we have

$$\nu_\mathfrak{p}(f(x)) = \nu_\mathfrak{p}(g(y)) = 1.$$

Let $\bar{R}$ be the $\mathfrak{p}$-adic completion of R. Since Σ is separable over K, $\bar{R}$ contains a field Σ_1 containing K and isomorphic with Σ, and $\bar{R}$ may be identified with the field of formal power series in $f(x)$ with coefficients in Σ_1 (III, §3). Moreover, D_x may be extended to a derivation $\bar{D}$ of $\bar{R}$ which is defined by the formula $\bar{D}(\sum_{k=r}^{\infty} s_k(f(x))^k) = (D_x f(x))(\sum_{k=r}^{\infty} k s_k(f(x))^{k-1})$ (where $s_k \in \Sigma_1$, $r \leqq k < \infty$). On the other hand, it is clear that $D_x f(x) = f'(x)$, where f' is the derivative of the polynomial f. If we expand $g(y)$ in $\bar{R}$, the expansion will be of the form $g(y) = z_1 f(x) + \sum_{k=2}^{\infty} s_k(f(x))^k$ (z_1, $s_k \in \Sigma_1$), with $z_1 \neq 0$. It follows that $\mathfrak{p}$ is not a pole of $D_x g(y)$, and that the residue class of $D_x g(y)$ is $\zeta f'(\xi)$, where ζ is the residue class of z_1 modulo $\mathfrak{p}$. But we have $D_x g(y) = g'(y) D_x y$, and, since η is separable over K, $g'(\eta) \neq 0$; it follows that $\nu_\mathfrak{p}(g'(y)) = 0$, whence $\nu_\mathfrak{p}(D_x y) \geqq 0$. Moreover, the residue class of $D_x y$ modulo $\mathfrak{p}$ is $\zeta f'(\xi)/g'(\eta)$. This being said, let $\mathfrak{x}$ be a repartition such that $\nu_\mathfrak{p}(\mathfrak{x}) \geqq -1$, whence $\nu_\mathfrak{p}(f(x)\mathfrak{x}(\mathfrak{p})) \geqq 0$, $\nu_\mathfrak{p}(g(y)\mathfrak{x}(\mathfrak{p})) \geqq 0$. Let φ and ψ be the residue classes of $f(x)\mathfrak{x}(\mathfrak{p})$ and $g(y)\mathfrak{x}(\mathfrak{p})$ respectively modulo $\mathfrak{p}$; then we have, by Formula (1), §3

$$(dx)^\mathfrak{p}(\mathfrak{x}) = \mathrm{Sp}_{\Sigma/K} \frac{\varphi}{f'(\xi)}, \quad (dy)^\mathfrak{p}(\mathfrak{x}) = \mathrm{Sp}_{\Sigma/K} \frac{\psi}{g'(\eta)}.$$

We have $((D_x y)\, dx)^\mathfrak{p}(\mathfrak{x}) = (dx)^\mathfrak{p}((D_x y)\mathfrak{x})$, and the residue class modulo $\mathfrak{p}$ of $f(x)(D_x y)\mathfrak{x}(\mathfrak{p})$ is $\zeta\varphi f'(\xi)/g'(\eta)$, whence

$$((D_x y) dx)^\mathfrak{p}(\mathfrak{x}) = \mathrm{Sp}_{\Sigma/K} \frac{\zeta\varphi}{g'(\eta)}.$$

On the other hand, we have $g(y)\mathfrak{x}(\mathfrak{p}) = (g(y)/f(x))f(x)\mathfrak{x}(\mathfrak{p})$, and $g(y)/f(x)$ belongs to the residue class ζ modulo $\mathfrak{p}$. It follows that $\psi = \zeta\varphi$, whence

$$((D_x y) dx)^\mathfrak{p}(\mathfrak{x}) = (dy)^\mathfrak{p}(\mathfrak{x}).$$

Theorem 9 is thereby proved.

COROLLARY. *The notation being as in Theorem* 15, *let* $y_1, \cdots, y_m$ *be elements of* R, *and let* F *be a polynomial with coefficients in the field of constants* K *of* R *in* m *letters* $Y_1, \cdots, Y_m$. *Then we have*

$$dF(y_1, \cdots, y_m) = \sum_{i=1}^{m} (\partial F/\partial Y_i)(y_1, \cdots, y_m)\, dy_i .$$

It is easily seen by direct computation that

$$D_x F(y_1, \cdots, y_m) = \textstyle\sum_{i=1}^{m} (\partial F/\partial Y_i)(y_1, \cdots, y_m) D_x y_i ;$$

the corollary follows immediately.

In particular, we have the formulas

$$d(y_1 + y_2) = dy_1 + dy_2,$$
$$d(y_1 y_2) = (dy_1)y_2 + y_1(dy_2).$$

§6. Extension of the notion of cotrace

Let R and S be fields of algebraic functions of one variable such that R is a subfield of S. We denote by K and L the fields of constants of R and S respectively, and we assume as usual that $K = L \cap R$. In the case where $K = L$, we have defined in §2 an operation of cotrace which assigns to every differential ω in R a differential $\mathrm{Cosp}_{R/S}\omega$ in S. We propose now to do the same thing in the case where $K \neq L$; however, in that case, we shall have to assume that R is separably generated. Let then u be a separating variable in R. Every differential ω of R can then be written in the form $x(du)_R$, with some $x \in R$. Set $\Omega(\omega) = x(du)_S$. It is clear that $\Omega(y\omega) = y\Omega(\omega)$ for every $y \in R$. We shall prove that we have $\Omega(\omega) = \mathrm{Cosp}_{R/S}\omega$ in the case where $K = L$, and that, in the general case, the differential $\Omega(\omega)$ does not depend on the choice of the separating variable u.

Assuming that $K = L$, we have, in virtue of Formula (6), §2, $\mathrm{Cosp}_{R/S}(du)_R = \mathrm{Cosp}_{R/S}\,(\mathrm{Cosp}_{K\langle u\rangle/R}(du)_{K\langle u\rangle}) = \mathrm{Cosp}_{K\langle u\rangle/S}(du)_{K\langle u\rangle} = (du)_S$. From this and from Formula (5), §2, it follows that $\Omega(\omega) = \mathrm{Cosp}_{R/S}\omega$ if $K = L$.

Now we shall prove that, in the general case, $\Omega((dv)_R) = (dv)_S$ for every element v of R. Let $F(U, V)$ be an irreducible polynomial with coefficients in K such that $F(u, v) = 0$. Then we have (Corollary to Theorem 9, §5),

$$(\partial F/\partial U)(u, v)(du)_R + (\partial F/\partial V)(u, v)(dv)_R = 0$$

and $(\partial F/\partial U)(u, v)(du)_S + (\partial F/\partial V)(u, v)(dv)_S = 0$. On the other hand, since v is separable over $K\langle u\rangle$, we have $(\partial F/\partial V)(u, v) \neq 0$. Thus we have

$$\Omega((dv)_R) = -\frac{\dfrac{\partial F}{\partial U}(u, v)}{\dfrac{\partial F}{\partial V}(u, v)}(du)_S = (dv)_S .$$

Considering in particular the case where v is itself a separating variable in R, we see that the operation $\omega \rightarrow \Omega(\omega)$ does not depend on the choice of the separating variable u.

The differential $\Omega(\omega)$ will be called the *cotrace of* ω, *taken from* R *to* S, and will be denoted by $\mathrm{Cosp}_{R/S}\, \omega$.

If T is an intermediary field between R and S and is itself a separably generated field of algebraic functions of one variable (with $L \cap T$ as its field of constants), we have, for every differential ω in R, $\mathrm{Cosp}_{R/S}\omega = \mathrm{Cosp}_{T/S}(\mathrm{Cosp}_{R/T}\omega)$. It is clearly sufficient to check the validity of this formula when $\omega = (du)_R$, u being a separating variable in R. In that case, the right side is equal to

$$\mathrm{Cosp}_{T/S}(du)_T = (du)_S = \mathrm{Cosp}_{R/S}(du)_R .$$

It follows in particular from our formula that, if u is a separating variable in S, it is also a separating variable in T.

We have proved

THEOREM 10. *Let* R *and* S *be fields of algebraic functions of one variable. Assume that* R *is separably generated, that* S *is an overfield of* R, *and that the field of constants of* R *is the intersection of* R *with the field of constants* L *of* S. *Then there exists a mapping* $\omega \rightarrow \mathrm{Cosp}_{R/S}\, \omega$ *of the set of differentials of* R *into the set of differentials of* S *which satisfies the following conditions:* 1) *if* $v \in R$, *we have* $\mathrm{Cosp}_{R/S}(dv)_R = (dv)_S$; 2) *the formulae* (4), (5), (6) *of* §2 *are valid* (*in* (6), T *is taken to be an intermediary field between* R *and* S *which is a separably generated field of algebraic functions of one variable whose field of constants is* $L \cap T$).

Now we prove

THEOREM 11. *Let* R *be a separably generated field of algebraic functions of one variable, and let* L *be an overfield of the field of constants* K *of* R; *set* $S = R\langle L\rangle$. *Let* ω *be a differential of* R, *and denote by* Ω *the differential* $\mathrm{Cosp}_{R/S}\, \omega$ *of* S. *Then we have, for any repartition* $\mathfrak{x}$ *in* R, $\omega(\mathfrak{x}) = \Omega(\mathrm{Cosp}_{R/S}\, \mathfrak{x})$. *Let* $\mathfrak{p}$ *be a place of* R *and* $\mathfrak{P}$ *a place of* S *above* $\mathfrak{p}$; *if we identify the residue field of* $\mathfrak{p}$ *with a subfield of that of* $\mathfrak{P}$, *we have* $\mathrm{res}_{\mathfrak{P}}\, \Omega = \mathrm{res}_{\mathfrak{p}}\, \omega$. *If* $\omega \neq 0$, *the divisor* $\mathfrak{d}(\Omega)$ *of* Ω *divides* $\mathrm{Con}_{R/S}\mathfrak{d}(\omega)$, *and is equal to it in the case where* R *and* S *have the same genus.*

We have $\omega(\mathfrak{x}) = \sum_{\mathfrak{q}} \omega^{\mathfrak{q}}(\mathfrak{x})$, the sum being extended over all places $\mathfrak{q}$ of R, and $\Omega\, (\mathrm{Cosp}_{R/S}\mathfrak{x}) = \sum_{\mathfrak{Q}} \Omega^{\mathfrak{Q}}\, (\mathrm{Cosp}_{R/S}\mathfrak{x})$, the sum being extended over all places $\mathfrak{Q}$ of S. In order to prove the first assertion of Theorem 11, it will therefore be sufficient to prove that $\omega^{\mathfrak{q}}(\mathfrak{x}) = \sum_{k=1}^{h} \Omega^{\mathfrak{Q}_k}\, (\mathrm{Cosp}_{R/S}\mathfrak{x})$, where $\mathfrak{q}$ is an arbitrary place of R and $\mathfrak{Q}_1, \cdots, \mathfrak{Q}_h$ the distinct places of S above $\mathfrak{q}$. Let (z_n) be a sequence of elements of R which converges to $\mathfrak{x}(\mathfrak{q})$ in the $\mathfrak{q}$-adic completion of R. Then this sequence converges to $(\mathrm{Cosp}_{R/S}\, \mathfrak{x})(\mathfrak{Q}_k)$ in the $\mathfrak{Q}_k$-adic completion of S $(1 \leqq k \leqq h)$; from a certain n on, we have $\omega^{\mathfrak{q}}(\mathfrak{x}) = \omega^{\mathfrak{q}}(z_n)$, $\Omega^{\mathfrak{Q}_k}\, (\mathrm{Cosp}_{R/S}\mathfrak{x}) = \Omega^{\mathfrak{Q}_k}(z_n)$ $(1 \leqq k \leqq h)$, and it will be sufficient to prove that, for every $z \in R$, $\omega^{\mathfrak{q}}(z) = \sum_{k=1}^{h} \Omega^{\mathfrak{Q}_k}(z)$. We first establish

LEMMA 1. *Let $\mathfrak{q}$ be a place of a separably generated field R of algebraic functions of one variable. Then there exists in R a separating variable which has $\mathfrak{q}$ as its unique pole.*

Let t be any separating variable in R, and let $\mathfrak{n}$ be the divisor of poles of t. If g is the genus of R, select an integer $m > 0$ such that $md(\mathfrak{q}) - d(\mathfrak{n})$ is greater than both $2g - 2$ and $g - 1$. Then no differential $\neq 0$ of R is multiple of $\mathfrak{q}^m\mathfrak{n}^{-1}$, whence, by the theorem of Riemann-Roch, $l(\mathfrak{q}^{-m}\mathfrak{n}) = md(\mathfrak{q}) - d(\mathfrak{n}) - g + 1 > 0$. This means that R contains an element $x_1 \neq 0$ which is $\equiv 0 \pmod{\mathfrak{q}^{-m}\mathfrak{n}}$. The place $\mathfrak{q}$ is the unique pole of either one of the elements x_1 and tx_1. Since

$$d(tx_1) = t\,dx_1 + x_1\,dt$$

and $dt \neq 0$, one at least of x_1 and tx_1 has a differential $\neq 0$ and is therefore a separating variable (cf. Theorem 4, §3).

This being said, we may return to the proof of the formula

$$\omega^{\mathfrak{q}}(z) = \sum_{k=1}^{h} \Omega^{\mathfrak{Q}_k}(z).$$

Let x be a separating variable in R which has $\mathfrak{q}$ as its unique pole; then we may write $\omega = y(dx)_R$, $y \in R$, whence $\Omega = y(dx)_S$, and we have to prove that

$$((dx)_R)^{\mathfrak{q}}(yz) = \sum_{k=1}^{h} ((dx)_S)^{\mathfrak{Q}_k}(yz).$$

Let $\mathfrak{q}_0$ be the pole of x in $K\langle x\rangle$; since $\mathfrak{q}$ is the only place of R above $\mathfrak{q}_0$, the left side of our formula is equal to $((dx)_{K\langle x\rangle})^{\mathfrak{q}_0}(\mathrm{Sp}_{R/K\langle x\rangle}yz)$. Let $\mathfrak{Q}_0$ be the pole of x in $L\langle x\rangle$; then $\mathfrak{Q}_0$ is the place of $L\langle x\rangle$ above $\mathfrak{q}_0$, and $\mathfrak{Q}_1, \cdots, \mathfrak{Q}_h$ are the places of S above $\mathfrak{Q}_0$. Let $\mathfrak{z}$ be the repartition of S which assigns yz to each of the places $\mathfrak{Q}_1, \cdots, \mathfrak{Q}_h$ and 0 to every other place. Then the right side of the formula to be proved is $(dx)_S(\mathfrak{z}) = (dx)_{L\langle x\rangle}(\mathrm{Sp}_{S/L\langle x\rangle}\mathfrak{z}) = ((dx)_{L\langle x\rangle})^{\mathfrak{Q}_0}(\mathrm{Sp}_{S/L\langle x\rangle}yz)$, in virtue of the Corollary 1 to Theorem 6, IV, §5. By Corollary 2 to Theorem 1, V, §4, we have $\mathrm{Sp}_{S/L\langle x\rangle}yz = \mathrm{Sp}_{R/K\langle x\rangle}yz$, and we are reduced to prove that $((dx)_{K\langle x\rangle})^{\mathfrak{q}_0}(u) = ((dx)_{L\langle x\rangle})^{\mathfrak{Q}_0}(u)$ for every $u \in K\langle x\rangle$. We may expand u in a power series in x^{-1} in the $\mathfrak{q}_0$-adic completion of $K\langle x\rangle: u = \sum_{i=-t}^{+\infty} c_i x^{-1}$. Then it follows from Lemma 4, §1 that both sides of the formula to be proved are equal to $-c_1$. This completes the proof of the first assertion of Theorem 11.

Denote by $\Sigma_s(\mathfrak{p})$ the field of elements of the residue field of $\mathfrak{p}$ which are separable over K, by $\mathfrak{P}_1 = \mathfrak{P}, \cdots, \mathfrak{P}_h$ the distinct places of S above $\mathfrak{p}$, and by $\Sigma_s(\mathfrak{P}_i)$ the field of elements of the residue field of $\mathfrak{P}_i$ which are separable over L $(1 \leqq i \leqq h)$. We know (Corollary 3 to Theorem 3, V, §5) that the algebra $(\Sigma_s(\mathfrak{p}))_L$ is the direct sum of h fields $Z_1', \cdots, Z_h'$ which are respectively isomorphic to the fields $\Sigma_s(\mathfrak{P}_1), \cdots, \Sigma_s(\mathfrak{P}_h)$; moreover there is an isomorphism φ_i' of Z_i' with $\Sigma_s(\mathfrak{P}_i)$ which induces on $\Sigma_s(\mathfrak{p})$ the natural isomorphism of this field into $\Sigma_s(\mathfrak{P}_i)$. Let γ be an arbitrary element of $\Sigma_s(\mathfrak{P}_1)$, and let ζ be the element of Z_1' which is mapped on γ by φ_1'. We may represent ζ in the form

$$\sum_{k=1}^{r} a_k \otimes \xi_k,$$

with $a_k \in L$, $\xi_k \in \Sigma_s(\mathfrak{p})$ $(1 \leqq k \leqq r$; the sign $\otimes$ represents the multiplication in $(\Sigma_s(\mathfrak{p}))_L)$. Denote by $\overline{\Sigma_s(\mathfrak{p})}$ the system of representatives of $\Sigma_s(\mathfrak{p})$ in the $\mathfrak{p}$-adic completion $\bar{R}$ of R which is a field containing K (Theorem 1, III, §3) and by $\overline{\Sigma_s(\mathfrak{P}_i)}$ the system of representatives of $\Sigma_s(\mathfrak{P}_i)$ in the $\mathfrak{P}_i$-adic completion $\bar{S}_i$ of S which is a field containing L. It is clear that, if we identify $\bar{R}$ with a subfield of $\bar{S}_i$, $\overline{\Sigma_s(\mathfrak{p})}$ is a subfield of $\overline{\Sigma_s(\mathfrak{P}_i)}$. Let $\bar{\xi}_k$ be the representative of ξ_k in $\overline{\Sigma_s(\mathfrak{p})}$, and denote by $\bar{\gamma}_i$ the element $\sum_{k=1}^h a_k\bar{\xi}_k$, the operations being performed in $\bar{S}_i$; then $\bar{\gamma}_i$ belongs to $\overline{\Sigma_s(\mathfrak{P}_i)}$. It is clear that $\bar{\gamma}_1$ is the representative of γ in $\overline{\Sigma_s(\mathfrak{P}_1)}$. On the other hand, since $\zeta \in Z_1'$, the element $\sum_{k=1}^h a_k\varphi_i'(\xi_k)$ of $\Sigma_s(\mathfrak{P}_i)$ is 0 if $i > 1$, from which it follows that $\bar{\gamma}_i = 0$ if $i > 1$. Denote by $\mathfrak{c}_k$ the repartition of R which assigns ξ_k to $\mathfrak{p}$ and 0 to every other place, and by $\mathfrak{c}$ the repartition of S which assigns $\bar{\gamma}_1$ to $\mathfrak{P}$ and 0 to every other place. Then it follows from what we have said that $\mathfrak{c} = \sum_{k=1}^h a_k \operatorname{Cosp}_{R/S} \mathfrak{c}_k$. Thus we have, in virtue of the first assertion of Theorem 11, $\Omega^{\mathfrak{P}}(\bar{\gamma}_1) = \Omega(\mathfrak{c}) = \sum_{k=1}^h a_k\omega(\mathfrak{c}_k) = \sum_{k=1}^h a_k \operatorname{Sp}_{\Sigma_s(\mathfrak{p})/K} \xi_k \operatorname{res}_{\mathfrak{p}} \omega$. The element $\operatorname{Sp}_{\Sigma_s(\mathfrak{p})/K} \xi_k \operatorname{res}_{\mathfrak{p}} \omega$ is the trace of the endomorphism of $\Sigma_s(\mathfrak{p})$ (considered as a vector space over K) produced by the multiplication by $\xi_k \operatorname{res}_{\mathfrak{p}} \omega$; this is obviously equal to the trace of the endomorphism of $(\Sigma_s(\mathfrak{p}))_L$ (considered as a vector space over L) produced by the multiplication by $\xi_k \operatorname{res}_{\mathfrak{p}} \omega$. It follows that $\Omega^{\mathfrak{P}}(\bar{\gamma}_1)$ is equal to the trace θ of the endomorphism of $(\Sigma_s(\mathfrak{p}))_L$ produced by the multiplication by $\zeta \operatorname{res}_{\mathfrak{p}} \omega$. The space $(\Sigma_s(\mathfrak{p}))_L$ is the direct sum of

$$Z_1', \cdots, Z_h',$$

and the multiplication by ζ maps Z_i' upon 0 if $i > 1$. It follows that $\theta = \operatorname{Sp}_{Z_1'/L} \zeta \cdot \operatorname{res}_{\mathfrak{p}} \omega = \operatorname{Sp}_{\Sigma_s(\mathfrak{P}_1)/L}\gamma \operatorname{res}_{\mathfrak{p}} \omega$. Since $\Omega^{\mathfrak{P}}(\bar{\gamma}_1) = \operatorname{Sp}_{\Sigma_s(\mathfrak{P}_1)/L}\gamma \operatorname{res}_{\mathfrak{p}} \omega$ for every $\gamma \in \Sigma_s(\mathfrak{P}_1)$, we have $\operatorname{res}_{\mathfrak{P}} \Omega = \operatorname{res}_{\mathfrak{p}} \omega$, which proves the second assertion of Theorem 11.

Next, we wish to prove that $\mathfrak{d}(\Omega)$ divides $\operatorname{Con}_{R/S} \mathfrak{d}(\omega)$. Write ω in the form $y(dx)_R$, where x is any separating variable in R, and $y \in R$; then

$$\mathfrak{d}(\omega) = \mathfrak{d}_R(y)\mathfrak{d}((dx)_R), \qquad \mathfrak{d}(\Omega) = \mathfrak{d}_S(y)\mathfrak{d}((dx)_S)$$

($\mathfrak{d}_R(y)$ and $\mathfrak{d}_S(y)$ represent the divisors of y in R and S respectively). We know that $\mathfrak{d}_S(y) = \operatorname{Con}_{R/S}\mathfrak{d}_R(y)$ (cf. IV, §7); this shows that we may assume without loss of generality that $\omega = (dx)_R$. Denote by $\mathfrak{D}$ the different of R with respect to $K\langle x\rangle$ and by $\mathfrak{E}$ the different of S with respect to $L\langle x\rangle$. Then we have, by Theorem 7, §3, $\mathfrak{d}(\omega) = \mathfrak{D}\mathfrak{n}_R^{-2}$, $\mathfrak{d}(\Omega) = \mathfrak{E}\mathfrak{n}_S^{-2}$, where $\mathfrak{n}_R$ and $\mathfrak{n}_S$ are the divisors of poles of x in R and S respectively. It follows immediately from the definitions that $\mathfrak{n}_S = \operatorname{Con}_{R/S} \mathfrak{n}_R$. Thus we have to prove that $\mathfrak{E}$ divides $\operatorname{Con}_{R/S} \mathfrak{D}$.

Let $\mathfrak{q}_0$ be any place of $K\langle x\rangle$; denote by E the set of places of R above $\mathfrak{q}_0$ and by F the set of places of S above $\mathfrak{q}_0$. If $\mathfrak{Q} \in F$, denote by $m(\mathfrak{Q})$ the ramification index of $\mathfrak{Q}$ with respect to $L\langle x\rangle$. Let y be an element of S such that

$$\nu_{\mathfrak{Q}}(y) = -m(\mathfrak{Q})$$

for all $\mathfrak{Q} \in F$ (cf. Corollary to Theorem 3, I, §6) and let z be any element of R such that $\nu_{\mathfrak{q}}(z) \geqq 0$ for all $\mathfrak{q} \in E$. Then we have also $\nu_{\mathfrak{Q}}(z) \geqq 0$ for all $\mathfrak{Q} \in F$; making use of Lemma 4, IV, §8, we see that $\operatorname{Sp}_{S/L\langle x\rangle} yz$ is integral at all places of

$L\langle x\rangle$ which lie above $\mathfrak{q}_0$. We also observe that, if y' is an element of R, we have $\mathrm{Sp}_{S/L\langle x\rangle}\, y'z = \mathrm{Sp}_{R/K\langle x\rangle}\, y'z$ in virtue of the Corollary 2 to Theorem 1, V, §4. This being said, consider first the case where L is purely transcendental over K. In this case, above any place $\mathfrak{q} \in E$ there lies only one place $\mathfrak{Q} \in F$, which is unramified with respect to R (Corollary 2 to Theorem 3, V, §5); we may then take for y an element of R. Making use again of Lemma 4, IV, §8, we see that

$$\nu_{\mathfrak{Q}}(y) \geqq -m(\mathfrak{q})$$

for every $\mathfrak{q} \in E$, where $m(\mathfrak{q})$ is the differential exponent of $\mathfrak{q}$ with respect to $K\langle x\rangle$; it follows that $m(\mathfrak{Q}) \leqq m(\mathfrak{q})$, and therefore that $\mathfrak{E}$ divides $\mathrm{Con}_{R/S}\mathfrak{D}$. This shows that, in order to prove our assertion in the general case, it will be sufficient to consider the case where L is algebraic over K. This being the case, y belongs to a field which can be obtained from R by adjunction of a finite number of elements of L and may therefore be represented in the form

$$y = \sum_{i=1}^{n} a_i y_i$$

with $y_i \in R$ $(1 \leqq i \leqq n)$, $a_1, \cdots, a_n$ being elements of L which we may assume to be linearly independent over K (cf. Lemma 2, IV, §1). We have $\mathrm{Sp}_{S/L\langle x\rangle} yz = \sum_{i=1}^{n} a_i\, \mathrm{Sp}_{R/K\langle x\rangle} y_i z$. This element is integral at every place $\mathfrak{L}$ of $L\langle x\rangle$ which lies above $\mathfrak{q}_0$. From this, we shall deduce that the elements $u_i = \mathrm{Sp}_{R/K\langle x\rangle} y_i z$ are all integral at $\mathfrak{q}_0$. Replacing if necessary x by $1/x$, we may assume that $\nu_{\mathfrak{q}_0}(x) \geqq 0$. Let G be an irreducible polynomial with coefficients in K such that $\nu_{\mathfrak{q}_0}(G(x)) > 0$. Write u_i in the form $(P_i(x))/(Q(x))$, where $P_1, \cdots, P_n, Q$ are polynomials with coefficients in K and have no common factor of degree >0. We wish to prove that Q is not divisible by G. Write $G = cG_1^{e_1} \cdots G_h^{e_h}$, where $c \in L$ and $G_1, \cdots, G_h$ are distinct irreducible polynomials with coefficients in L and leading coefficient 1. To each G_k there corresponds a place $\mathfrak{L}_k$ of $L\langle x\rangle$ above $\mathfrak{q}_0$ for which $\nu_{\mathfrak{L}_k}(G_k(x)) = 1$, whence $\nu_{\mathfrak{L}_k}(G(x)) = e_k$. Were Q divisible by G, then we would have $\nu_{\mathfrak{L}_k}(\sum_{i=1}^{n} a_i P_i) \geqq e_k$ $(1 \leqq k \leqq h)$, and $\sum_{i=1}^{n} a_i P_i$ would be divisible by each $G_k^{e_k}$, and therefore also by G. But, $a_1, \cdots, a_n$ being linearly independent over K, it is easily seen that this would imply that $P_1, \cdots, P_n$ are all divisible by G, which is impossible. Thus we see that whenever $z \in R$ is integral at all places $\mathfrak{q} \in E$, the elements $\mathrm{Sp}_{R/K\langle x\rangle}\, y_i z$ are integral at $\mathfrak{q}_0$. This means that

$$\nu_{\mathfrak{q}}(y_i) \geqq -m(\mathfrak{q}) \qquad (1 \leqq i \leqq n)$$

for all places $\mathfrak{q} \in E$. If $e(\mathfrak{Q})$ is the ramification index with respect to R of a place $\mathfrak{Q} \in F$, we have $\nu_{\mathfrak{Q}}(y_i) \geqq -e(\mathfrak{Q})m(\mathfrak{q})$, whence $-m(\mathfrak{Q}) = \nu_{\mathfrak{Q}}(y) \geqq -e(\mathfrak{Q})m(\mathfrak{q})$, which proves that $\mathfrak{E}$ divides $\mathrm{Con}_{R/S}\, \mathfrak{D}$.

If R and S have the same genus g, then $\mathfrak{d}(\Omega)$ and $\mathfrak{d}(\omega)$ have the same degree $2g - 2$. Making use of the Corollary to Theorem 2, V, §3, we see that $\mathrm{Con}_{R/S}\, \mathfrak{d}(\omega)$ is of degree $2g - 2$ equal to the degree of $\mathfrak{d}(\Omega)$. It follows immediately that we then have $\mathfrak{d}(\Omega) = \mathrm{Con}_{R/S}\, \mathfrak{d}(\omega)$. Theorem 11 is now completely proved.

COROLLARY 1. *The notation being as in Theorem* 11, *the genus of S is at most equal to the genus of R.*

Let g and G be the genera of R and S respectively, and let ω be a differential $\neq 0$ of R. Then we have seen at the end of the proof of Theorem 11 that $\mathrm{Con}_{R/S}\mathfrak{d}(\omega)$ is of degree $2g - 2$, while $\mathfrak{d}(\Omega)$ is of degree $2G - 2$; Corollary 1 therefore follows immediately from the third assertion of Theorem 11. Thus we see that Theorem 11 yields a new proof of the first assertion of Theorem 5, V, §5.

COROLLARY 2. *The notation being as in Theorem* 11, *every differential of the first kind of S is a linear combination with coefficients in S of cotraces (from R to S) of differentials of the first kind of R.*

Let Ω be a differential of the first kind of S. If x is a separating variable in R, we may write $\Omega = y(dx)_S$, with $y \in S$. Since $\mathfrak{d}((dx)_S)$ divides $\mathrm{Con}_{R/S}\mathfrak{d}((dx)_R)$, $(dx)_S$ has no variable zero with respect to R, from which it follows that y has no variable pole with respect to R. Let $(a_i)_{i\in I}$ be a base of L with respect to K (I being some set of indices); then we may write y in the form $\sum_{i\in I} a_i y_i$, with $y_i \in R$ (cf. the remark which follows the Corollary 2 to Theorem 5, V, §5). Set $\omega_i = y_i(dx)_R$; then we have $\Omega = \sum_{i\in I} a_i \,\mathrm{Cosp}_{R/S}\omega_i$. Let $\mathfrak{x}$ be any repartition of R which has no pole; then $\mathrm{Cosp}_{R/S}\mathfrak{x}$ has no pole in S, and we have $0 = \Omega(\mathrm{Cosp}_{R/S}\mathfrak{x}) = \sum_{i\in I} a_i\omega_i(\mathfrak{x})$. Since $\omega_i(\mathfrak{x}) \in K$ for all $i \in I$, we conclude that $\omega_i(\mathfrak{x}) = 0$ for all $i \in I$. This means that each ω_i is a differential of the first kind.

COROLLARY 3. *Let the notation be as in Theorem* 11, *and assume that the field L is perfect. If a place $\mathfrak{P}$ of S is not ramified with respect to R, then the residue field of the place $\mathfrak{p}$ of R which lies below $\mathfrak{P}$ is separable over K.*

Let x be a uniformizing variable at $\mathfrak{p}$ in R, and therefore also at $\mathfrak{P}$ in S. Since L is perfect, the residue field of $\mathfrak{P}$ is separable over L; making use of Theorem 5, §3 and of Theorem 11, we have $\mathrm{res}_{\mathfrak{p}} x^{-1}(dx)_R = \mathrm{res}_{\mathfrak{P}}\, x^{-1}(dx)_S = 1 \neq 0$, and our assertion follows from Theorem 5, §3.

COROLLARY 4. *Let $\mathfrak{p}$ be a place of a separably generated field R of algebraic functions of one variable. Then there exists an overfield of L the field of constants K of R which is algebraic of finite degree and purely unseparable over K and which has the property that the residue field of the place $\mathfrak{P}$ of $R\langle L\rangle$ which lies above $\mathfrak{p}$ is separable over L.*

Let $\bar{K}$ be an algebraically closed overfield of K, and let L_1 be the smallest perfect subfield of $\bar{K}$ containing K. Then L_1 is purely unseparable over K; denote by $\mathfrak{P}_1$ the place of $R\langle L_1\rangle$ above $\mathfrak{p}$ and by x a uniformizing variable at $\mathfrak{P}_1$ in $R\langle L_1\rangle$. Then x belongs to a field which can be obtained by adjunction to R of a finite number of elements $a_1, \cdots, a_m$ of L_1. It follows easily from Corollary 3 that the field $L = K\langle a_1, \cdots, a_m\rangle$ has the required properties.

REMARK. Using the notation of the proof of the third assertion of Theorem 11, we see that $\mathfrak{d}(\Omega)(\mathrm{Con}_{R/S}\mathfrak{d}(\omega))^{-1} = \mathfrak{d}((dx)_S)(\mathrm{Con}_{R/S}\mathfrak{d}(dx)_R)^{-1}$. The fields R and S being given, the left side depends only on ω, while the right side depends only on x. It follows that the divisor $\mathfrak{M} = \mathfrak{d}(\Omega)\,(\mathrm{Con}_{R/S}\mathfrak{d}(\omega))^{-1}$ depends only on R and S. The formula $\mathfrak{d}(\Omega) = \mathfrak{M}\,\mathrm{Con}_{R/S}\mathfrak{d}(\omega)$ shows that $\mathfrak{M}$ plays a role analogous to that of the different of a separable extension of finite degree (cf. Theorem 2, §2); however, while the different is an integral divisor, $\mathfrak{M}$ is the inverse of an integral divisor.

THEOREM 12. *Let R be a separably generated field of algebraic functions of one variable and let S be an overfield of R which is of finite degree over R. Denote by ω a differential of R and by z an element of S. Then we have* $\mathrm{Sp}_{S/R}(z\ \mathrm{Cosp}_{R/S}\omega) = (\mathrm{Sp}_{S/R}z)\omega$.

Let x be a separating variable in R. Write $\omega = y(dx)_R$, with $y \in R$, whence $\mathrm{Cosp}_{R/S}\omega = y(dx)_S$. Since $\mathrm{Sp}_{S/R}yz = y\ \mathrm{Sp}_{S/R}z$, we see that it will be sufficient to prove Theorem 12 in the case where $\omega = (dx)_R$.

Let $\mathfrak{x}$ be any repartition in R. Then $(\mathrm{Sp}_{S/R}(z(dx)_S))(\mathfrak{x}) = \mathrm{Sp}_{L/K}((dx)_S(z\mathrm{Cosp}_{R/S}\mathfrak{x}))$, where K and L are the fields of constants of R and S respectively. Since $(dx)_S = \mathrm{Cosp}_{L\langle x\rangle/S}(dx)_{L\langle x\rangle}$, we have

$$(1)\qquad (dx)_S(z\mathrm{Cosp}_{R/S}\mathfrak{x}) = (dx)_{L\langle x\rangle}(\mathrm{Sp}_{S/L\langle x\rangle}z\mathrm{Cosp}_{R/S}\mathfrak{x})$$

(we make use here of the first definition of the cotrace; cf. §2). We set $\mathfrak{y} = \mathrm{Sp}_{S/L\langle x\rangle}(z\ \mathrm{Cosp}_{R/S}\mathfrak{x})$. Then we have $(\mathrm{Sp}_{S/R}z(dx)_S)(\mathfrak{x}) = \mathrm{Sp}_{L/K}((dx)_{L\langle x\rangle}(\mathfrak{y}))$. We shall prove that the right side of this last formula is equal to $(dx)_{K\langle x\rangle}(\mathrm{Sp}_{L\langle x\rangle/K\langle x\rangle}\mathfrak{y})$. Denote by $\mathfrak{p}$ the pole of x in $K\langle x\rangle$, and by $\mathfrak{P}$ its pole in $L\langle x\rangle$. It follows immediately from Theorem 2, II, §5 that there exists an element $u \in L\langle x\rangle$ such that $\nu_{\mathfrak{Q}}(\mathfrak{y} - u) \geqq 0$ for all places $\mathfrak{Q} \neq \mathfrak{P}$ of $L\langle x\rangle$ (we denote here by the same symbol u an element of $L\langle x\rangle$ and the repartition of $L\langle x\rangle$ which is associated with this element). If $\mathfrak{y}' = \mathfrak{y} - u$, we have $(dx)_{L\langle x\rangle}(\mathfrak{y}) = (dx)_{L\langle x\rangle}(\mathfrak{y}')$. Set $\mathfrak{y}'(\mathfrak{P}) = \sum_{k=-m}^{1} c_k x^{-k} + x^{-2}w$, where w is an integral element of the $\mathfrak{P}$-adic completion of $L\langle x\rangle$. Then we have $(dx)_{L\langle x\rangle}(\mathfrak{y}') = -c_1$ (Lemma 4, §1), whence $\mathrm{Sp}_{L/K}((dx)_{L\langle x\rangle}(\mathfrak{y})) = -\mathrm{Sp}_{L/K}c_1$. On the other hand, we have $\mathrm{Sp}_{L\langle x\rangle/K\langle x\rangle}\mathfrak{y} = \mathrm{Sp}_{L\langle x\rangle/K\langle x\rangle}\mathfrak{y}' + \mathrm{Sp}_{L\langle x\rangle/K\langle x\rangle}u$, whence $(dx)_{K\langle x\rangle}(\mathrm{Sp}_{L\langle x\rangle/K\langle x\rangle}\mathfrak{y}) = (dx)_{K\langle x\rangle}(\mathrm{Sp}_{L\langle x\rangle/K\langle x\rangle}\mathfrak{y}')$. If $\mathfrak{q}$ is any place $\neq \mathfrak{p}$ of $K\langle x\rangle$, then $\nu_{\mathfrak{Q}}(\mathfrak{y}') \geqq 0$ for every place $\mathfrak{Q}$ of $L\langle x\rangle$ above $\mathfrak{q}$, whence $\nu_{\mathfrak{q}}(\mathrm{Sp}_{L\langle x\rangle/K\langle x\rangle}\mathfrak{y}') \geqq 0$ (cf. Corollary 1 to Theorem 5, IV, §5). Therefore, $(dx)_{K\langle x\rangle}(\mathrm{Sp}_{L\langle x\rangle/K\langle x\rangle}\mathfrak{y}') = ((dx)_{K\langle x\rangle})^{\mathfrak{p}}(v)$, where v is the $\mathfrak{p}$-component of $\mathrm{Sp}_{L\langle x\rangle/K\langle x\rangle}\mathfrak{y}'$. Denote by $\overline{K\langle x\rangle}$ the $\mathfrak{p}$-adic completion of $K\langle x\rangle$ and by $\overline{L\langle x\rangle}$ the $\mathfrak{p}$-adic completion of $L\langle x\rangle$. Since $\mathfrak{P}$ is the only place of $L\langle x\rangle$ above $\mathfrak{p}$, we have

$$v = \sum_{k=-m}^{1} (\mathrm{Sp}_{\overline{L\langle x\rangle}/\overline{K\langle x\rangle}}c_k)x^{-k} + x^{-2}\ \mathrm{Sp}_{\overline{L\langle x\rangle}/\overline{K\langle x\rangle}}w.$$

We know that $\mathrm{Sp}_{\overline{L\langle x\rangle}/\overline{K\langle x\rangle}}w$ is integral in $\overline{K\langle x\rangle}$. On the other hand, any base of L with respect to K is also a base of $L\langle x\rangle$ with respect to $K\langle x\rangle$ (Lemmas 2 and 3, IV, §1) and therefore also of $\overline{L\langle x\rangle}$ with respect to $\overline{K\langle x\rangle}$ (cf. Theorem 4, IV, §5); it follows immediately that $\mathrm{Sp}_{\overline{L\langle x\rangle}/\overline{K\langle x\rangle}}c_k = \mathrm{Sp}_{L/K}c_k$. Making use again of Lemma 4, §1, we have $((dx)_{K\langle x\rangle})^{\mathfrak{p}}(v) = -\mathrm{Sp}_{L/K}c_1$. The formula

$\mathrm{Sp}_{L/K}((dx)_{L\langle x\rangle}(\mathfrak{y})) = (dx)_{K\langle x\rangle}(\mathrm{Sp}_{L\langle x\rangle/K\langle x\rangle}\mathfrak{y})$ is thereby proved. Coming back to formula (1), and making use of Lemma 3, IV, §7, we conclude that

$$(dx)_S(z\ \mathrm{Cosp}_{R/S}\mathfrak{x}) = ((dx)_{K\langle x\rangle})\ (\mathrm{Sp}_{S/K\langle x\rangle}z\ \mathrm{Cosp}_{R/S}\mathfrak{x})$$
$$= (dx)_{K\langle x\rangle}\ (\mathrm{Sp}_{R/K\langle x\rangle}\ (\mathrm{Sp}_{S/R}z\mathrm{Cosp}_{R/S}\mathfrak{x})).$$

Since $(dx)_R = \mathrm{Cosp}_{R/K\langle x\rangle}(dx)_{K\langle x\rangle}$, this is equal to $(dx)_R(\mathrm{Sp}_{S/R}z\ (\mathrm{Cosp}_{R/S}\mathfrak{x})) = (dx)_R\ ((\mathrm{Sp}_{S/R}z)\mathfrak{x}) = ((\mathrm{Sp}_{S/R}z)(dx)_R(\mathfrak{x})$. The left side of (1) being equal to $(\mathrm{Sp}_{S/R}z(dx)_S)(\mathfrak{x})$, Theorem 12 is proved.

§7. Derivations of the field of constants

Let R be a field of algebraic functions of one variable which is separable over its field of constants K, and let D be a derivation of the field K. Then we can (in infinitely many ways) extend D to a derivation of R. For, if x is any separating variable in R, there exists a unique derivation (which we shall denote by D^x) of R which coincides with D on K and which maps x upon 0 (Lemma 7, §4). Let ω be any differential in R; then we may represent ω in the form $y\ dx$, with $y \in R$. We shall define $D^x\omega$ to be the differential

$$D^x\omega = (D^xy)\ dx.$$

We obtain in this way a certain mapping D^x (which is not linear) of the space of differentials of R into itself. If z is any element of R, we have

$$D^x(z\omega) = (D^xz)\ \omega + z(D^x\omega).$$

In fact, the left side is $(D^x(zy))\ dx = ((D^xz)y + z(D^xy))\ dx$, which is equal to the right side.

Lemma 1. *Let R be a separably generated field of algebraic functions of one variable and let D be a derivation of the field of the field of constants K of R. Denote by x a separating variable in R and by D^x the derivation of R which extends D and which maps x upon 0. Then D^x commutes with the derivation with respect to x in R (i.e., with the derivation D_x which maps every element of K upon 0 and x upon 1).*

The operation $\Delta = D^xD_x - D_xD^x$ is a derivation of R (§4). We see immediately that Δ maps upon 0 the elements of K and the element x, and therefore all elements of $K\langle x\rangle$ (Lemma 1, §4). Making use of the uniqueness assertion in Lemma 2, §4, we conclude that $\Delta = 0$.

Lemma 2. *The notation being as in Lemma 1, let z be any element of R. Then we have $D^x(dz) = d(D^xz)$.*

We have $dz = (D_xz)\ dx$ (Theorem 9, §5), whence, by Lemma 1, $D^x(dz) = (D^xD_xz)\ dx = (D_xD^xz)\ dx = d(D^xz)$.

Lemma 3. *The notation being as in Lemma 1, let x' be any other separating variable in R. Then, if ω is any differential of R, $D^x\omega - D^{x'}\omega$ is an exact differential.*

Write $\omega = y dx = y' dx'$, with y and y' in R. The operation $D^{x'} - D^x - (D^{x'}x)D_x$ is a derivation of R which maps x and every element of K upon 0 and which is therefore the zero derivation (cf. proof of Lemma 2); thus $D^{x'}y' = D^x y' + (D^{x'}x)D_x y'$ and $0 = D^{x'}x' = D^x x' + (D^{x'}x)D_x x'$. We have $D^{x'}\omega = (D^{x'}y')dx'$, and $dx' = (D_x x')dx$ (Theorem 9, §5), whence $D^{x'}\omega = (D^x y')(D_x x')\,dx - (D^x x')(D_x y')\,dx$. On the other hand, we have $y = y'(D_x x')$, whence $D^x\omega = (D^x y)\,dx = (D^x y')(D_x x')\,dx + y'(D^x D_x x')\,dx = (D^x y')(D_x x')\,dx + y'(D_x D^x x')\,dx$, by Lemma 1. It follows that $D^x\omega - D^{x'}\omega = y'(D_x D^x x')\,dx + (D^x x')(D_x y')dx = (D_x(y' \cdot D^x x'))\,dx$, which proves Lemma 3.

LEMMA 4. *The notation being as in Lemma 1, let also ω be a differential of R, $\mathfrak{p}$ a place of R, and u an element of the $\mathfrak{p}$-adic completion $\bar{R}$ of R. Then we have $D(\omega^{\mathfrak{p}}(u)) = (D^x\omega)^{\mathfrak{p}}(u) + \omega^{\mathfrak{p}}(D^x u)$.*

In this statement, $D^x u$ represents the effect on u of the continuous derivation of $\bar{R}$ which extends D^x (cf. Lemma 5, §4).

Set $\omega = y\,dx$, with $y \in R$. Then the formula to be proved becomes $D((dx)^{\mathfrak{p}}(yu)) = (dx)^{\mathfrak{p}}((D^x y)u + y(D^x u))$, or, if we set $yu = v$, $D((dx)^{\mathfrak{p}}(v)) = (dx)^{\mathfrak{p}}(D^x v)$. Let $\mathfrak{p}_0$ be the place of $K\langle x\rangle$ below $\mathfrak{p}$, and let w be the trace of v, taken from $\bar{R}$ to the $\mathfrak{p}_0$-adic completion of $K\langle x\rangle$. Making use of Lemma 3, §4, our formula becomes $D(((dx)_{K\langle x\rangle})^{\mathfrak{p}_0}(w)) = ((dx)_{K\langle x\rangle})^{\mathfrak{p}_0}(D^x w)$. Assume first that $\mathfrak{p}_0$ is the pole $\mathfrak{q}_0$ of x in $K\langle x\rangle$. Then we may expand w in the form $w = \sum_{k=-m}^{+\infty} c_k x^{-k}$, with $c_k \in K$. Making use of Formula (5), §4, we have $D^x w = \sum_{k=-m}^{+\infty} (Dc_k)x^{-k}$, and we see that both sides of the formula to be proved are equal to $-Dc_1$. Assume now that $\mathfrak{p}_0 \neq \mathfrak{q}_0$. Then we can find an element $z \in K\langle x\rangle$ which has no other poles than possibly $\mathfrak{p}_0$ and $\mathfrak{q}_0$ and which is such that $\nu_{\mathfrak{p}_0}(w - z) \geqq 0$. Then we have $((dx)_{K\langle x\rangle})^{\mathfrak{p}_0}(w) = ((dx)_{K\langle x\rangle})^{\mathfrak{p}_0}(z) = -((dx)_{K\langle x\rangle})^{\mathfrak{q}_0}(z)$. On the other hand, if $\mathfrak{r}_0$ is any place $\neq \mathfrak{q}_0$ of $K\langle x\rangle$, then D^x maps into itself the ring $\mathfrak{o}$ of $\mathfrak{r}_0$. In fact, the elements of $\mathfrak{o}$ are the fractions $P(x)/Q(x)$ where P and Q are polynomials with coefficients in K and $\nu_{\mathfrak{r}_0}(Q(x)) = 0$. Since $D^x(P(x)/Q(x))$ is the quotient of a polynomial in x by $Q^2(x)$, we see that $D^x(\mathfrak{o}) \subset \mathfrak{o}$. It follows that $D^x z$ has no other poles than possibly $\mathfrak{p}_0$ and $\mathfrak{q}_0$ and that $D^x w - D^x z$ is integral in the $\mathfrak{p}_0$-adic completion of $K\langle x\rangle$. We conclude that $((dx)_{K\langle x\rangle})^{\mathfrak{p}_0}(D^x w) = -((dx)_{K\langle x\rangle})^{\mathfrak{q}_0}(D^x z)$. Since we know already that $D(((dx)_{K\langle x\rangle})^{\mathfrak{q}_0}(z)) = ((dx)_{K\langle x\rangle})^{\mathfrak{q}_0}(D^x z)$, our formula is completely proved.

THEOREM 13. *Let R be a separably generated field of algebraic functions of one variable, and let D be a derivation of the field of constants K of R. Denote by x a separating variable in R and by D^x the derivation of R which extends D and which maps x upon 0. Let ω be a differential of R and $\mathfrak{p}$ a place of R. Let Σ_s be the field of elements of the residue field of $\mathfrak{p}$ which are separable over K, and denote again by D the derivation of Σ_s which extends the derivation D of K. Then we have $\mathrm{res}_{\mathfrak{p}} D^x\omega = D(\mathrm{res}_{\mathfrak{p}}\omega)$.*

We form the $\mathfrak{p}$-adic completion $\bar{R}$ of R, and we denote by $\bar{\Sigma}_s$ the system of representatives of Σ_s in $\bar{R}$ which is a field containing K (Theorem 1, III, §3). The derivation D^x may be extended to a continuous derivation of $\bar{R}$, which we denote again by D^x (Lemma 5, §4). The derivation D of K may be extended to a derivation of $\bar{\Sigma}_s$ and it follows from Lemma 2, §4 that this derivation is the restriction of D^x to $\bar{\Sigma}_s$. It is clear that, if $\bar{\xi}$ is the representative in $\bar{\Sigma}_s$ of an element $\xi \in \Sigma_s$, then $D^x\bar{\xi}$ is the representative of $D\xi$. This being said, let γ be any element of Σ_s, and let $\bar{\gamma}$ be its representative in $\bar{\Sigma}_s$. If we denote by ρ the residue of ω at $\mathfrak{p}$, we have $D(\omega^{\mathfrak{p}}(\bar{\gamma})) = D(\mathrm{Sp}_{\Sigma_s/K}\gamma\rho)$. The right side of this formula is equal, in virtue of Lemma 3, §4, to $\mathrm{Sp}_{\Sigma_s/K}D(\gamma\rho) = \mathrm{Sp}_{\Sigma_s/K}(D\gamma)\rho + \mathrm{Sp}_{\Sigma_s/K}\gamma(D\rho) = \omega^{\mathfrak{p}}(D^x\bar{\gamma}) + \mathrm{Sp}_{\Sigma_s/K}\gamma(D\rho)$. On the other hand, we have, by Lemma 4, $D(\omega^{\mathfrak{p}}(\bar{\gamma})) = (D^x\omega)^{\mathfrak{p}}(\bar{\gamma}) + \omega^{\mathfrak{p}}(D^x\bar{\gamma})$. We conclude that $(D^x\omega)^{\mathfrak{p}}(\bar{\gamma}) = \mathrm{Sp}_{\Sigma_s/K}\gamma(D\rho)$. This being true for every element γ of Σ_s, we have $\mathrm{res}_{\mathfrak{p}}D^x\omega = D\rho$.

§8. Differentials of the second kind

In this section, we shall limit ourselves to the consideration of fields of characteristic 0.

We have defined a differential of a field of algebraic functions of one variable to be of the second kind if its residues are all equal to 0; in particular, every differential of the first kind is of the second kind. It follows from Theorem 5, §3 that every exact differential is of the second kind. It is clear that the differentials of the second kind form a vector space over the field of constants of the field under consideration; one of the main results of this chapter will consist in the evaluation of the dimension of the factor space of this space by the subspace of exact differentials.

Let R and S be fields of algebraic functions of one variable of characteristic 0; we assume that R is a subfield of S and that the field of constants K of R is the intersection of R with the field of constants L of S. Then, if ω is a differential of the second kind of R, $\mathrm{Cosp}_{R/S}\omega$ is of the second kind in S. For, let T be the field $R\langle L\rangle$; then $\theta = \mathrm{Cosp}_{R/T}\omega$ is of the second kind in virtue of Theorem 11, §6 and $\mathrm{Cosp}_{R/S}\omega = \mathrm{Cosp}_{T/S}\theta$ is of the second kind in virtue of Theorem 3, §2. On the other hand, if S is of finite degree over R, and if Ω is a differential of the second kind in S, then $\mathrm{Sp}_{S/R}\Omega$ is of the second kind in R, as follows from Theorem 1, §2.

Theorem 14. *Let R be a field of algebraic functions of one variable of characteristic 0, and let ω be a differential of R. If $\mathfrak{p}$ is a place of R, there exists an $x \in R$ such that $\nu_{\mathfrak{p}}(\omega - dx) \geqq -1$. In order that there should exist an $x \in R$ such that $\nu_{\mathfrak{p}}(\omega - dx) \geqq 0$, it is necessary and sufficient that* $\mathrm{res}_{\mathfrak{p}}\omega = 0$. *If this condition is satisfied, then there exists for every integer n an element $x_n \in R$ such that*

$$\nu_{\mathfrak{p}}(\omega - dx_n) \geqq n.$$

Write $\omega = y\,dt$, where t is a uniformizing variable at $\mathfrak{p}$, and $y \in R$. Denote by Σ the residue field of $\mathfrak{p}$, by $\bar{R}$ the $\mathfrak{p}$-adic completion of R, and by $\bar{\Sigma}$ the system of representatives of Σ in $\bar{R}$ which is a field containing the field of constants of R (Theorem 1, III, §3). Then $\bar{R}$ may be identified with the field of formal power series in t with coefficients in $\bar{\Sigma}$ (III, §3), and we may expand y in the form

$\sum_{k=-m}^{\infty} c_k t^k$ (where m is some integer > 0). If n is in integer $\geqq -m$, we can find an $x_n \in R$ whose expansion is of the form $\sum_{k=1-m}^{\infty} c'_k t^k$ with $c'_k \in \bar{\Sigma}$ and $kc'_k = c_{k-1}$ for $1 - m \leqq k \leqq n + 1, k \neq 0$. If D_t is the derivation with respect to t in R, we have, by formula (5), §4, $D_t x_n = \sum_{k=1-m}^{\infty} kc'_k t^{k-1}$. Thus we have $\nu_\mathfrak{p}(y - D_t x_n - c_{-1}t^{-1}) \geqq n$, whence, since obviously $\nu_\mathfrak{p}(dt) \geqq 0$, $\nu_\mathfrak{p}(\omega - dx_n - c_{-1}t^{-1}dt) \geqq n$. In particular, we have $\nu_\mathfrak{p}(\omega - dx_{-1}) \geqq -1$. Taking now n to be $\geqq 0$, we see that $\text{res}_\mathfrak{p}(\omega - dx_n - c_{-1}t^{-1}dt) = 0$, whence, since $\text{res}_\mathfrak{p}\, dx_n = 0$, $\text{res}_\mathfrak{p}\omega = c_{-1} \cdot \text{res}_\mathfrak{p}\, t^{-1}dt = c_{-1}$, by Theorem 5, §3. Thus, if $\text{res}_\mathfrak{p}\omega = 0$, we have $\nu_\mathfrak{p}(\omega - dx_n) \geqq n$. Conversely, if there exists an $x \in R$ such that $\nu_\mathfrak{p}(\omega - dx) \geqq 0$, we have $\text{res}_\mathfrak{p}\omega = \text{res}_\mathfrak{p}\, dx = 0$. Theorem 14 is thereby proved.

LEMMA 1. *Let $\mathfrak{p}$ be a place of a field R of algebraic functions of one variable of characteristic* 0, *and let x be an element of R. If $\nu_\mathfrak{p}(x) \geqq 0$, we have $\nu_\mathfrak{p}(dx) \geqq 0$. If $\nu_\mathfrak{p}(x) \neq 0$, we have $\nu_\mathfrak{p}(dx) = \nu_\mathfrak{p}(x) - 1$.*

The first assertion, which is true independently of any assumption on the characteristic of R, follows immediately from Theorem 7, §3. If $\nu_\mathfrak{p}(x) \neq 0$, then the ramification index of $\mathfrak{p}$ with respect to $K\langle x\rangle$ is $|\nu_\mathfrak{p}(x)|$, and it follows from Theorem 7, IV, §8 that the differential exponent of $\mathfrak{p}$ with respect to $K\langle x\rangle$ is $|\nu_\mathfrak{p}(x)| - 1$. The exponent of $\mathfrak{p}$ in the divisor of poles of x is 0 if $\nu_\mathfrak{p}(x) > 0$ and $-\nu_\mathfrak{p}(x)$ if $\nu_\mathfrak{p}(x) < 0$; in any case, this exponent is $(1/2)(|\nu_\mathfrak{p}(x)| - \nu_\mathfrak{p}(x))$. The second assertion of Lemma 1 then follows immediately from Theorem 7, §3.

LEMMA 2. *Let $\mathfrak{p}_1, \cdots, \mathfrak{p}_m$ be distinct places of a field R of algebraic functions of one variable of characteristic* 0. *Let $\mathfrak{M}$ be a vector space (over the field of constants K of R) of differentials of R, and let $\mathfrak{N}$ be the subspace of $\mathfrak{M}$ composed of those differentials in $\mathfrak{M}$ whose residues at $\mathfrak{p}_1, \cdots, \mathfrak{p}_m$ are equal to* 0. *Assume that $\mathfrak{M}$ contains every differential which is a multiple of $(\mathfrak{p}_1 \cdots \mathfrak{p}_m)^{-1}$. Then $\mathfrak{M}/\mathfrak{N}$ is a vector space of finite dimension equal to* $\max\{0, (\sum_{i=1}^{m} d(\mathfrak{p}_i)) - 1\}$.

Denote by Σ_i the residue field of $\mathfrak{p}_i$ $(1 \leqq i \leqq m)$, and form the vector space $\prod_{i=1}^{m} \Sigma_i$ over K. To every $\omega \in \mathfrak{M}$ we assign the element $\Phi(\omega)$ of $\prod_{i=1}^{m} \Sigma_i$ whose Σ_i-coordinate is, for each i, $\text{res}_{\mathfrak{p}_i}\omega$. We obtain in this way a linear mapping ϕ of $\mathfrak{M}$ into $\prod_{i=1}^{m} \Sigma_i$ whose kernel is obviously $\mathfrak{N}$. It follows immediately from Theorem 4, III, §5 that $\Phi(\mathfrak{M})$ is the subspace of $\prod_{i=1}^{m} \Sigma_i$ composed of the elements $(\rho_1, \cdots, \rho_m)$ which satisfy the condition $\sum_{i=1}^{m} \text{Sp}_{\Sigma_i/K}\rho_i = 0$. The mapping $(\rho_1, \cdots, \rho_m) \to \sum_{i=1}^{m} \text{Sp}_{\Sigma_i/K}\rho_i$ is a linear function on $\prod_{i=1}^{m} \Sigma_i$ which is not identically 0 if $m \neq 0$. It follows immediately that $\dim(\mathfrak{M}/\mathfrak{N}) = \dim \Phi(\mathfrak{M}) = \max\{0, (\sum_{i=1}^{m} d(\mathfrak{p}_i)) - 1\}$.

THEOREM 15. *Let R be a field of algebraic functions of one variable of characteristic* 0 *and let P and Q be two disjoint finite sets of places of R. Denote by $\mathfrak{D}$ the space of differentials of R having no pole in the set P and whose residues at all places not in Q are* 0, *and let $\mathfrak{E}$ be the space of differentials of elements of R which have the places of P as zeros. Then the factor space $\mathfrak{D}/\mathfrak{E}$ is of finite dimension equal to $2g + p + q$, where g is the genus of R, $p = \max\{0, (\sum_{\mathfrak{p}\in P} d(\mathfrak{p})) - 1\}$, and $q = \max\{0, (\sum_{\mathfrak{q}\in Q} d(\mathfrak{q})) - 1\}$.*

Denote by $\mathfrak{D}'$ the space of differentials belonging to $\mathfrak{D}$ and whose residues at all places $\mathfrak{q} \in Q$ are 0 (i.e., $\mathfrak{D}'$ is the space of differentials of the second kind in $\mathfrak{D}$) and by $\mathfrak{E}'$ the space of exact differentials having no pole in the set P. Then it follows from Lemma 2 that $\mathfrak{D}/\mathfrak{D}'$ is of finite dimension equal to q.

In order to evaluate the dimension of $\mathfrak{D}'/\mathfrak{E}'$, we introduce the following notion. An integral divisor $\mathfrak{a}$ of a field R of algebraic functions of one variable is called *special* or *non special* according as to whether or not there exists a differential $\neq 0$ of R which is multiple of $\mathfrak{a}$. If g is the genus of R, any integral divisor of degree $>2g - 2$ is clearly non special.

LEMMA 3. *Let $\mathfrak{a}$ be any non special integral divisor of a field R of algebraic functions of one variable of characteristic* 0, *and let ω be a differential of the second kind in R. Then there exists in R an exact differential dx such that $\omega - dx \equiv 0 \pmod{\mathfrak{a}^{-2}}$.*

Let $\mathfrak{r}_1, \cdots, \mathfrak{r}_s$ be the distinct poles of ω. Since ω is of the second kind, it follows from Theorem 14 that we can find for each i $(1 \leqq i \leqq s)$ an element $x_i \in R$ such that $\nu_{\mathfrak{r}_i}(\omega - dx_i) \geqq 0$. Let $\mathfrak{x}$ be the repartition of R which assigns x_i to $\mathfrak{r}_i$ $(1 \leqq i \leqq s)$ and 0 to every other place. Making use of Theorem 2, II, §5, we see that there exists an $x \in R$ such that $x - \mathfrak{x} \equiv 0 \pmod{\mathfrak{a}^{-1}}$. Set $\mathfrak{a} = \prod_{\mathfrak{p}} \mathfrak{p}^{a(\mathfrak{p})}$. We have $\nu_{\mathfrak{r}_i}(x - x_i) \geqq -a(\mathfrak{r}_i)$, whence $\nu_{\mathfrak{r}_i}(dx - dx_i) \geqq -2a(\mathfrak{r}_i)$ (for, if $a(\mathfrak{r}_i) > 0$, then, by Lemma 1, $\nu_{\mathfrak{r}_i}(dx - dx_i) \geqq -a(\mathfrak{r}_i) - 1 \geqq -2a(\mathfrak{r}_i)$, while, if $a(\mathfrak{r}_i) = 0$, $\nu_{\mathfrak{r}_i}(dx - dx_i) \geqq 0$); if $\mathfrak{p}$ does not occur among $\mathfrak{r}_1, \cdots, \mathfrak{r}_s$, then $\nu_{\mathfrak{p}}(x) \geqq -a(\mathfrak{p})$, whence, as above, $\nu_{\mathfrak{p}}(dx) \geqq -2a(\mathfrak{p})$. It follows immediately that we have in any case $\nu_{\mathfrak{p}}(\omega - dx) \geqq -2a(\mathfrak{p})$, and Lemma 4 is proved.

This being said, let us return to the proof of Theorem 15. Since R has infinitely many places, we can find a finite number of distinct places $\mathfrak{s}_1, \cdots, \mathfrak{s}_t$ of R, none of which belongs to P, such that the divisor $\mathfrak{a} = \mathfrak{s}_1 \cdots \mathfrak{s}_t$ is non special. Let $\mathfrak{D}''$ be the space of differentials belonging to $\mathfrak{D}'$ which are $\equiv 0 \pmod{\mathfrak{a}^{-2}}$. If $\omega \in \mathfrak{D}'$, we can find an $x \in R$ such that $\omega - dx \equiv 0 \pmod{\mathfrak{a}^{-2}}$. A place $\mathfrak{p} \in \mathfrak{P}$, not being a pole of ω and not occurring in $\mathfrak{a}$, cannot be a pole of dx, which proves that $dx \in \mathfrak{E}'$ and that $\omega - dx \in \mathfrak{D}''$. This means that we have $\mathfrak{D}' = \mathfrak{E}' + \mathfrak{D}''$, and therefore that $\mathfrak{D}'/\mathfrak{E}'$ is isomorphic with $\mathfrak{D}''/\mathfrak{E}''$, where $\mathfrak{E}'' = \mathfrak{E}' \cap \mathfrak{D}''$. Now, let $\mathfrak{M}$ be the space of all differentials which are $\equiv 0 \pmod{\mathfrak{a}^{-2}}$. Then $\mathfrak{D}''$ is the space of differentials of the second kind belonging to $\mathfrak{M}$, and, for a differential belonging to $\mathfrak{M}$ to be in $\mathfrak{D}''$, it is sufficient that its residues at $\mathfrak{s}_1, \cdots, \mathfrak{s}_t$ be equal to 0. We may assume that $t > 0$; making use of Lemma 2, we see that $\mathfrak{M}/\mathfrak{D}''$ is of dimension $(\sum_{j=1}^{t} d(\mathfrak{s}_j)) - 1 = d(\mathfrak{a}) - 1$. On the other hand, the theorem of Riemann-Roch gives $0 = l(\mathfrak{a}^2) = -2d(\mathfrak{a}) - g + 1 + \dim \mathfrak{M}$, whence $\dim \mathfrak{M} = 2d(\mathfrak{a}) + g - 1$. Thus we see that $\mathfrak{D}''$ is a vector space of finite dimension equal to $2d(\mathfrak{a}) + g - 1 - (d(\mathfrak{a}) - 1) = d(\mathfrak{a}) + g$. We have now to evaluate the dimension of $\mathfrak{E}''$. It follows immediately from Lemma 1 that, in order for an element y of R to be such that $dy \equiv 0 \pmod{\mathfrak{a}^{-2}}$, it is necessary and sufficient that $y \equiv 0 \pmod{\mathfrak{a}^{-1}}$. Since $\mathfrak{a}$ is non special, it follows from the theorem of Riemann-Roch that the dimension of the space $\mathfrak{F}$ of elements of R which are $\equiv 0 \pmod{\mathfrak{a}^{-1}}$ is $d(\mathfrak{a}) - g + 1$. The mapping $y \to dy$ maps $\mathfrak{F}$ linearly onto $\mathfrak{E}''$, and the kernel of this mapping is the space of constants, which is of dimension 1. Thus we have $\dim \mathfrak{E}'' = \dim \mathfrak{F} - 1 = d(\mathfrak{a}) - g$. We see that we have proved that $\mathfrak{D}'/\mathfrak{E}'$ is a vector space of finite dimension equal to $(d(\mathfrak{a}) + g) - (d(\mathfrak{a}) - g) = 2g$.

Finally, we have to estimate the dimension of $\mathfrak{E}'/\mathfrak{E}$. Let $\mathfrak{G}'$ be the space of all elements of R which have no pole in the set P, and let $\mathfrak{G}$ be the space of elements which admit the places of P as zeros. The mapping $z \to dz$ maps $\mathfrak{G}'$ linearly onto $\mathfrak{E}'$, and $\mathfrak{G}$ onto $\mathfrak{E}$. Let $\mathfrak{C}$ be the space of constants; then $\mathfrak{C}$ is the kernel of our mapping, from which it follows easily that $\mathfrak{E}'/\mathfrak{E}$ is isomorphic with $\mathfrak{G}'/(\mathfrak{G} + \mathfrak{C})$. Denote by $\Sigma_1, \cdots, \Sigma_n$ the residue fields of the places $\mathfrak{p}_1, \cdots, \mathfrak{p}_n$ belonging to P, and form the vector space $\prod_{i=1}^n \Sigma_i$ over K. To every $z \in \mathfrak{G}'$ associate the element $\Theta(z) \in \prod_{i=1}^n \Sigma_i$ whose Σ_i-coordinate is, for each i, the value taken by z at $\mathfrak{p}_i$. Then it follows from Theorem 3, I, §6 that Θ maps $\mathfrak{G}'$ onto $\prod_{i=1}^n \Sigma_i$. The kernel of Θ being $\mathfrak{G}$, we see that $\mathfrak{G}'/\mathfrak{G}$ is a vector space of finite dimension equal to $\dim \prod_{i=1}^n \Sigma_i = \sum_{\mathfrak{p} \in P} d(\mathfrak{p})$. If P is not empty, then $\mathfrak{C}$ has only 0 in common with $\mathfrak{G}$, whence, since $\dim \mathfrak{C} = 1$, $\dim \mathfrak{E}'/\mathfrak{E} = (\sum_{\mathfrak{p} \in P} d(\mathfrak{p})) - 1 = p$; if P is empty, then $\mathfrak{E}' = \mathfrak{E}$, whence $\dim \mathfrak{E}'/\mathfrak{E} = 0 = p$. Theorem 15 is now completely proved.

COROLLARY 1. *Let R be a field of algebraic functions of one variable of characteristic 0 and of genus g. Then the factor space of the space of differentials of the second kind of R by the space of exact differentials is of finite dimension equal to $2g$.*

This follows immediately from Theorem 15 by taking P and Q to be empty.

COROLLARY 2. *Let R be a field of algebraic functions of one variable of characteristic 0, and let L be an overfield of the field of constants K of R. Then every differential of the second kind of $R\langle L\rangle$ can be written as the sum of an exact differential and a linear combination with coefficients in L of cotraces of differentials of the second kind of R.*

Let g be the genus of R. Then it follows from Corollary 1 that we can find $2g$ differentials $\omega_1, \cdots, \omega_{2g}$ of the second kind of R which are linearly independent (over K) modulo the space of exact differentials of R. Set $\Omega_i = \mathrm{Cosp}_{R/R\langle L\rangle}\omega_i$ $(1 \leqq i \leqq 2g)$. Then we know that the differentials Ω_i are of the second kind in $R\langle L\rangle$. We shall see that they are linearly independent (over L) modulo the space of exact differentials of $R\langle L\rangle$. We let $a_1, \cdots, a_{2g}$ be elements of L such that $\sum_{i=1}^{2g} a_i\Omega_i = dx$, with $x \in R\langle L\rangle$. The differential on the left side of this formula has no variable pole with respect to R, as follows from Theorem 11, §6, if we observe that $R\langle L\rangle$ has the same genus as R (Theorem 5, V, §6); making use of Lemma 1, we conclude that x has no variable pole with respect to R and can therefore be written in the form of a linear combination with coefficients in L of elements of R (Corollary 2 to Theorem 4, V, §6). Let $\{b_j\}_{j \in J}$ be a base of L with respect to K; then $x = \sum_{j \in J} b_j x_j$, $x_j \in R$. Write $a_i = \sum_{j \in J} c_{ij} b_j$, and set $\eta_j = \sum_{i=1}^{2g} c_{ij}\omega_i$. Then we have $\sum_{j \in J} b_j((dx_j)_{R\langle L\rangle} - \mathrm{Cosp}_{R/R\langle L\rangle}\eta_j) = 0$. Let t be a non constant element of R, and set $(dx_j)_R - \eta_j = u_j(dt)_R$, with $u_j \in R$. Then $\sum_{j \in J} u_j b_j = 0$, whence $u_j = 0$ for every j, by Theorem 1, V, §4. Since

$\sum_{i=1}^{2g} c_{ij}\omega_i = (dx_j)_R$, we have $c_{ij} = 0$ for every i and j, whence $a_i = 0$ ($1 \leqq i \leqq 2g$), which proves our assertion. The field $R\langle L\rangle$ being of the same genus g as R, it follows from Corollary 1 that every differential of the second kind in $R\langle L\rangle$ is congruent modulo the space of exact differentials to a linear combination of $\Omega_1, \cdots, \Omega_{2g}$, which proves Corollary 2.

REMARK. It follows from Corollary 2 and from the corollary 2 to Theorem 4, §6 that every differential of the second kind of $R\langle L\rangle$ which has no variable pole with respect to R is a linear combination with coefficients in L of cotraces of differentials of the second kind of R.

We shall now consider fields of algebraic functions of one variable over algebraically closed fields of characteristic 0.

LEMMA 4. *Let R be a field of algebraic functions of one variable over an algebraically closed field K, and let E be any infinite set of places of R. Then, if g is the genus of R, we can find g places $\mathfrak{p}_1, \cdots, \mathfrak{p}_g$ in E such that the divisor $\mathfrak{p}_1 \cdots \mathfrak{p}_g$ is non special.*

We may assume that $g > 0$. Let ω_1 be any differential of the first kind $\neq 0$ in R; then we can find a place $\mathfrak{p}_1 \in E$ which is not a zero of ω_1, and it is clear that the space of differentials which are $\equiv 0 \pmod{\mathfrak{p}_1}$ is of dimension $\leqq g - 1$. Let h be any integer such that $1 \leqq h \leqq g$. Assume that we have already found $h - 1$ places $\mathfrak{p}_1, \cdots, \mathfrak{p}_{h-1}$ in E such that the space of differentials which are $\equiv 0 \pmod{\mathfrak{p}_1 \cdots \mathfrak{p}_{h-1}}$ is of dimension $\leqq g - h + 1$. If this space is $\neq \{0\}$, take a differential $\omega \neq 0$ in it, and let $\mathfrak{p}_h$ be a place in E which is not a zero of ω; if this space is $\{0\}$, let $\mathfrak{p}_h$ be any place of E distinct from $\mathfrak{p}_1, \cdots, \mathfrak{p}_{h-1}$. In the first case, the space of differentials of R which are multiples of $\mathfrak{p}_1 \cdots \mathfrak{p}_h$ has a dimension less than the space of differentials which are $\equiv 0 \pmod{\mathfrak{p}_1 \cdots \mathfrak{p}_{h-1}}$, and this dimension is therefore $\leqq g - h$; this conclusion is also true in the second case. At the end of this procedure, we obtain a set of g places of E with the required property.

The notation being as in Lemma 4, set $\mathfrak{a} = \mathfrak{p}_1 \cdots \mathfrak{p}_g$. Then it follows from Lemma 3 that every differential of the second kind is congruent modulo the space of exact differentials to a differential which is $\equiv 0 \pmod{\mathfrak{a}^{-2}}$. For each i ($1 \leqq i \leqq g$), let x_i be a uniformizing variable at $\mathfrak{p}_i$. If ω is a differential of the second kind which is $\equiv 0 \pmod{\mathfrak{a}^{-2}}$, write $\omega = u_i\, dx_i$, with $u_i \in R$. Making use of the Corollary to Theorem 6, §3, we see that the coefficient of the term in x_i^{-1} in the expansion of u_i in a power series in x_i in the $\mathfrak{p}_i$-adic completion of R is 0. We may therefore write

$$\omega = (a_i(\omega)x_i^{-2} + b_i(\omega) + x_i v_i)dx_i$$

with v_i integral at $\mathfrak{p}_i$. Let $\mathfrak{D}$ be the space of differentials of the second kind which are $\equiv 0 \pmod{\mathfrak{a}^{-2}}$. Then $a_1, \cdots, a_g, b_1, \cdots, b_g$ are $2g$ linear functions on $\mathfrak{D}$. If $a_i(\omega) = b_i(\omega) = 0$ ($1 \leqq i \leqq g$) for some $\omega \in \mathfrak{D}$, then ω admits $\mathfrak{p}_1, \cdots, \mathfrak{p}_g$ as zeros and is therefore $\equiv 0 \pmod{\mathfrak{a}}$, whence $\omega = 0$. On the other hand, $\mathfrak{D}$ cannot contain any exact differential $\neq 0$; for, if $x \in R$ is such that $dx \in \mathfrak{D}$, then it follows from Lemma 1 that $x \equiv 0 \pmod{\mathfrak{a}^{-1}}$, and the theorem of Riemann-Roch gives $l(\mathfrak{a}^{-1}) = d(\mathfrak{a}) - g + 1 = 1$, which proves that the only multiples of $\mathfrak{a}^{-1}$ are the constants. This being the case, it follows immediately from the Corollary 1 to Theorem 15 that $\mathfrak{D}$ is of dimension $2g$. The functions $a_i(\omega)$, $b_i(\omega)$ are therefore linearly independent. Thus we have proved

THEOREM 16. *Let R be a field of algebraic functions of one variable over an algebraically closed field $K$of characteristic* 0, *and let g be the genus of R. Let $\mathfrak{p}_1, \cdots, \mathfrak{p}_g$ be g distinct places of R such that the divisor $\mathfrak{a} = \mathfrak{p}_1 \cdots \mathfrak{p}_g$ is non special; for each i let x_i be a uniformizing variable at $\mathfrak{p}_i$. Being given $2g$ elements a_i, b_i of K $(1 \leqq i \leqq g)$, there exists a uniquely determined differential ω of the second kind of R which is such that $\nu_{\mathfrak{p}_i}(\omega - (a_i x_i^{-2} + b_i)dx_i) > 0$ $(1 \leqq i \leqq g)$.*

CHAPTER VII

THE RIEMANN SURFACE

Throughout this chapter, C will denote the field of complex numbers and R a field of algebraic functions of one variable over C. We shall denote by Σ the Riemann sphere, obtained by adjunction of a point ∞ to C; Σ is therefore a compact topological space, homeomorphic to a 2-dimensional sphere in euclidean 3-space.

§1. Definition of the Riemann surface

Let S be the set of all places of the field R. If $x \in R$ and if $\mathfrak{p}$ is a place of R, then either $\mathfrak{p}$ is a pole of x, in which case we say that x takes the value ∞ at $\mathfrak{p}$, or this is not the case, and then, since C is algebraically closed, the value $x(\mathfrak{p})$ taken by x at $\mathfrak{p}$ is some complex number. Thus we can associate to every $x \in R$ a mapping $\mathfrak{p} \rightarrow x(\mathfrak{p})$ of S into Σ. If x and x' are distinct elements of R, there exists a place $\mathfrak{p}$ which is neither a zero nor a pole of $x' - x$, and which is not a pole of x, which proves that the mappings $\mathfrak{p} \rightarrow x(\mathfrak{p})$ and $\mathfrak{p} \rightarrow x'(\mathfrak{p})$ of S into Σ are distinct. We shall henceforth use the same symbol x to denote an element of R and the mapping of S into Σ which is associated with it.

We shall introduce a topology in S with respect to which the mappings x $(x \in R)$ shall be continuous. It is well known that, being given a set of mappings of a set X into a topological space E, there is a weakest topology on X which makes these mappings continuous. We shall henceforth consider S as a topological space whose topology is the weakest with respect to which all mappings x $(x \in R)$ are continuous. The open sets of S are those which can be represented as unions of sets of the form

$$(1) \qquad x_1^{-1}(U_1) \cap x_2^{-1}(U_2) \cap \cdots \cap x_k^{-1}(U_k)$$

where $x_1, \cdots, x_k$ are in R, while $U_1, \cdots, U_k$ are open subsets of Σ ($x^{-1}(U)$ denotes the set of places $\mathfrak{p}$ such that $x(\mathfrak{p}) \in U$).

This topology can be defined in an alternative manner. Let us form, for each $x \in R$, a copy Σ_x of the space Σ, and let P be the Cartesian product of the spaces Σ_x for all $x \in R$. If we assign to every $\mathfrak{p} \in S$ the point $\Phi(\mathfrak{p})$ of P whose Σ_x-coordinate is, for every x, $x(\mathfrak{p})$, we obtain a mapping Φ of S onto a subset S' of P. This mapping is one-to-one. For, if $\mathfrak{p}$ and $\mathfrak{q}$ are distinct places of R, there exists an $x \in R$ which takes distinct values at $\mathfrak{p}$ and $\mathfrak{q}$ (by Theorem 3, I, §6), whence $\Phi(\mathfrak{p}) \neq \Phi(\mathfrak{q})$. The set (1) is the inverse image under Φ of the set $V(x_1, U_1; \cdots; x_k, U_k)$ composed of those points of P whose Σ_{x_i}-coordinate is in U_i $(1 \leqq i \leqq k)$. The set $V(x_1, U_1; \cdots; x_k, U_k)$ is open in P, and every open set of P is a union of sets of this form. It follows immediately that Φ is a homeomorphism of S with the subspace of P carried by S'. In particular, since the Hausdorff separation axiom holds in P, and therefore in S', this axiom holds in S, that is, S is a Hausdorff space.

Now, we shall prove that S *is compact*. Since P is a product of compact spaces, it is compact; it will therefore be sufficent to prove that S' is closed in P. Let $\mathfrak{u} = (u_x)_{x \in R}$ be a point of P which is adherent to S'. We shall denote by $\mathfrak{o}$ the set of elements $x \in R$ such that $u_x \neq \infty$. We shall prove that $\mathfrak{o}$ is a V-ring in R and that $\mathfrak{u} = \phi(\mathfrak{p})$, where $\mathfrak{p}$ is the place which corresponds to $\mathfrak{o}$.

If c is a constant, then $c(\mathfrak{p}) = c$ for every place $\mathfrak{p}$. This means that all points of S' have the same Σ_c-coordinate, viz. c. Since $\mathfrak{u}$ is adherent to S', it follows immediately that $u_c = c$; in particular, we see that $C \subset \mathfrak{o}$. Let a be a point of Σ and ε a real number >0. If $a \neq \infty$, denote by $\delta(a, \varepsilon)$ the set of points $a' \neq \infty$ of Σ such that $| a' - a | < \varepsilon$; by $\delta(\infty, \varepsilon)$ we shall represent the set composed of ∞ and of the points $a' \neq \infty$ of Σ such that $| a' | > \varepsilon^{-1}$. In any case, $\delta(a, \varepsilon)$ is an open subset of Σ. Let x and y be elements of R. Denote by V_ε the set of points $\mathfrak{v} = (v_z)_{z \in R}$ of P such that the inclusion $v_z \in \delta(u_z, \varepsilon)$ holds whenever z is one of the elements x, y, $x - y$ or xy. Then V_ε is an open subset of P and contains $\mathfrak{u}$; it follows that V_ε contains a point $\mathfrak{v}_\varepsilon = (v_{z,\varepsilon})_{z \in R}$ of S'. Set $\mathfrak{v}_\varepsilon = \Phi(\mathfrak{p}_\varepsilon)$, $\mathfrak{p}_\varepsilon \in S$. If x and y are in $\mathfrak{o}$, then u_x and u_y are $\neq \infty$, and therefore also $v_{x,\varepsilon}$ and $v_{y,\varepsilon}$. Since $\mathfrak{v}_\varepsilon = \phi(\mathfrak{p}_\varepsilon)$, we have $v_{x,\varepsilon} = x(\mathfrak{p}_\varepsilon)$, $v_{y,\varepsilon} = y(\mathfrak{p}_\varepsilon)$, $v_{x-y,\varepsilon} = (x - y)(\mathfrak{p}_\varepsilon) = v_{x,\varepsilon} - v_{y,\varepsilon}$ and, similarly, $v_{xy,\varepsilon} = v_{x,\varepsilon}v_{y,\varepsilon}$. On the other hand, it is clear that $u_z = \lim_{\varepsilon \to 0} v_{z,\varepsilon}$, whenever z is one of the elements x, y, $x - y$ or xy. We conclude that u_{x-y} and u_{xy} are respectively equal to $u_x - u_y$ and to u_{xy}. Thus we see that $\mathfrak{o}$ is a ring and that the mapping $x \to u_x$ $(x \in \mathfrak{o})$ is a homomorphism of $\mathfrak{o}$ into C which maps the elements of C upon themselves. Now, assume that $x \notin \mathfrak{o}$ and that $y = x^{-1}$. If $v_{x,\varepsilon} = \infty$, then $\mathfrak{p}_\varepsilon$ is a pole of x and therefore a zero of y, whence $v_{y,\varepsilon} = 0$; if $v_{x,\varepsilon} \neq \infty$, then $v_{y,\varepsilon} = v_{x,\varepsilon}^{-1}$ and $| v_{x,\varepsilon} | > \varepsilon^{-1}$ whence $| v_{y,\varepsilon} | < \varepsilon$. In any case, we have $| v_{y,\varepsilon} | < \varepsilon$. This being true for every $\varepsilon > 0$, $u_y = 0$, whence $x^{-1} \in \mathfrak{o}$. Finally, we have $\mathfrak{o} \neq R$. For, if x is an element of $\mathfrak{o}$ not in C and $x' = x - u_x$, then $u_{x'} = 0$, but $x' \neq 0$, and x'^{-1} is an element of R which cannot be in $\mathfrak{o}$ since otherwise we would have $u_{x'^{-1}}u_{x'} = 1$. Thus we have proved that $\mathfrak{o}$ is a V-ring. Let $\mathfrak{p}$ be the corresponding place. The elements of $\mathfrak{p}$ are clearly the elements x of $\mathfrak{o}$ such that $u_x = 0$; if x is any element of $\mathfrak{o}$, we have $x - u_x \in \mathfrak{p}$, whence $u_x = x(\mathfrak{p})$. It follows that $\mathfrak{u} = \Phi(\mathfrak{p})$, and our assertion that S' is closed is now proved.

We shall prove that, if $\mathfrak{p}$ is any place of S, then any uniformizing variable x at $\mathfrak{p}$ maps some neighbourhood of $\mathfrak{p}$ in S topologically onto a neighbourhood of the point 0 in Σ. Denote by $\mathfrak{p}_1 = \mathfrak{p}, \cdots, \mathfrak{p}_h$ the distinct zeros of x, and set $e_i = \nu_{\mathfrak{p}_i}(x)$ $(1 \leqq i \leqq h)$. Then we have $e_1 = 1$ and $\sum_{i=1}^{h} e_i = [R:C\langle x\rangle]$ (Theorem 4, I, §8); we shall denote this last number by n. Let $a_1, \cdots, a_h$ be h distinct complex numbers; we know (by the corollary to Theorem 3, I, §6) that there exists an element $y \in R$ such that $\nu_{\mathfrak{p}_i}(y - a_i) = 1$ $(1 \leqq i \leqq h)$. Let $F(X, Y)$ be an irreducible polynomial with complex coefficients such that $F(x, y) = 0$. Since $F(x, Y)$ is irreducible in the ring $C\langle x\rangle[Y]$, F is of degree $\leqq n$ in Y. Since $y(\mathfrak{p}_i) = a_i$ $(1 \leqq i \leqq h)$, we have $F(0, a_i) = 0$. We assert that a_i is a root of multiplicity e_i of the equation $F(0, Y) = 0$. For, write $F(0, Y) = (Y - a_i)^{m_i}G_i(Y)$, where G_i

is a polynomial which is not divisible by $Y - a_i$, whence $\nu_{\mathfrak{p}_i}(F(0, y)) = m_i$. We can write $F(X, Y) - F(0, Y) = XH(X, Y)$, where H is a polynomial, whence $F(0, y) = -xH(x, y)$ and therefore $m_i \geqq \nu_{\mathfrak{p}_i}(x) = e_i$. On the other hand, $\sum_{i=1}^h m_i$ is at most equal to the degree of $F(0, Y)$, which is itself $\leqq n = \sum_{i=1}^h e_i$. It follows that $m_i = e_i$ $(1 \leqq i \leqq h)$; moreover we see that F is of degree n in Y, whence $R = C\langle x, y\rangle$. This being said, we shall make use of the continuity of the roots of an algebraic equation. Let s be a positive real number smaller than all the quantities $(1/2) \mid a_i - a_j \mid$, for $1 \leqq i \leqq j \leqq h$. Then we can find a number $r > 0$ with the following properties: if ξ is any complex number $\neq 0$ such that $\mid \xi \mid \leqq r$, then, for each i $(1 \leqq i \leqq h)$ the equation $F(\xi, Y) = 0$ has exactly e_i distinct roots η_{ik} $(1 \leqq k \leqq e_i)$ such that $\mid \eta_{ik} - a_i \mid \leqq s$. Since $\mid a_i - a_j \mid > 2s$ for $i \neq j$, we see that the numbers η_{ik} are all distinct; there are n of these numbers. It follows from Corollary 2 to Theorem 1, I, §4 that there exists for each η_{ik} a place $\mathfrak{q}_{ik}$ of R which is a common zero of $x - \xi$ and $y - \eta_{ik}$. The n places $\mathfrak{q}_{ik}$ are mutually distinct; since $[R:C\langle x - \xi\rangle] = [R:C\langle x\rangle] = n$, it follows from Theorem 4, I, §8 that the places $\mathfrak{q}_{ik}$ are all the zeros of $x - \xi$ in R, and that these zeros are of order 1. Let N be the set of all places $\mathfrak{q}$ such that $|x(\mathfrak{q}) \mid \leqq r$, $\mid y(\mathfrak{q}) - a_1 \mid \leqq s$. Then it is clear that N is a closed neighbourhood of $\mathfrak{p}$ in the space S. Since $e_1 = 1$, it follows from what we have proved above that, for every $\xi \neq 0$ such that $\mid \xi \mid \leqq r$, N contains exactly one zero of $x - \xi$ (it is the place which was denoted above by $\mathfrak{q}_{11}$). This shows that x maps N in a one-to-one way onto the closed circular disc of Σ composed of the points $\xi \in \Sigma$ such that $\mid \xi \mid \leqq r$. This mapping is continuous; since N is closed in S, it is compact, which proves that our mapping is topological.

Thus we see that every point of S has a neighbourhood which is homeomorphic to a neighbourhood of a point in the plane. It is therefore appropriate to call S a surface. We shall say that S is the *Riemann surface* of the field R. We have proved the following results:

THEOREM 1. *Let S be the Riemann surface of the field R. Then every element x of R induces a continuous mapping $\mathfrak{p} \to x(\mathfrak{p})$ of S into the Riemann sphere, and the topology of S is the weakest with respect to which all these mappings are continuous. If x is a uniformizing variable at a point $\mathfrak{p} \in S$, then there exists a neighbourhood of $\mathfrak{p}$ in S which is mapped topologically by x onto a neighbourhood of* 0 *in the Riemann sphere.*

Let Δ be a subset of S which has the following property: there exist an open subset U of S containing Δ and an element $x \in R$ such that x maps U topologically onto a neighbourhood of 0 in Σ and Δ onto a closed circular disc of center 0 in Σ (of radius $r > 0$). Then we say that Δ is a *closed disc* in S, or, if more precision is needed, a *closed x-disc*. An open subset V of S is called an *open disc* (or an *open x-disc*) if its adherence in S is a closed disc. The point of an x-disc (open or closed) which is mapped by x upon 0 is called the *x-center* of the disc. The radius of the image of the disc under x is called the *x-radius* of the disc. The boundary of an x-disc (open or closed) is mapped by x upon a circumference in Σ; this boundary is called the *circumference* of the disc. It follows easily from Theorem 1 that, if x is a uniformizing variable at a point $\mathfrak{p}$ of S, then the closed x-discs of x-center $\mathfrak{p}$ form a fundamental system of neighbourhoods of $\mathfrak{p}$; the same is of course true of the open x-discs of x-center $\mathfrak{p}$. On the other hand, any closed x-disc of x-center $\mathfrak{p}$ is contained in an open x-disc of x-center $\mathfrak{p}$, and every open x-disc of x-center $\mathfrak{p}$ in a closed x-disc of x-center $\mathfrak{p}$.

EXAMPLES OF RIEMANN SURFACES.

1) Assume that $R = C\langle x\rangle$, with x transcendental over C. For each $a \in C$, $x - a$ has a single zero $\mathfrak{p}_a$; on the other hand, x has a unique pole $\mathfrak{p}_\infty$. We know that there are no other places than $\mathfrak{p}_\infty$ and the places $\mathfrak{p}_a$ ($a \in C$). We conclude that, in this case, x maps S in a one-to-one way upon Σ. This mapping being continuous and S being compact, we see that S is homeomorphic with Σ.

2) Assume that R is the field of elliptic functions of one variable with the periods ζ and ζ' (cf. II, §8). Then we have seen that to every complex number a there corresponds a place $\mathfrak{p}_a$ of R, which is the set of elliptic functions in the field which have a as a zero; moreover, all places of R are of the form $\mathfrak{p}_a$, and a necessary and sufficent condition for $\mathfrak{p}_a$ and $\mathfrak{p}_{a'}$ to be identical is for a and a' to belong to the same coset of the additive group G of complex numbers modulo the subgroup H generated by ζ and ζ'. Thus we see that S is in this case in a one-to-one correspondence with the group G/H. If $\mathfrak{p}(\mathfrak{z})$ is the place which corresponds to a $\mathfrak{z} \in G/H$, then, for every $x \in R$, $x(\mathfrak{p}(\mathfrak{z})) = x(z)$, where z is a complex number belonging to the coset $\mathfrak{z}$ modulo H. Now, the function $x(z)$ is meromorphic and therefore continuous (as a mapping of C into Σ). It follows easily that the mapping $\mathfrak{z} \to x(\mathfrak{p}(\mathfrak{z}))$ of G/H into Σ is continuous. This being true for every $x \in R$, and the topology of S being the weakest to make the mappings $\mathfrak{p} \to x(\mathfrak{p})$ continuous, we conclude that the mapping $\mathfrak{z} \to \mathfrak{p}(\mathfrak{z})$ of G/H into S is continuous. Since G/H is compact, our mapping is a homeomorphism. Thus we see that S is in this case homeomorphic with a 2-dimensional torus.

In the first example given above, the field R was of genus 0 and its Riemann surface was homeomorphic with a sphere, which is a surface of topological genus 0. In the second example, R was of genus 1 and its Riemann surface was homeomorphic with a torus, i.e., with a surface of topological genus 1. We shall prove later that the topological genus of the Riemann surface of R is always equal to the genus of R.

§2. MEROMORPHIC FUNCTIONS ON THE RIEMANN SURFACE

Let S be the Riemann surface of R, and let $\mathfrak{p}$ be a point of S. Let x be a uniformizing variable at $\mathfrak{p}$ and N an open x-disc of x-center $\mathfrak{p}$; set $D = x(N)$. If $\xi \in D$, let $\mathfrak{p}_\xi$ be the point of N which is mapped by x upon ξ. Let y be any element of R. If we set $f(\xi) = y(\mathfrak{p}_\xi)$, we obtain a mapping f of D into Σ. We shall prove that this mapping is a meromorphic function on D.

Let $F(X, Y)$ be an irreducible polynomial with complex coefficients such that $F(x, y) = 0$. Then we have $(\partial F/\partial Y)(x, y) \neq 0$, and the element $(\partial F/\partial Y)(x, y)$ has only a finite number of zeros. On the other hand, since y has only a finite number of poles, there are only a finite number of $\xi \in D$ for which $f(\xi) = \infty$. Thus we see that there exists a subset D_1 of D, which can be obtained from D by removing a finite number of points and which has the following property: if $\xi \in D_1$, we have $f(\xi) \neq \infty$, $F(\xi, f(\xi)) = 0$, $(\partial F/\partial Y)(\xi, f(\xi)) \neq 0$. Making use of the theorem on implicit functions, we conclude that f is holomorphic at every point of D_1. On the other hand, f is clearly continuous on D; since D_1 differs from D only by a finite number of points (which are therefore isolated points in D), it follows from a well known theorem in the theory of functions of complex variables that f is meromorphic on D.

We can find in D an open disc D' of center 0 and radius $r' > 0$ in which f has no pole except perhaps at 0; inside D', we may represent f by a Laurent series

$$f(\xi) = \sum_{k=m}^{\infty} a_k \xi^k$$

which converges everywhere on D' except perhaps at 0. On the other hand, the $\mathfrak{p}$-adic completion $\bar{R}$ of R may be identified with the field of formal power series in x with complex coefficients (III, §3), and y admits the representation $y = \sum_{k=m'}^{\infty} a'_k x^k$ in $\bar{R}$. Assuming that $y \neq 0$ and that $a_m \neq 0$, $a'_{m'} \neq 0$, we shall prove that $m = m'$ and $a_k = a'_k$ for every $k \geqq m$.

When the element $y \in R$ and the meromorphic function f are in the relation described above, we shall say that f is the representing function of y in D'. If f and g are representing functions in D' of elements y and z of R, it is clear that $f - g$ and fg are representing functions in D' of $y - z$ and yz respectively (by $f - g$ we mean the unique meromorphic function on D' whose value at any point $\xi \in D'$ which is not a pole of either f or g is $f(\xi) - g(\xi)$; similar definition for fg). On the other hand, if h is any integer, the representing function of x^h in D' is obviously ξ^h. This being said, the integer m is characterized by the fact that the function $\xi^{-m}f(\xi)$ takes a finite value $\neq 0$ at 0, viz. a_m. It follows that $x^{-m}y - a_m$ has $\mathfrak{p}$ as a zero, whence $m' = m$, $a'_{m'} = a_m$. Assume that it is true that $a_k = a'_k$ for $m \leqq k < n$, where n is an integer $> m$. The representing function of $x^{-n}(y - \sum_{m \leqq k < n} a_k x^k)$ is the function $\xi^{-n}(f(\xi) - \sum_{m \leqq k < n} a_k \xi^k)$. But the element $x^{-n}(y - \sum_{m \leqq k < n} a_k x^k)$ takes the value a'_n at $\mathfrak{p}$, while the function $\xi^{-n}(f(\xi) - \sum_{m \leqq k < n} a_k \xi^k)$ takes the value a_n at 0; it follows that $a_n = a'_n$. Our assertion is thereby proved.

Thus we have proved

THEOREM 2. *Let $\mathfrak{p}$ be a place of R and x a uniformizing variable at $\mathfrak{p}$. Let y be an element of R, and let $y = \sum_{k=m}^{\infty} a_k x^k$ be the formal power series in x which represents y in the $\mathfrak{p}$-adic completion of R. Then there exists a neighbourhood N of $\mathfrak{p}$ in S on which x has no pole and such that, for any $\mathfrak{q} \in N$ except perhaps* 0, *the series $\sum_{k=m}^{\infty} a_k(x(\mathfrak{q}))^k$ converges to $y(\mathfrak{q})$.*

We can make use of Theorem 2 to study the local nature of the mapping of S into Σ given by an arbitrary element of R.

THEOREM 3. *Let x be a non constant element of R and let $\mathfrak{p}$ be a place of R. Denote by e the ramification index of $\mathfrak{p}$ with respect to $C\langle x \rangle$. Then every neighbourhood of $\mathfrak{p}$ on the Riemann surface of R contains an open neighbourhood N of $\mathfrak{p}$ which has the following properties: x maps N onto an open subset $x(N)$ of Σ; $\mathfrak{p}$ is the only point of N which is mapped upon $x(\mathfrak{p})$ by x; if ξ is any point of $x(N)$ other than $x(\mathfrak{p})$, then there are exactly e points of N which are mapped upon ξ by x.*

If $x(\mathfrak{p}) \neq \infty$, we set $y = x - x(\mathfrak{p})$; if $x(\mathfrak{p}) = \infty$, we set $y = x^{-1}$. In any case, we have $\nu_{\mathfrak{p}}(y) = e$. Let t be a uniformizing variable at $\mathfrak{p}$, and let N_0 be any given neighbourhood of $\mathfrak{p}$. Then it follows from Theorem 2 that there exists a neighbourhood N_1 of $\mathfrak{p}$ which has the following properties: N_1 is contained in N_0; t maps N_1 topologically onto a disc Δ_1 of center 0 in the complex plane; if $\mathfrak{q} \in N_1$, then $y(\mathfrak{q}) = f(t(\mathfrak{q}))$, where f is a holomorphic function on Δ_1 having 0 as a zero of order e. From the elementary theory of functions of a complex variable, it follows that there exist open neighbourhoods U and Δ_2 of 0 in the complex plane with the following properties: Δ_2 is contained in Δ_1; 0 is the only zero of f in Δ_2; if ξ is any point $\neq 0$ in U, then there are exactly e points of Δ_2 which are mapped upon ξ by f. If N is the set of points $\mathfrak{q} \in N_1$ such that $y(\mathfrak{q}) \in U$, then N is an open neighbourhood of $\mathfrak{p}$ contained in N_0 and has the properties stated in Theorem 3.

We shall now define the notion of *meromorphic function* at a point $\mathfrak{p}$ of the Riemann surface S of R. Let f be a mapping into Σ of some neighbourhood of $\mathfrak{p}$ in S, and let x be a uniformizing variable at $\mathfrak{p}$. It follows from Theorem 2 that there exists a mapping f_x into Σ of some neighbourhood of the origin in the complex plane and a neighbourhood N of $\mathfrak{p}$ in S which satisfy the following conditions: f is defined on N; f_x is defined on $x(N)$; we have $f(\mathfrak{q}) = f_x(x(\mathfrak{q}))$ for all $\mathfrak{q} \in N$. We shall say that f belongs to the class $\mathfrak{M}(\mathfrak{p}, x)$ if, for a suitable choice of N, f_x is meromorphic on $x(N)$. Although the function f_x clearly depends on the choice of the uniformizing variable x, we shall see that the class $\mathfrak{M}(\mathfrak{p}, x)$ does not. Let x' be any other uniformizing variable at $\mathfrak{p}$, and let the neighbourhood N' and the function $f_{x'}$ have, with respect to x', the same properties as N and f_x with respect to x. On the other hand, replacing the consideration of f by that of the function x, we obtain a function $x_{x'}$ defined on a neighbourhood of 0 in the complex plane. It is clear that the equality $f_{x'}(\xi') = f_x(x_{x'}(\xi'))$ has a meaning and is true whenever $|\xi'|$ is sufficently small. Since $x \in R$, the function $x_{x'}$ is meromorphic on some neighbourhood of 0, and, since $x(\mathfrak{p}) = 0$, $x_{x'}(0) = 0$. Thus we see that, if f_x is meromorphic on some neighbourhood of 0, then the same is true of $f_{x'}$, i.e., we have $\mathfrak{M}(\mathfrak{p}, x) \subset \mathfrak{M}(\mathfrak{p}, x')$. We would prove in the same way that $\mathfrak{M}(\mathfrak{p}, x') \subset \mathfrak{M}(\mathfrak{p}, x)$. Our assertion that the class $\mathfrak{M}(\mathfrak{p}, x)$ does not depend on the choice of the uniformizing variable x is thereby proved. We shall set $\mathfrak{M}(\mathfrak{p}) = \mathfrak{M}(\mathfrak{p}, x)$; any mapping belonging to the class $\mathfrak{M}(\mathfrak{p})$ is said to be a *meromorphic function at* $\mathfrak{p}$. If f is a mapping of an open subset U of S into Σ which is meromorphic at every point of U, then f is said to be *meromorphic on* U. If furthermore $f(U)$ does not contain the point ∞, then f is said to be *holomorphic* on U. It is clear that any meromorphic mapping of U is also a continuous mapping. Every $x \in R$ provides a meromorphic mapping into Σ of the whole Riemann surface S; we shall see that these are the only meromorphic mappings of S.

THEOREM 4. *The Riemann surface S of the field R is connected, and every meromorphic mapping of S into Σ is produced by some element of R.*

Let f be a meromorphic mapping of some open subset U of S. Then a place $\mathfrak{p}$ is called a zero of f if $f(\mathfrak{p}) = 0$, a pole of f if $f(\mathfrak{p}) = \infty$. The poles of f are isolated, and, if U is connected, the same is true of the zeros of f unless $f(U) = \{0\}$. For, let $\mathfrak{p}$ be a point of U, and let x be a uniformizing variable at $\mathfrak{p}$. Then there exists a neighbourhood N of $\mathfrak{p}$, contained in U, which is mapped topologically by x onto a circular disc Δ of center 0 in Σ, and we have, for $\mathfrak{q} \in N, f(\mathfrak{q}) = f^*(x(\mathfrak{q}))$, where f^* is a function defined on Δ and meromorphic on some neighbourhood of 0. If $\mathfrak{p}$ is a pole of f, then 0 is a pole of f^*, and there is a neighbourhood of 0 in Δ in which f^* has no other pole than 0, which proves that $\mathfrak{p}$ is an isolated pole of f. On the other hand, if every neighbourhood of $\mathfrak{p}$ contains a zero of f distinct from $\mathfrak{p}$, then every neighbourhood of 0 in Δ contains a zero of f^* distinct from 0, from which it follows that f^* is constant and equal to 0 on some neighbourhood of 0, i.e., that f maps some neighbourhood of $\mathfrak{p}$ in U upon $\{0\}$. This shows first that the set of zeros of f in U is relatively closed in U and secondly that the set of non isolated zeros of f in U is open. If U is connected, the set of non isolated zeros of f in U, which is relatively closed and open in U, coincides with the whole of U, which proves our assertion.

Now, let f and g be meromorphic mappings of the same open connected subset U of S. If $g(U) = \{0\}$, then we set $f - g = f$. Assume now that $g(U) \neq \{0\}$, and denote by U_1 the set of points of U which are poles of neither f nor g and which are not zeros of g; U_1 is therefore an open set. If $\mathfrak{p} \in U_1$, then we set $u(\mathfrak{p}) = f(\mathfrak{p}) - g(\mathfrak{p})$, $v(\mathfrak{p}) = f(\mathfrak{p})(g(\mathfrak{p}))^{-1}$. It follows easily from the definitions that u and v are meromorphic mappings of U_1. Let now $\mathfrak{p}$ be a point of U not in U_1; we select a uniformizing variable x at $\mathfrak{p}$ and a neighbourhood N of $\mathfrak{p}$ in U which has the following property: $\mathfrak{p}$ is the only point of N not in U_1; x maps N topologically onto an open disc Δ of center 0 in Σ. If $\mathfrak{q} \in N$, we set $f(\mathfrak{q}) = f^*(x(\mathfrak{q}))$, $g(\mathfrak{q}) = g^*(x(\mathfrak{q}))$; the functions f^* and g^* are then meromorphic on some neighbourhood of 0 in Δ. We may therefore write, for $\xi \neq 0$ in Δ, $f^*(\xi) = \xi^h f_1^*(\xi)$, $g^*(\xi) = \xi^k g_1^*(\xi)$, where h and k are integral exponents, while f_1^* and g_1^* are functions defined on Δ which are holomorphic on some neighbourhood of 0 and take values $\neq 0$ at 0. Let Δ' be a disc of center 0 in which f_1^* and g_1^* are everywhere holomorphic and $\neq 0$. If ξ is any point $\neq 0$ of Δ', we set

$$u_1^*(\xi) = f^*(\xi) - g^*(\xi) = \xi^l(\xi^{h-l}f_1^*(\xi) - \xi^{k-l}g_1^*(\xi)),$$
$$v_1^*(\xi) = f^*(\xi)(g^*(\xi))^{-1} = \xi^{h-k}f_1^*(\xi)(g_1^*(\xi))^{-1}$$

where $l = \max\{h, k\}$. We see immediately that the functions u_1^*, v_1^* defined in this way can be extended to functions u^*, v^* defined on Δ' and meromorphic on some neighbourhood of 0. We set $\alpha = u^*(0)$, $\beta = v^*(0)$ (with α and β in Σ), and we extend the domain of definition of the functions u and v, previously defined on U_1, by setting $u(\mathfrak{p}) = \alpha$, $v(\mathfrak{p}) = \beta$. It is clear that the extended functions defined in this way are meromorphic at $\mathfrak{p}$. Doing the same thing for all points of $U - U_1$, we extend u and v to meromorphic mappings of U. The meromorphic mappings of U which extend u and v are uniquely determined, as follows immediately from the continuity of a meromorphic mapping and from the fact that U_1 is everywhere dense in U. We shall denote our extended mappings by $f - g$ and f/g respectively. If 0_U denotes the mapping which associates 0 to every point of U and 1_U the mapping which associates 1 to every point of U, then we define $-g$ by $-g = 0_U - g$ and (if $g \neq 0_U$) g^{-1} by $(1_U)/g$. We define $f + g$ to be $f - (-g)$ and fg to be $f(g^{-1})^{-1}$ if $g \neq 0_U$ and 0_U if $g = 0_U$. Thus we have two laws of composition, addition and multiplication, in the set $\mathfrak{M}(U)$ of meromorphic functions on U. It is easy to see that these laws of composition satisfy the axioms for fields. Let us for instance check the associativity of addition. Let f_1, f_2, and f_3 be elements of $\mathfrak{M}(U)$. It is easy to check on the base of our definitions that $(f_1 + f_2)(\mathfrak{p}) = f_1(\mathfrak{p}) + f_2(\mathfrak{p})$ for any $\mathfrak{p} \in U$ which is not a pole of either f_1 or f_2. Thus $(f_1 + f_2) + f_3$ and $f_1 + (f_2 + f_3)$ coincide on the set U_2 of points of U which are not poles of any of f_1, f_2, or f_3. The set U_2 being everywhere dense in U, it follows from the continuity of meromorphic mappings that $(f_1 + f_2) + f_3 = f_1 + (f_2 + f_3)$. The other axioms would be verified in a similar manner.

It follows that the set $\mathfrak{M}(U)$ has the structure of a field; if we identify any $c \in C$ with the constant mapping which assigns c to every point of U, we see that C is a subfield of $\mathfrak{M}(U)$. On the other hand, every element x of R defines a function $\mathfrak{p} \to x(\mathfrak{p})$ belonging to $\mathfrak{M}(U)$, and we obtain in this way an isomorphism of R with a subfield of $\mathfrak{M}(U)$ which coincides with the identity on C.

From now on, we shall denote by U a component of S. Since S is clearly locally connected, U is open. On the other hand, since U is closed in S, it is compact. We shall see that *a meromorphic mapping f of U which has no pole is necessarily a constant mapping*. For, since U is compact, the continuous real valued function $|f|$ must reach its maximum at some point $\mathfrak{p}_0 \in U$. Let x_0 be a uniformizing variable at $\mathfrak{p}_0$, and let N_0 be an open x_0-disc of x_0-center $\mathfrak{p}_0$. If $\mathfrak{p} \in N_0$, we may write $f(\mathfrak{p}) = f^*(x(\mathfrak{p}))$, where f^* is holomorphic on a neighbourhood of 0 in $x_0(N_0)$. The function $|f^*|$ has a maximum at the point 0; it follows that f^* has a constant value c on some neighbourhood of 0. The point $\mathfrak{p}_0$ is a non isolated zero of $f - c$, from which it follows that $f - c$ maps the whole of U upon $\{0\}$.

Let x be a non constant element of R of which we shall furthermore assume that, if $S - U$ is not empty, x has at least one pole on this set. Let f be any meromorphic function on U. We denote by $\mathfrak{p}_1, \cdots, \mathfrak{p}_h$ the distinct points of U which are poles of either x or f. For each i let t_i be a uniformizing variable at $\mathfrak{p}_i$ $(1 \leqq i \leqq h)$. We can find integral exponents a_i and b_i such that $(t_i^{a_i}x)(\mathfrak{p}_i)$ and $(t_i^{b_i}f)(\mathfrak{p}_i)$ are both different from 0 and ∞. If k and l are arbitrary non negative integers, then $(t_i^{ka_i+lb_i}x^kf^l)(\mathfrak{p}_i)$ is $\neq \infty$. Let p and q be positive integers, and let

λ_{kl} $(0 \leqq k \leqq p, 0 \leqq l \leqq q)$ be arbitrary complex numbers; we set

$$z = \sum_{k=0}^{p} \sum_{l=0}^{q} \lambda_{kl} x^k f^l .$$

For $\mathfrak{p}$ in a suitable neighbourhood of $\mathfrak{p}_i$, we may write $z(\mathfrak{p}) = z_i^*(t_i(\mathfrak{p}))$, where z_i^* is a meromorphic function on a neighbourhood of 0. Moerover, it is clear that the Laurent expansion of z_i^* around 0 will be of the form

$$z_i^*(\xi) = \sum_{r=pa_i^*+qb_i^*}^{\infty} L_{ir}(\cdots, \lambda_{kl}, \cdots)\xi^r$$

where $a_i^* = \min \{0, a_i\}$, $b_i^* = \min \{0, b_i\}$, and where each L_{ir} is a linear form in $(p + 1)(q + 1)$ variables. From now on, we shall take q to be the number $-\sum_{i=1}^h a_i^*$ (this number is >0 because x is not constant on U and must therefore have at least one pole on U) and we shall take for p a positive integer which is $\geqq -q(\sum_{i=1}^h b_i^* + 1)$. Under these conditions, we have $(p + 1)(q + 1) > -p\sum_{i=1}^h a_i^* - q\sum_{i=1}^h b_i^*$, from which it follows that the system of homogeneous linear equations $L_{ir}(\cdots, \lambda_{kl}, \cdots) = 0$ $(1 \leqq i \leqq h, pa_i^* + qb_i^* \leqq r < 0)$ has a non trivial solution. Assume that the numbers λ_{kl} represent a solution of this system. Then it is clear that the function z has no pole on U. This function is therefore a constant, which proves that f is algebraic of degree $\leqq q$ over $C\langle x\rangle$. Thus we see that every element of $\mathfrak{M}(U)$ is algebraic over $C\langle x\rangle$, and the degree of an element of $\mathfrak{M}(U)$ over $C\langle x\rangle$ is bounded by the number q which depends only on x. Let n be the degree of R over $C\langle x\rangle$; then we know (Corollary to Theorem 4, I, §8) that n is the degree of the divisor of poles $\mathfrak{n}$ of x. Now, it is clear that $\mathfrak{n} = \prod_{i=1}^h \mathfrak{p}_i^{-a_i^*}\mathfrak{n}'$, where $\mathfrak{n}'$ is an integral divisor, whence $q \leqq n$; moreover, if $U \neq S$, then x has at least one pole not in U, whence $q < n$. Since R is of characteristic 0, there exists an element u of R such that $R = C\langle x, u\rangle$, and u is clearly of degree n over $C\langle x\rangle$. On the other hand, considering u as an element of $\mathfrak{M}(U)$, we see that it is of degree $\leqq q$ over $C\langle x\rangle$. Therefore we have $q \geqq n$, whence $q = n$, which proves that $U = S$. On the other hand, would $\mathfrak{M}(U)$ contain an element f which is not in R, then $R\langle f\rangle$ would be algebraic of finite degree $> n$ over $C\langle x\rangle$, and would therefore contain some element of degree $>n$ over $C\langle x\rangle$, which is impossible. Theorem 4 is thereby proved.

REMARK. The connectedness of S could be proved in a shorter way as follows. Assume for a moment that S would have a component $U \neq S$, and let $\mathfrak{q}$ be a point of S not in U. Making use of Lemma 1, IV, §1, we see that there would exist an $x \in R$ which admits $\mathfrak{q}$ as its only pole. Then x would induce on U a meromorphic function with no pole, and would therefore be constant on U. But this is impossible, because, x not being in the field of constants of R, for any $c \in C$, $x - c$ has only a finite number of zeros.

§3. ON SINGULAR HOMOLOGY THEORY

We shall give in this section a brief statement of the main definitions and results of singular homology theory. For more details and for the proofs of theorems, we refer the reader to the book of S. Eilenberg and N. Steenrod on homology theory, to be published shortly (cf. also the paper by S. Eilenberg, *Singular homology theory*, Ann. of Math., 1944, vol. 45, p. 407).

We denote by $\boldsymbol{R}^n$ the n-dimensional Cartesian space. By T^n we mean the smallest convex subset of $\boldsymbol{R}^{n+1}$ containing the points $\boldsymbol{e}_{1,n} = (1, 0, \cdots, 0)$, $\boldsymbol{e}_{2,n} = (0, 1, \cdots, 0), \cdots, \boldsymbol{e}_{n+1,n} = (0, 0, \cdots, 1)$; thus T^n is the set of points

$$(t_1, \cdots, t_{n+1})$$

of $\boldsymbol{R}^{n+1}$ such that $0 \leqq t_i \leqq 1$ $(1 \leqq i \leqq n + 1)$ and $\sum_{i=1}^{n+1} t_i = 1$.

Let X be a topological space. By an *n-simplex* in S is meant a continuous mapping σ of T^n into X. If we take all n-simplices in S as free generators of a free abelian group, then the elements of the group obtained in this way are called the *n-chains* on X. Any n-chain γ may be written in the form $\sum_{k=1}^{h} a_k\sigma_k$, where $\sigma_1, \cdots, \sigma_h$ are distinct n-simplices, and $a_1, \cdots, a_h$ are integers $\neq 0$ (if γ is the zero n-chain, then we set $h = 0$). We then say that $\sigma_1, \cdots, \sigma_h$ are the simplices which *occur* in γ, and that σ_k occurs with the coefficient a_k; a simplex which does not occur in γ is also said to occur with coefficient 0 in γ. The union of the sets of points $\sigma_k(T^n)$ $(1 \leqq k \leqq h)$ is called the *set of points of* γ, and is denoted by $|\gamma|$; this set is empty if $\gamma = 0$. The set of points of any n-chain is clearly a compact set. If $|\gamma|$ is contained on some subset Y of X, then we say that the *chain* γ *is on* Y.

The set T^0 consists of a single point. Thus, a 0-simplex σ is uniquely determined by giving the point $\mathfrak{p} = \sigma(T^0)$ of X. We shall therefore identify the 0-simplices with the points of X.

If $n \geqq 1$ and $1 \leqq i \leqq n + 1$, we denote by f_i the barycentric mapping of T^n into T^{n+1} which maps $\boldsymbol{e}_{j,n-1}$ upon $\boldsymbol{e}_{j,n}$ if $j < i$ and upon $\boldsymbol{e}_{j+1,n}$ if $j \geqq i$. Thus we have $f_i(t_1, \cdots, t_n) = \sum_{j<i} t_j\boldsymbol{e}_{j,n} + \sum_{j \geqq i} t_j\boldsymbol{e}_{j+1,n}$. If σ is an n-simplex, each of the mappings σf_i $(1 \leqq i \leqq n + 1)$ is an $(n - 1)$-simplex. We set

$$\partial\sigma = \sum_{i=1}^{n+1} (-1)^{i+1}(\sigma f_i)$$

and we extend ∂ to a homomorphism, also denoted by ∂, of the group of n-chains into the group of $(n - 1)$-chains. If γ is an n-chain, then $\partial\gamma$ is called the *boundary* of γ. If γ is a 0-chain, we make the convention that $\partial\gamma$ is the number 0.

It can be proved easily that $\partial(\partial\gamma) = 0$ for any n-chain γ (if $n \geqq 1$). Any n-chain whose boundary is 0 is called an *n-cycle*. If Y is a subset of X, an n-chain is said to be an *n-cycle modulo* Y if its boundary is on Y. The n-cycles are the same thing as the n-cycles modulo the empty set $\emptyset$.

An n-chain γ is said to *bound* on X if there exists an $(n + 1)$-chain φ such that $\partial\varphi = \gamma$. If Y is a subset of X, then γ is said to *bound modulo* Y if γ is of the form $\partial\varphi + \gamma_1$, where φ is an $(n + 1)$-chain while γ_1 is an n-chain on Y. Any n-chain which bounds modulo Y is an n-cycle modulo Y, but not conversely in general. To bound is the same thing as to bound modulo $\emptyset$.

Let Z be the group of n-cycles modulo Y, and let B be the group of n-chains which bound modulo Y. Then the factor group Z/B is called the *n-dimensional homology group of X modulo Y*, and is denoted by $H_n(X, Y)$. The group $H_n(X, \emptyset)$ is called the *n-dimensional homology group of X* and is denoted by $H_n(X)$. The elements of $H_n(X, Y)$ are called the *homology classes of X modulo Y* (*homology classes of X* if $Y = \emptyset$). Two cycles modulo Y which belong to the same homology class modulo Y are said to be *homologous to each other modulo Y* (*homologous to each other* if $Y = \emptyset$).

If $\gamma = \sum_{k=1}^{h} a_k \mathfrak{p}_k$ is a 0-cycle on X (each $\mathfrak{p}_k$ being a point of X), then a necessary condition for γ to bound is for the number $\sum_{k=1}^{h} a_k$ to be 0. This condition is also sufficent if X is arcwise connected. It follows that, if X is not empty and arcwise connected, then $H_0(X)$ is an infinite cyclic group. It follows from Theorem 1, §1 that any connected open subset of the Riemann surface of a field of algebraic functions of one variable over C is arcwise connected.

Let X and X' be topological spaces, and let Y and Y' be subsets of X and X' respectively. Assume that we have a continuous mapping f of X into X' which maps Y into Y'; then we say that f is a continuous mapping of the pair (X, Y) into the pair (X', Y'). Let σ be an n-simplex on X; then $f\sigma$ is an n-simplex on X'. The mapping $\sigma \to f\sigma$ can be extended to a homomorphism of the group of n-chains of X into that of X'; we shall also denote by f this homomorphism. Then f maps any cycle modulo Y in X upon a cycle modulo Y' in X', and, if the given cycle bounds modulo Y, then its image bounds modulo Y'. It follows that f defines in a natural way a homomorphism (also denoted by f) of $H_n(X, Y)$ into $H_n(X', Y')$.

Let g be a continuous mapping of X' into a space X'', and assume that g maps Y' into a subset Y'' of X''. Then the homomorphism of $H_n(X, Y)$ into $H_n(X'', Y'')$ which corresponds to the mapping gf can be obtained by first applying the homomorphism of $H_n(X, Y)$ into $H_n(X', Y')$ which corresponds to f, and then the homomorphism of $H_n(X', Y')$ into $H_n(X'', Y'')$ which corresponds to g.

Two continuous mappings f and f' of the pair (X, Y) into the pair (X', Y') are said to be *homotopic to each other* if there exists a continuous mapping F of the pair $(X \times [0, 1], Y \times [0, 1])$ into the pair (X', Y') such that $F(x, 0) = f(x)$, $F(x, 1) = f'(x)$ for all $x \in X$. If this is the case, then the homomorphisms of $H_n(X, Y)$ into $H_n(X', Y')$ which correspond to f and f' are identical with each other.

A subset Y of a space X is called a *deformation retract* of X if there exists a continuous mapping f of X into itself with the following properties: f is homotopic to the identity mapping (which maps every point upon itself); f coincides with the identity on Y; f maps X into Y. If this is the case, then the homomorphism of $H_n(Y)$ into $H_n(X)$ which corresponds to the identity mapping of Y into X is an isomorphism of $H_n(Y)$ onto $H_n(X)$. In particular, every n-cycle on X is homologous to an n-cycle on Y.

Let Y be a subset of a space X. Assume that Y contains an open subset U of X

such that $\bar{U} \subset V \subset Y$, where V is itself an open set ($\bar{U}$ denotes the adherence of U). Then the homomorphism of $H_n(X - U, Y - U)$ into $H_n(X, Y)$ which corresponds to the identity mapping is an isomorphism of the first of these groups onto the second. This statement is called the *excision theorem*.

Let Y be a subset of X, and let c be an element of $H_n(X, Y)$, with $n > 0$; if γ is an n-chain of the class c, then $\partial\gamma$ is an $(n - 1)$-cycle on Y whose homology class ∂c in $H_{n-1}(Y)$ depends only on c. The mapping $c \to \partial c$ is a homomorphism of $H_n(X, Y)$ into $H_{n-1}(Y)$ which is called the *boundary homomorphism.* If $n = 0$, then we set $\partial c = 0$ for any $c \in H_0(X, Y)$.

Let Y and Z be subsets of a space X, such that $Z \subset Y$. We consider the following sequence of groups:

$$\{0\}, H_0(X, Y), \cdots, H_{n-1}(X, Y), H_{n-1}(X, Z), H_{n-1}(Y, Z), H_n(X, Y), \cdots.$$

Then we have a series of homomorphisms, each mapping a group of the sequence (except the first one) on the previous one. These homomorphisms are defined as follows: a) the boundary homomorphism ∂ of $H_n(X, Y)$ into $H_{n-1}(Y, Z)$, obtained by first applying the boundary homomorphism of $H_n(X, Y)$ into $H_{n-1}(Y)$ and then the homomorphism of $H_{n-1}(Y)$ into $H_{n-1}(Y, Z)$ which corresponds to the identity mapping of $(Y, \emptyset)$ into (Y, Z); b) the homomorphism of $H_{n-1}(Y, Z)$ into $H_{n-1}(X, Z)$ which corresponds to the identity mapping of (Y, Z) into (X, Z); c) the homomorphism of $H_{n-1}(X, Z)$ into $H_{n-1}(X, Y)$ which corresponds to the identity mapping of (X, Z) into (X, Y). The composite object formed by our sequence of groups and by the homomorphisms defined above is called the *homology sequence* of the triplet (X, Y, Z). The *exactness theorem* asserts that this sequence is exact, i.e., that, if A, B, C are three consecutive terms of the sequence, then the kernel of the homomorphism of B into A coincides with the image of C under the homomorphism of C into B.

If $Z = \emptyset$, then we speak of the homology sequence of the pair (X, Y); the groups of this sequence are

$$\{0\}, H_0(X, Y), \cdots, H_{n-1}(X, Y), H_{n-1}(X), H_{n-1}(Y), H_n(X, Y), \cdots.$$

Assume that the space X is the union of a finite number of mutually disjoint open sets $X_1, \cdots, X_m$. Let Y be a subset of X; set $Y_i = Y \cap X_i$ $(1 \leqq i \leqq m)$. If $1 \leqq i \leqq m$, the identity mapping of the pair (X_i, Y_i) into (X, Y) defines a homomorphism η_i of $H_n(X_i, Y_i)$ into $H_n(X, Y)$. It can be proved that the homomorphisms η_i are isomorphisms, and that $H_n(X, Y)$ is the direct sum of the groups $\eta_i(H_n(X_i, Y_i))$ $(1 \leqq i \leqq m)$. Thus we see that, in that case, $H_n(X, Y)$ may be identified with the product $\prod_{i=1}^m H_n(X_i, Y_i)$.

To every n-chain γ on a space X it is possible to associate a new n-chain, the *barycentric subdivision* of γ, denoted by subd γ. We shall not give the definition of subd γ, but we shall give the main properties of the operation $\gamma \to$ subd γ.

a) This operation is a homomorphism into itself of the group of n-chains;
b) if γ is a 0-chain, then subd $\gamma = \gamma$;
c) if γ is a 1-simplex and $\partial\gamma = \mathfrak{b} - \mathfrak{a}$ (where $\mathfrak{a}$ and $\mathfrak{b}$ are points), then subd γ

is the sum of two 1-simplices, whose respective boundaries are $\mathfrak{b}_1 - \mathfrak{a}$ and $\mathfrak{b} - \mathfrak{b}_1$, $\mathfrak{b}_1$ being a point of $|\gamma|$ which is called the barycenter of γ;

d) for any n-chain γ, we have ∂ (subd γ) = subd $\partial\gamma$;

e) if γ lies on some subset Y of X, then subd γ lies on Y and is the same whether we consider γ as a chain on X or on Y;

f) if $\partial \gamma$ lies on some subset Y of X, then subd γ is homologous to γ modulo Y; in particular, we see that subd γ is homologous to γ on $|\gamma|$ modulo $|\partial\gamma|$;

g) if p is an integer $\geqq 0$, define $\text{subd}^p \gamma$ by induction by means of the formulas $\text{subd}^0 \gamma = \gamma$, $\text{subd}^{p+1} \gamma = \text{subd}(\text{subd}^p \gamma)$; then, if X is covered by a family $\mathcal{F}$ of open sets, given any chain γ, there is an integer $p \geqq 0$ such that every simplex occurring on $\text{subd}^p \gamma$ lies on some set of the family $\mathcal{F}$.

Let Δ be a closed disc in the complex plane, and let a be the center of Δ. Then $\{a\}$ is clearly a deformation retract of Δ, from which it follows easily that $H_n(\Delta) = \{0\}$ for every $n > 0$. Assume that $\Delta \neq \{a\}$, and let Γ be the circumference of Δ. Then it can be proved that $H_0(\Gamma)$ and $H_1(\Gamma)$ are both infinite cyclic, while $H_n(\Gamma) = \{0\}$ if $n > 1$. The groups $H_0(\Delta, \Gamma)$, $H_1(\Delta, \Gamma)$, and $H_n(\Delta, \Gamma)$ for $n > 2$, consist each of its zero element only, while $H_2(\Delta, \Gamma)$ is infinite cyclic. Since Γ is a deformation retract of $\Delta - \{a\}$, we see that $H_n(\Delta, \Delta - \{a\}) = \{0\}$ if $n \neq 2$, while $H_2(\Delta, \Delta - \{a\})$ is infinite cyclic.

If I is an interval, either open or closed, then $H_n(I) = \{0\}$ for every $n > 0$.

§4. Periods of differentials

We have learned in Chapter VI how to associate to every element x of R a differential dx of the field R; moreover, we know that, provided x is not constant, every differential of R may be written in the form $y\,dx$, with $y \in R$. If x and x' are non constant elements of R, then

$$(1) \qquad y\,dx = (y\,D_{x'}x)\,dx'$$

where $D_{x'}$ is the derivation with respect to x' in R.

Let U be an open set of the Riemann surface S of R. To every pair (f, x) formed by a meromorphic function f on U and a non constant element x of R, we shall associate a new object which we shall call a *meromorphic differential* on U, with the convention that the meromorphic differentials attached to the pairs (f, x) and (f', x') will be held identical to each other if and only if $f = f'D_x x'$, where D_x is the derivation with respect to x in R. If $U = S$, then f is an element of R (Theorem 4, §2), and we shall then identify the meromorphic differential attached to the pair (f, x) with the differential $f\,dx$ of R, which is permissible in view of Formula (1) above. In the case where U is an arbitrary open set, we shall still denote by $f\,dx$ the meromorphic differential attached to the pair (f, x). If $\omega = f\,dx$ and $\omega' = f'\,dx$ are meromorphic differentials on U, and if g is any meromorphic function on U, we set $g\omega = (gf)\,dx$, $\omega + \omega' = (f + f')\,dx$; it can be verified without difficulty that these definitions are coherent with our convention of identification.

Let $f\,dx$ be a meromorphic differential on U, and let $\mathfrak{p}$ be a point of U. Let

t be a uniformizing variable at $\mathfrak{p}$. Then we may write $f\,dx = (fD_tx)\,dt$. If $f\,D_t x$ has no pole at $\mathfrak{p}$, then we say that $f\,dx$ is *holomorphic* at $\mathfrak{p}$. In order to legitimate this definition, we have to show that the fact that $f\,D_tx$ has no pole at $\mathfrak{p}$ does not depend on which uniformizing variable t we select. Let t' be any other uniformizing variable at $\mathfrak{p}$; then $f\,dx = (f\,D_tx)\,dt = (f\,D_tx\,D_{t'}t)\,dt'$. If we represent t by a power series in t' in the $\mathfrak{p}$-adic completion of R, its expansion is of the form $t = \sum_{k=1}^{\infty} a_k t'^k$, whence $D_{t'}t = \sum_{k=1}^{\infty} ka_k t'^{k-1}$, which proves that $D_{t'}t$ has no pole at $\mathfrak{p}$, and therefore that, if $f\,D_tx$ has no pole at $\mathfrak{p}$, then the same is true of $f\,D_{t'}x = f\,D_tx\,D_{t'}t$. A meromorphic differential which is holomorphic at every point of U is said to be *holomorphic on* U.

Let g be a meromorphic function on the open subset U of S, and let x be a non constant element of R. Let $\mathfrak{p}$ be any point of U; denote by t a uniformizing variable at $\mathfrak{p}$ and by V an open t-disc of t-center $\mathfrak{p}$. Then we have, for $\mathfrak{q} \in V$, $g(\mathfrak{q}) = G(t(\mathfrak{q}))$, where G is a meromorphic function on the disc $t(V)$. The derivative G' of G is likewise meromorphic on $t(V)$; set, for $\mathfrak{q} \in V$, $g'(\mathfrak{q}) = G'(t(\mathfrak{q}))$. Then $D_xt\,g'$ is a meromorphic function on V, whose value at $\mathfrak{p}$ we shall denote by $f_{\mathfrak{p}}$. The point $f_{\mathfrak{p}} \in \Sigma$ does not depend on the choice of the uniformizing variable t at $\mathfrak{p}$ or of the set V. For, let t_1 be any uniformizing variable at $\mathfrak{p}$, and let V_1 be any open t_1-disc of t_1-center $\mathfrak{p}$. Set $g(\mathfrak{q}) = G_1(t_1(\mathfrak{q}))$, where G_1 is meromorphic on $t_1(V_1)$. If $\tau \in t(V \cap V_1)$, we have $G(\tau) = G_1(\theta(\tau))$, where $\theta(\tau)$ is defined by the condition that $\theta(t(\mathfrak{q})) = t_1(\mathfrak{q})$ for $\mathfrak{q} \in V \cap V_1$. It follows that $G'(\tau) = G_1'(\theta(\tau))\theta'(\tau)$, where θ' is the derivative of the holomorphic function θ. Set $g_1'(\mathfrak{q}) = G_1'(t_1(\mathfrak{q}))$ for $\mathfrak{q} \in V_1$; then, since $\theta'(t(\mathfrak{q})) = (D_tt_1)(\mathfrak{q})$ (as follows easily from Theorem 2, §2), we have $g'(\mathfrak{q}) = (g_1'\,D_tt_1)(\mathfrak{q})$ for $\mathfrak{q} \in V \cap V_1$, whence $(g'\,D_xt)(\mathfrak{q}) = (g_1'\,D_xt_1)(\mathfrak{q})$, which proves our assertion. It is easily seen that the mapping $\mathfrak{p} \to f_{\mathfrak{p}}$ is a meromorphic function f on U; we shall denote this function by dg/dx (it is easily seen that, if $g \in R$, then dg/dx, as defined here, coincides with the ratio of the differentials dg and dx, i.e. is equal to D_xg). We see immediately that, if x' is any other non constant element of R, then $dg/dx' = (dg/dx)\,D_{x'}x$. It follows that the differential $(dg/dx)dx$ does not depend on the choice of x. We shall denote this meromorphic differential by dg. We see easily that, if g is holomorphic on U, then the same is true of dg.

Let $f\,dx$ be a meromorphic differential on an open set U, and let U_1 be an open subset of U. If there exists a meromorphic function g on U_1 such that $dg/dx = f$ on U_1, then we say that g is a *primitive* of $f\,dx$ on U_1. *If U_1 is connected, then two primitives g, g' of $f\,dx$ on U_1 can differ from each other only by a constant.* For, we can find a point $\mathfrak{p}_0$ of U_1 at which g and g' are both holomorphic and which is such that $x(\mathfrak{p}_0) \neq \infty$ and that $x - x(\mathfrak{p}_0)$ is uniformizing at $\mathfrak{p}_0$. Using $x - x(\mathfrak{p}_0)$ for the t of the definition given above, and observing that $D_x(g' - g) = 0$, we see easily that $g' - g$ coincides with a constant a on some neighbourhood of $\mathfrak{p}_0$. The function $g' - g - a$ being meromorphic on U_1 and having a non isolated zero, it follows that this function is identically zero on U_1, which proves our assertion.

Now, let ω be a differential which is holomorphic on some open set U. We propose to define the *integral* $\int_\gamma \omega$ *of* ω on any 1-chain γ on U.

LEMMA 1. *Let ω be a differential which is holomorphic on an open disc V. Then ω has a primitive on V.*

Assume that V is an x-disc, with x in R. Then it follows immediately from Theorem 3, §2 that $x - x(\mathfrak{p})$ is uniformizing at $\mathfrak{p}$ for every $\mathfrak{p} \in V$. Write $\omega = f\,dx$, and, for $\mathfrak{p} \in V$, $f(\mathfrak{p}) = F(x(\mathfrak{p}))$, where F is holomorphic on $x(V)$. Then it is well known that there exists a holomorphic function G on $x(V)$ whose derivative is F. If we set $g(\mathfrak{p}) = G(x(\mathfrak{p}))$ (for $\mathfrak{p} \in V$), then g is a primitive of ω.

This being said, we shall first define $\int_\gamma \omega$ in the case where γ is a 1-simplex on some open disc V contained in U. The 0-chain $\partial\gamma$ is of the form $\mathfrak{b} - \mathfrak{a}$, where $\mathfrak{a}$ and $\mathfrak{b}$ are points of V. Let g be a primitive of ω on V. We shall prove that the number $g(\mathfrak{b}) - g(\mathfrak{a})$ does not depend on the choices of either V or g. Let V' be any open disc contained in U and containing $|\gamma|$, and let g' be a primitive of ω on V'. Since $|\gamma|$ is obviously connected, it is contained in some component W of $V \cap V'$, and we have seen above that $g' - g$ is constant on W, whence $g'(\mathfrak{b}) - g'(\mathfrak{a}) = g(\mathfrak{b}) - g(\mathfrak{a})$, which proves our assertion. In this case we set

$$\int_\gamma \omega = g(\mathfrak{b}) - g(\mathfrak{a}).$$

Let $\mathfrak{b}_1$ be the barycenter of the 1-simplex γ. Then subd γ is of the form $\gamma_1 + \gamma_2$, where γ_1 and γ_2 are 1-simplices such that $\partial\gamma_1 = \mathfrak{b}_1 - \mathfrak{a}$, $\partial\gamma_2 = \mathfrak{b} - \mathfrak{b}_1$. It follows that we have $\int_\gamma \omega = \int_{\gamma_1} \omega + \int_{\gamma_2} \omega$.

Now, let γ be a 1-chain such that every 1-simplex occurring in γ lies on some open disc. If $\gamma = \sum_{k=1}^h a_k\sigma_k$, where $\sigma_1, \cdots, \sigma_h$ are the simplices occuring in γ, we set

$$\int_\gamma \omega = \sum_{k=1}^h a_k \int_{\sigma_k} \omega.$$

It follows immediately from the definitions that we have

$$\int_\gamma \omega = \int_{\mathrm{subd}\,\gamma} \omega.$$

Finally, let γ be an arbitrary 1-chain on U. The open discs contained in U form a covering of U by open sets; it follows that, for p large enough, every simplex occurring in $\mathrm{subd}^p\,\gamma$ lies on some open disc contained in U. Since we have, for $p' > p$, $\mathrm{subd}^{p'}\gamma = \mathrm{subd}^{p'-p}(\mathrm{subd}^p\,\gamma)$, the number $\int_{\mathrm{subd}^p\gamma} \omega$ does not depend on the choice of the integer p satisfying the condition stated above. It is this number which we call the *integral* on γ of ω, and which we denote by $\int_\gamma \omega$.

It is clear that the mapping $\gamma \to \int_\gamma \omega$ is a homomorphism of the group of 1-chains on U into the additive group of complex numbers, and that, for γ fixed, the mapping $\omega \to \int_\gamma \omega$ is a linear function on the vector space (over C) of holomorphic differentials on U. Moreover, we have $\int_\gamma \omega = \int_{\mathrm{subd}^p\gamma} \omega$ for any 1-chain γ on U and any integer $p \geqq 0$.

LEMMA 2. *Let ω be a holomorphic differential on an open set U of the Riemann surface, and let γ be a 1-chain on U. Set $\partial\gamma = \sum_{k=1}^{h} a_k\mathfrak{p}_k$, where $\mathfrak{p}_1, \cdots, \mathfrak{p}_h$ are points of U. Assume that ω admits a primitive g on U. Then we have*

$$\int_\gamma \omega = \sum_{k=1}^{h} a_k\, g(\mathfrak{p}_k).$$

We select an integer $p \geqq 0$ such that every simplex occurring in $\text{subd}^p\gamma$ lies on some open disc contained in U, and we set $\text{subd}^p\,\gamma = \sum_{l=1}^{m} b_l\sigma_l$, where $\sigma_1, \cdots, \sigma_m$ are the simplices occurring in $\text{subd}^p\,\gamma$. If $\partial\sigma_l = \mathfrak{b}_l - \mathfrak{a}_l$, we have $\int_\gamma \omega = \int_{\text{subd}^p\gamma} \omega = \sum_{l=1}^{m} b_l(g(\mathfrak{b}_l) - g(\mathfrak{a}_l))$. On the other hand we have $\sum_{l=1}^{m} b_l(\mathfrak{b}_l - \mathfrak{a}_l) = \partial\,(\text{subd}^p\,\gamma) = \partial\gamma = \sum_{k=1}^{h} a_k\mathfrak{p}_k$; Lemma 2 follows immediately from this.

THEOREM 5 (CAUCHY'S THEOREM). *Let ω be a holomorphic differential on an open set U of the Riemann surface of the field R, and let γ be a 1-cycle which bounds on U. Then we have $\int_\gamma \omega = 0$.*

Set $\gamma = \partial\varphi$, where φ is a 2-chain on U. There exists an integer $p \geqq 0$ such that every 2-simplex occurring in $\text{subd}^p\,\varphi$ lies on some open disc contained in U. We have $\int_\gamma \omega = \int_{\text{subd}^p\gamma} \omega$ and $\text{subd}^p\,\gamma = \partial(\text{subd}^p\,\varphi)$. It follows immediately that it is sufficient to prove Theorem 5 in the case where γ is the boundary of a 2-simplex φ contained in some open disc V contained in U. Since ω admits a primitive on V (by Lemma 1) and $\partial\gamma = 0$, the result follows immediately from Lemma 2 in that case.

LEMMA 3. *Let U be an open subset of the Riemann surface of R, and let Q be a finite subset of U. Let γ and γ' be 1-cycles on U modulo Q which represent the same element of $H_1(U, Q)$. Then $\gamma' - \gamma$ is a cycle homologous to 0 on U.*

We consider the following four terms of the homology sequence attached to the pair (U, Q):

$$H_1(Q) \rightarrow H_1(U) \rightarrow H_1(U, Q) \rightarrow H_0(Q).$$

Since γ and γ' represent the same element of $H_1(U, Q)$, $\partial\gamma$ and $\partial\gamma'$ represent the same element of $H_0(Q)$, whence $\partial\gamma = \partial\gamma'$ since Q is a finite set. It follows that $\gamma' - \gamma$ is a cycle whose homology class in $H_1(U)$ belongs to the kernel of the homomorphism of $H_1(U)$ into $H_1(U, Q)$. By the exactness theorem, this kernel is the image of $H_1(Q)$, which is easily seen to contain only its zero element, since Q is finite. Lemma 3 is thereby proved.

The notation being as in Lemma 3, let ω be a holomorphic differential on U, and let c be an element of $H_1(U, Q)$. Let γ be a 1-chain belonging to c. Then it follows from Cauchy's theorem and from Lemma 3 that $\int_\gamma \omega$ depends only on c and ω. This number is called the *integral of ω on c* and is denoted by $\int_c \omega$. If Q is empty, i.e. if $c \in H_1(U)$, then $\int_c \omega$ is also called the *period of ω relative to c.* If ω and Q are kept fixed, the mapping $c \rightarrow \int_c \omega$ is a homomorphism of $H_1(U, Q)$ into the additive group of complex numbers. If c is fixed, the mapping $\omega \rightarrow \int_c \omega$ is a linear function on the vector space (over C) of holomorphic differentials on U. Let U_1 be an open subset of U and Q_1 a subset of $Q \cap U_1$; denote by c_1 an element of $H_1(U_1, Q_1)$ and by c the element of $H_1(U, Q)$ which corresponds to c_1 by the identity mapping of (U_1, Q_1) into (U, Q). Then it is clear that $\int_{c_1} \omega = \int_c \omega$.

Now, let us consider the case of a differential ω which is meromorphic but not holomorphic on U. Any point $\mathfrak{p} \in U$ at which ω is not holomorphic will be called a *pole* of ω. To simplify, we shall consider only the case where ω has only a finite number of poles; this is certainly the case when ω is a differential of the field R. Let then P be the set of poles of ω (or any finite set containing it), and set $U_1 = U - P$. Let Q be a finite subset of U_1. Then $\int_c \omega$ is defined for every $c \in H_1(U_1, Q)$. Now, the identity mapping of U_1 into U defines a certain homomorphism of $H_1(U_1, Q)$ into $H_1(U, Q)$; we propose to study more closely this homomorphism.

In what follows, P is to denote an arbitrary finite subset of the open set U and Q an arbitrary subset of $U - P$. We consider the following four terms of the homology sequence attached to the triplet $(U, U - P, Q)$:

$$H_2(U, U-P) \underset{\eta_1}{\rightarrow} H_1(U-P, Q) \underset{\eta}{\rightarrow} H_1(U, Q) \underset{\eta_2}{\rightarrow} H_1(U, U-P)$$

(the letters η_1, η, η_2 denoting the homomorphisms of the homology sequence). We know that $\eta(H_1(U - P, Q))$ is the kernel of η_2 and that the kernel of η is $\eta_1(H_2(U, U - P))$.

Since P is finite, it is easily seen by induction on the number of elements of P that we can assign to every $\mathfrak{p} \in P$ a closed disc $D_\mathfrak{p}$ containing $\mathfrak{p}$ in its interior, contained in U, in such a way that $D_\mathfrak{p} \cap D_\mathfrak{q} = \emptyset$ for $\mathfrak{p} \neq \mathfrak{q}$. Let W be the union of the sets $D_\mathfrak{p}$, $\mathfrak{p} \in P$. Then the adherence of $U - W$ in U is contained in an open set contained in $U - P$; therefore, it follows from the excision theorem that $H_2(U, U - P)$ is isomorphic with $H_2(W, W - P)$ and $H_1(U, U - P)$ to $H_1(W, W - P)$. The sets $D_\mathfrak{p}$ being mutually disjoint, $H_2(W, W - P)$ is isomorphic with the product $\prod_{\mathfrak{p} \in P} H_2(D_\mathfrak{p}, D_\mathfrak{p} - \{\mathfrak{p}\})$, and $H_1(W, W - P)$ with $\prod_{\mathfrak{p} \in P} H_1(D_\mathfrak{p}, D_\mathfrak{p} - \{\mathfrak{p}\})$. Now, we consider the following seven terms of the homology sequence attached to the pair $(D_\mathfrak{p}, D_\mathfrak{p} - \{\mathfrak{p}\})$:

$$H_2(D_\mathfrak{p}) \rightarrow H_2(D_\mathfrak{p}, D_\mathfrak{p} - \{\mathfrak{p}\}) \underset{\zeta}{\rightarrow} H_1(D_\mathfrak{p} - \{\mathfrak{p}\}) \rightarrow H_1(D_\mathfrak{p}) \underset{\zeta_1}{\rightarrow} H_1(D_\mathfrak{p}, D_\mathfrak{p} - \{\mathfrak{p}\})$$

$$\underset{\zeta_2}{\rightarrow} H_0(D_\mathfrak{p} - \{\mathfrak{p}\}) \rightarrow H_0(D_\mathfrak{p})$$

(the greek letters denoting the homomorphisms of the homology sequence). We know that $H_2(D_\mathfrak{p}) = \{0\}$, $H_1(D_\mathfrak{p}) = \{0\}$; the exactness theorem then shows that ζ is an isomorphism of $H_2(D_\mathfrak{p}, D_\mathfrak{p} - \{\mathfrak{p}\})$ with $H_1(D_\mathfrak{p} - \{\mathfrak{p}\})$. Let $C_\mathfrak{p}$ be the circumference of $D_\mathfrak{p}$; we know that $H_1(C_\mathfrak{p})$ is infinite cyclic. Let $\gamma_\mathfrak{p}$ be a 1-cycle on $C_\mathfrak{p}$ whose homology class generates $H_1(C_\mathfrak{p})$. It is clear that $C_\mathfrak{p}$ is a deformation retract of $D_\mathfrak{p} - \{\mathfrak{p}\}$; therefore, $H_1(D_\mathfrak{p} - \{\mathfrak{p}\})$ is infinite cyclic and generated by the homology class of $\gamma_\mathfrak{p}$ in $H_1(D_\mathfrak{p} - \{\mathfrak{p}\})$. It follows that $H_2(D_\mathfrak{p}, D_\mathfrak{p} - \{\mathfrak{p}\})$ is infinite cyclic; furthermore, since ζ is the boundary homomorphism, $H_2(D_\mathfrak{p},$

$D_\mathfrak{p} - \{\mathfrak{p}\})$ is generated by the homology class of a 2-chain $\delta_\mathfrak{p}$ on $D_\mathfrak{p}$ such that $\partial\delta_\mathfrak{p} = \gamma_\mathfrak{p}$. This means that $H_2(W, W - P)$ is generated by the homology classes of the chains $\delta_\mathfrak{p}$ for all $\mathfrak{p} \in P$; the same can therefore be said of $H_2(U, U - P)$. The mapping η_1 being the boundary homomorphism, we see that the kernel of η in $H_1(U - P, Q)$ is generated by the homology classes in this group of the chains $\gamma_\mathfrak{p}$.

Now, let $\mathfrak{q}$ be any point $\neq \mathfrak{p}$ in $D_\mathfrak{p}$. Since $D_\mathfrak{p}$ and $D_\mathfrak{p} - \{\mathfrak{p}\}$ are arcwise connected, the homology groups $H_0(D_\mathfrak{p})$ and $H_0(D_\mathfrak{p} - \{\mathfrak{p}\})$ are generated by the homology classes of $\mathfrak{q}$ in these groups. It follows that the image of $H_1(D_\mathfrak{p}, D_\mathfrak{p} - \{\mathfrak{p}\})$ under ζ_2 is $\{0\}$ and therefore that $H_1(D_\mathfrak{p}, D_\mathfrak{p} - \{\mathfrak{p}\}) = \zeta_1(H_1(D_\mathfrak{p})) = \{0\}$. This proves $H_1(U, U - P) = \{0\}$, and therefore that η maps $H_1(U - P, Q)$ onto $H_1(U, Q)$. Thus we have proved

LEMMA 4. *Let U be an open subset of the Riemann surface of R, P a finite subset of U, and Q a subset of $U - P$. Assign to each $\mathfrak{p} \in P$ a closed disc $D_\mathfrak{p} \subset U$ containing $\mathfrak{p}$ in its interior in such a way that $D_\mathfrak{p} \cap D_\mathfrak{q} = \emptyset$ if $\mathfrak{p} \neq \mathfrak{q}$. Denote by $C_\mathfrak{p}$ the circumference of $D_\mathfrak{p}$ and by $\gamma_\mathfrak{p}$ a 1-cycle on $C_\mathfrak{p}$ whose homology class generates $H_1(C_\mathfrak{p})$. Then the identity mapping of $(U - P, Q)$ into (U, Q) defines a homomorphism of $H_1(U - P, Q)$ onto $H_1(U, Q)$ whose kernel is generated by the homology classes of the cycles $\gamma_\mathfrak{p}$ in $H_1(U - P, Q)$. Let $\delta_\mathfrak{p}$ be a 2-chain on $D_\mathfrak{p}$ such that $\partial\delta_\mathfrak{p} = \gamma_\mathfrak{p}$, and let $d_\mathfrak{p}$ be the homology class of $\delta_\mathfrak{p}$ in $H_2(U, U - P)$. Then $H_2(U, U - P)$ is a free abelian group in which the elements $d_\mathfrak{p}$ form a system of free generators.*

COROLLARY. *The notation being as in Lemma 4, let γ be a 1-chain on U such that $|\partial\gamma| \cap P = \emptyset$; then there is a 1-chain γ' on $U - P$ such that $\gamma' - \gamma$ is a cycle homologous to 0 on U.*

Set $Q = |\partial\gamma|$; then it follows from Lemma 4 that the homology class c of γ in $H_1(U, Q)$ can be represented by a 1-chain γ' on $U - P$. Since Q is finite, $\gamma' - \gamma$ is a cycle homologous to 0 on U in virtue of Lemma 3.

This being said, let us return to the study of the meromorphic differential ω and to the notation introduced above. If z_1 is any element of $H_1(U - P)$ which is mapped upon 0 by the homomorphism of $H_1(U - P)$ into $H_1(U)$ which corresponds to the identity mapping, then $\int_{z_1} \omega$ is called a *logarithmic period* of ω. The cycles $\gamma_\mathfrak{p}$ being defined as in Lemma 4, we see that every logarithmic period of ω is a linear combination with integral coefficients of the numbers $\int_{\gamma_\mathfrak{p}} \omega$. We shall show how these numbers may be computed.

Let then $\mathfrak{p}$ be any point of P, and let D be a closed disc containing $\mathfrak{p}$ in its interior and such that $D \cap P = \{\mathfrak{p}\}$. We can find a uniformizing variable x at $\mathfrak{p}$ such that D is an x-disc of x-center $\mathfrak{p}$. If we write $\omega = f\,dx$, there will exist an open x-disc V of x-center $\mathfrak{p}$ on which f is defined and meromorphic with no pole except perhaps at $\mathfrak{p}$. It follows that, for $\mathfrak{q} \in V$, $\mathfrak{q} \neq \mathfrak{p}$, $f(\mathfrak{q})$ may be represented in the form $f(\mathfrak{q}) = \sum_{i=-r}^{+\infty} c_i(x(\mathfrak{q}))^i$ (with $r > 0$). Denote by f_1 the meromorphic function defined on V by $f_1(\mathfrak{q}) = f(\mathfrak{q}) - c_{-1}(x(\mathfrak{q}))^{-1}$ (for $\mathfrak{q} \in V$, $\mathfrak{q} \neq \mathfrak{p}$). Then $f_1\,dx$ has a primitive on V. For it is known that $\sum_{-r \leq i < \infty, i \neq -1}(i + 1)^{-1}c_i(x(\mathfrak{q}))^{i+1}$ converges for every $\mathfrak{q} \neq \mathfrak{p}$ in V; if we denote its sum by $g(\mathfrak{q})$, the mapping

$\mathfrak{q} \to g(\mathfrak{q})$ of $V - \{\mathfrak{p}\}$ into Σ can be extended to a meromorphic function (also denoted by g) on V, and we have $dg = f_1\, dx$. We may assume that $D \subset V$. Let C be the circumference of D, and let γ be a 1-cycle on C whose homology class generates $H_1(C)$. Then we have $\int_\gamma f_1\, dx = 0$ by Lemma 2, whence

$$\int_\gamma \omega = c_{-1} \int_\gamma x^{-1}\, dx.$$

Incidentally, we observe that, if $c_{-1} = 0$, there exists an open disc containing $\mathfrak{p}$ on which ω has a primitive.

Now, let $\mathfrak{a}_1, \mathfrak{a}_2, \mathfrak{a}_3$ be three distinct points of C. Denote by V_i the set of points $\mathfrak{q} \in V$ such that $x(\mathfrak{q})$ is not of the form $\rho\, x(\mathfrak{a}_i)$, with ρ real $\geqq 0$. Then $x^{-1}\, dx$ admits on each V_i a primitive. For, if $\mathfrak{q} \in V_i$, $x(\mathfrak{q})$ may be written in the form $x(\mathfrak{a}_i) r(\mathfrak{q}) \exp((-1)^{1/2}\theta_i(\mathfrak{q}))$, where $r(\mathfrak{q})$ and $\theta_i(\mathfrak{q})$ are continuous real functions on V_i, with $r(\mathfrak{q}) > 0$, $0 < \theta_i(\mathfrak{q}) < 2\pi$; the function Λ_i defined on V_i by $\Lambda_i(\mathfrak{q}) = \log r(\mathfrak{q}) + (-1)^{1/2}\theta_i(\mathfrak{q})$ is then a primitive of $x^{-1}\, dx$ on V_i. If i and j are two distinct indices between 1 and 3, we can find a 1-simplex σ_{ij} on C such that $\partial\sigma_{ij} = \mathfrak{a}_j - \mathfrak{a}_i$ and $\mathfrak{a}_k \notin |\sigma_{ij}|$, where k is the number between 1 and 3 which is distinct from i and j. Then $\gamma_1 = \sigma_{12} + \sigma_{23} + \sigma_{31}$ is a cycle on C, and we have

$$\frac{1}{(-1)^{1/2}} \int_{\gamma_1} x^{-1}\, dx = (\theta_3(\mathfrak{a}_2) - \theta_3(\mathfrak{a}_1)) + (\theta_1(\mathfrak{a}_3) - \theta_1(\mathfrak{a}_2)) + (\theta_2(\mathfrak{a}_1) - \theta_2(\mathfrak{a}_3)).$$

If $i \neq j$, we have $x(\mathfrak{a}_j)/x(\mathfrak{a}_i) = \exp((-1)^{1/2}\theta_i(\mathfrak{a}_j))$; since $\theta_i(\mathfrak{a}_j)$ and $\theta_j(\mathfrak{a}_i)$ are both between 0 and 2π, their sum is 2π, whence

$$((-1)^{1/2})^{-1} \int_{\gamma_1} x^{-1}\, dx = 6\pi - 2(\theta_1(\mathfrak{a}_2) + \theta_2(\mathfrak{a}_3) + \theta_3(\mathfrak{a}_1)).$$

Since $(x(\mathfrak{a}_2)/x(\mathfrak{a}_1))(x(\mathfrak{a}_3)/x(\mathfrak{a}_2)(x(\mathfrak{a}_1)/x(\mathfrak{a}_3) = 1$, the sum $\theta_1(\mathfrak{a}_2) + \theta_2(\mathfrak{a}_3) + \theta_3(\mathfrak{a}_1)$ is a multiple of 2π. Since $0 < \theta_i(\mathfrak{a}_j) < 2\pi$, the value of this sum is either 2π or 4π, which proves that $\int_{\gamma_1} x^{-1}\, dx = \pm 2\pi(-1)^{1/2}$. The chain $\sigma_{21} + \sigma_{32} + \sigma_{13}$ is also a cycle γ_1', and we see immediately that $\int_{\gamma_1'} x^{-1}\, dx = -\int_{\gamma_1} x^{-1}\, dx$.

We can find an integer $p \geqq 0$ such that every simplex occurring in $\text{subd}^p\gamma$ lies on one of the sets V_i; let then $\tau_1, \cdots, \tau_h$ be these simplices, and set $\text{subd}^p\gamma = \sum_{l=1}^h a_l\tau_l$. If $\partial\tau_l = \mathfrak{b}_l - \mathfrak{c}_l$, we see immediately that

$$\exp\left(\int_{\tau_l} x^{-1}\, dx\right) = x(\mathfrak{b}_l)/x(\mathfrak{c}_l).$$

Since γ is a cycle, we have $\sum_{l=1}^h a_l(\mathfrak{b}_l - \mathfrak{c}_l) = 0$, whence $\exp(\int_\gamma x^{-1}\, dx) = 1$, which prove that $\int_\gamma x^{-1}\, dx$ is a multiple of $2\pi(-1)^{1/2}$. Since the homology class of γ is a generator of $H_1(C)$, γ_1 is homologous on C to $r\gamma$, r being an integer, whence $2\pi(-1)^{1/2} = r \int_\gamma x^{-1}\, dx$. It follows immediately that $r = \pm 1$. Replacing if necessary γ by $-\gamma$, we shall assume from now on that γ has been selected in such a way that $\int_\gamma x^{-1}\, dx = 2\pi(-1)^{1/2}$. This condition determines entirely the homology class z of γ in $H_1(C)$ when D and x are given. We shall see that z actually does not depend on the choice of x, x being such that D is an x-disc. In fact, let x' be any element of R such that D is an x'-disc. Then x' is uniformizing at some interior point $\mathfrak{p}'$ of D. Set $x(\mathfrak{p}') = a$; if $a = 0$, then we have (by Schwarz's lemma) $x' = bx$, where b is a constant, whence $\int_\gamma x'^{-1}\,dx' = \int_\gamma x^{-1}\,dx$. If $a \neq 0$, denote by l the set of points $\mathfrak{q} \in D$ such that $x(\mathfrak{q})$ is of the form ta, with t real and $0 \leqq t \leqq 1$. Then it is easily seen that $x^{-1}\,dx - (x - a)^{-1}\,dx$ has a primitive on $V - l$, whence $\int_\gamma x^{-1}\,dx = \int_\gamma (x - a)^{-1}\,dx$ by Lemma 2. On the other hand, there exists an open set containing D on which $(x - a)^{-1}x'$ is holomorphic and has no zero. If we denote by u this function, we have $0 = \int_\gamma u^{-1}\,du = \int_\gamma x'^{-1}\,dx' - \int_\gamma (x - a)^{-1}\,dx$, whence $\int_\gamma x'^{-1}\,dx' = \int_\gamma x^{-1}\,dx$. Thus we see that we may attach to any closed disc D a definite generator z of $H_1(C)$, where C is the circumference of D. We shall say that z is the *positive generator* of $H_1(C)$. It is characterized by the condition that $\int_z x^{-1}\,dx = 2\pi(-1)^{1/2}$ whenever D is an x-disc.

In the case where ω is a differential of the field R, the number which we have denoted above by c_{-1} is the residue of ω at $\mathfrak{p}$ (cf. corollary to Theorem 6, VI, §3); this number does not depend on the choice of x. We shall see that this is still true in the present case. Using the same notation as above, we observe that $\omega - (\sum_{i=-r}^{-1} c_i x^i\,dx)$ has no pole at $\mathfrak{p}$. Let x' be any other uniformizing variable at $\mathfrak{p}$, and write $\omega = f'\,dx'$; then $f' - (\sum_{i=-r}^{-1} c_i x^i)D_x x'$ has no pole at $\mathfrak{p}$, from which our assertion follows immediately if we observe that $(\sum_{i=-r}^{-1} c_i x^i)\,dx = (\sum_{i=-r}^{-1} c_i x^i)D_x x' dx'$ is a differential of R. The number c_{-1} is called the *residue of* ω *at* $\mathfrak{p}$. Thus we have proved

THEOREM 6. *Let ω be a meromorphic differential on an open set U of the Riemann surface of the field R. Let $\mathfrak{p}$ be a point of U and let D be a closed disc containing $\mathfrak{p}$ in its interior and on which ω has no pole except perhaps at $\mathfrak{p}$. Denote by C the circumference of D and by z the positive generator of $H_1(C)$. Then we have $\int_z \omega = 2\pi(-1)^{1/2}\rho$, where ρ is the residue of ω at $\mathfrak{p}$.*

Moreover, using the same notation as above, we see that the homology classes of the cycles γ_1 and γ_1' in $H_1(C)$ do not depend on the choice of the simplices σ_{ij}, and that one and only one of these homology classes is the positive generator of $H_1(C)$. If γ_1 belongs to the positive generator, then we say that $\mathfrak{a}_1$, $\mathfrak{a}_2$, and $\mathfrak{a}_3$, in this order, determine the *positive orientation* on C; if not, then we say that $\mathfrak{a}_1$, $\mathfrak{a}_2$, and $\mathfrak{a}_3$ determine the *negative orientation* on C. In the first case, we set $\varepsilon(\mathfrak{a}_1, \mathfrak{a}_2, \mathfrak{a}_3) = 1$, in the second case $\varepsilon(\mathfrak{a}_1, \mathfrak{a}_2, \mathfrak{a}_3) = -1$. The following properties of this symbol are easily established: $\varepsilon(\mathfrak{a}_1, \mathfrak{a}_2, \mathfrak{a}_3)$ does not change if we perform an even permutation of $\mathfrak{a}_1$, $\mathfrak{a}_2$, $\mathfrak{a}_3$, but changes its sign if we perform an odd permutation; if C' and C'' are the two components of $C - (\{\mathfrak{a}_1\} \cup \{\mathfrak{a}_2\})$, then $\varepsilon(\mathfrak{a}_1, \mathfrak{a}_2, \mathfrak{a})$ is constant on each of the sets C' and C'', but does not have the same value on these two sets.

Let ω be a meromorphic differential on an open set U; assume that ω has only a finite number of poles on U. Then it follows immediately from Lemma 4 and Theorem 6 that the logarithmic periods of ω are the linear combinations with integral coefficients of the numbers $2\pi(-1)^{1/2}\rho$, ρ running over the residues of ω. If the residues of ω are all 0, then the same is true of its logarithmic periods.

Assume that this is the case, and let P be the set of poles of ω on U. Let Q be a finite subset of $U - P$. If $c \in H_1(U, Q)$, then c is the image of some $c_1 \in H_1(U - P, Q)$ under the homomorphism which corresponds to the identity mapping of the pair $(U - P, Q)$ into (U, Q) (this by Lemma 4), and, since the logarithmic periods of ω are 0, the number $\int_{c_1} \omega$ depends only on c, not on the choice of c_1. We shall denote this number by $\int_c \omega$.

THEOREM 7. *Let ω be a meromorphic differential on an open set U of the Riemann surface of the field R. We assume that ω has only a finite number of poles on U and that the residues of ω at these poles are all* 0. *Then a necessary and sufficient condition for the existence of a primitive of ω on U is that $\int_z \omega = 0$ for every $z \in H_1(U)$.*

We may obviously assume without loss of generality that U is connected. The necessity of the condition follows immediately from Lemma 2. Now, assume that the condition is satisfied. Let P be the set of poles of ω on U. Then $U - P$ is open and is easily seen to be connected, and therefore also arcwise connected. Let $\mathfrak{p}_0$ be any fixed point of $U - P$. Then, for any $\mathfrak{p} \in U - P$, we can select a 1-chain $\gamma_\mathfrak{p}$ on $U - P$ such that $\partial\gamma_\mathfrak{p} = \mathfrak{p} - \mathfrak{p}_0$. Set $g(\mathfrak{p}) = \int_{\gamma_\mathfrak{p}} \omega$. We shall see that g is holomorphic on $U - P$. Let $\mathfrak{p}$ be any point of $U - P$, and let V be an open disc contained in $U - P$ and containing $\mathfrak{p}$. Then ω admits a primitive g_1 on V. If $\mathfrak{q}$ is any point of V, we can find a 1-chain $\gamma_{\mathfrak{p}\mathfrak{q}}$ on V such that $\partial\gamma_{\mathfrak{p}\mathfrak{q}} = \mathfrak{q} - \mathfrak{p}$. Then $\gamma_\mathfrak{p} + \gamma_{\mathfrak{p}\mathfrak{q}} - \gamma_\mathfrak{q}$ is a 1-cycle on $U - P$, and it follows from our assumption on ω that the integral of ω on this cycle is 0, whence $g(\mathfrak{q}) = g(\mathfrak{p}) + \int_{\gamma_{\mathfrak{p}\mathfrak{q}}} \omega = g(\mathfrak{p}) + g_1(\mathfrak{q}) - g_1(\mathfrak{p})$. This formula shows that $g - g_1$ is constant on V, and therefore that g is a primitive of ω on V. Now, let $\mathfrak{a}$ be a pole of ω. Then, the residue of ω at $\mathfrak{a}$ being 0, we have seen that there exists an open disc $W \subset U$ containing $\mathfrak{a}$ on which ω admits a primitive g_2. The set $W - W \cap P$ being obviously connected, $g_2 - g$ is constant on this set, from which it follows that g can be extended to a function which is meromorphic on W. This being true for every pole of ω, we see that g can be extended to a meromorphic function on U, which is obviously a primitive of ω.

§5. THE BILINEAR FUNCTION $j(\omega, \omega')$

Let ω and ω' be differentials of the field R, and let $\mathfrak{p}$ be a place of R on which we shall assume that, if it is a pole of one of the differentials ω or ω', then the residue of the other at this place is 0. Under this condition we shall attach to the pair (ω, ω') and to the place $\mathfrak{p}$ a number $j_\mathfrak{p}(\omega, \omega')$.

We first consider the case where $\text{res}_\mathfrak{p}\, \omega = 0$. Then we know that, for any integer m, we can find an $f \in R$ such that $\nu_\mathfrak{p}(\omega - df) \geqq m$ (Theorem 14, VI, §8). We take $f \in R$ such that $\nu_\mathfrak{p}(\omega - df) \geqq -\nu_\mathfrak{p}(\omega') - 1$, with the further specification that, if $\nu_\mathfrak{p}(\omega) \geqq 0$, then $\nu_\mathfrak{p}(f) > 0$ (this is obviously always possible). Then we shall see that the number $\text{res}_\mathfrak{p}\, f\omega'$ does not depend on the choice of f. For, let f' be any element of R which satisfies the same conditions as f. Then we have $\nu_\mathfrak{p}(d(f' - f)) \geqq -\nu_\mathfrak{p}(\omega') - 1$ and $\nu_\mathfrak{p}(f' - f) > 0$ in case $\nu_\mathfrak{p}(\omega) \geqq 0$. Set $g = f' - f$; if $\nu_\mathfrak{p}(g) \neq 0$, we have $\nu_\mathfrak{p}(dg) = \nu_\mathfrak{p}(g) - 1$ (Lemma 1, VI, §8), whence

$\nu_{\mathfrak{p}}(g\omega') \geqq 0$ and $\mathrm{res}_{\mathfrak{p}}\, g\omega' = 0$. If $\nu_{\mathfrak{p}}(g) = 0$, then $\nu_{\mathfrak{p}}(dg) = \nu_{\mathfrak{p}}(g - g(\mathfrak{p})) - 1$ and $\mathrm{res}_{\mathfrak{p}}\, ((g - g(\mathfrak{p}))\omega') = 0$ whence $\mathrm{res}_{\mathfrak{p}}\, g\omega' = g(\mathfrak{p})\, \mathrm{res}_{\mathfrak{p}}\, \omega'$; but, if $\mathrm{res}_{\mathfrak{p}}\, \omega' \neq 0$, then $\nu_{\mathfrak{p}}(\omega) \geqq 0$, whence $g(\mathfrak{p}) = 0$. Thus we have $\mathrm{res}_{\mathfrak{p}}\, g\omega' = 0$ in every case, which proves our assertion. We shall define (if $\mathrm{res}_{\mathfrak{p}}\, \omega = 0$) the number $j_{\mathfrak{p}}(\omega, \omega')$ to be $\mathrm{res}_{\mathfrak{p}}\, f\omega'$. It will be observed that it follows from our discussion that, if $\nu_{\mathfrak{p}}(\omega) \geqq 0$, $\mathrm{res}_{\mathfrak{p}}\, \omega' = 0$, then the condition $\nu_{\mathfrak{p}}(f) > 0$ is unnecessary.

Before defining $j_{\mathfrak{p}}(\omega, \omega')$ in the case where $\mathrm{res}_{\mathfrak{p}}\, \omega \neq 0$, we shall first prove that $j_{\mathfrak{p}}(\omega', \omega) = -j_{\mathfrak{p}}(\omega, \omega')$ in the case where $\mathrm{res}_{\mathfrak{p}}\, \omega = \mathrm{res}_{\mathfrak{p}}\, \omega' = 0$. We can select elements f and f' in R such that $\nu_{\mathfrak{p}}(\omega - df) \geqq \max\{-\nu_{\mathfrak{p}}(\omega') - 1, \nu_{\mathfrak{p}}(\omega) + 1\}$, $\nu_{\mathfrak{p}}(\omega' - df') \geqq \max\{-\nu_{\mathfrak{p}}(\omega) - 1, \nu_{\mathfrak{p}}(\omega') + 1\}$. Then we have $j_{\mathfrak{p}}(\omega, \omega') = \mathrm{res}_{\mathfrak{p}}\, f\omega' = \mathrm{res}_{\mathfrak{p}}\, f\, df' + \mathrm{res}_{\mathfrak{p}}\, f(\omega' - df')$. Since $\nu_{\mathfrak{p}}(\omega - df) \geqq \nu_{\mathfrak{p}}(\omega) + 1$, we have $\nu_{\mathfrak{p}}(df) = \nu_{\mathfrak{p}}(\omega)$. If this number is <0, we have $\nu_{\mathfrak{p}}(f) = \nu_{\mathfrak{p}}(\omega) + 1$, whence $\nu_{\mathfrak{p}}(f(\omega' - df')) \geqq 0$ and $\mathrm{res}_{\mathfrak{p}}\, f(\omega' - df') = 0$. If $\nu_{\mathfrak{p}}(\omega) \geqq 0$, we have $\nu_{\mathfrak{p}}(f) \geqq 0$, $\nu_{\mathfrak{p}}(\omega' - df') \geqq -1$; but we have $\mathrm{res}_{\mathfrak{p}}\, (\omega' - df') = 0$, and a differential which is of order $\geqq -1$ at a place where its residue is 0 has no pole at this place; it follows that $\nu_{\mathfrak{p}}(\omega' - df') \geqq 0$, whence $\mathrm{res}_{\mathfrak{p}}\, f(\omega' - df') = 0$. Thus we have in any case $j_{\mathfrak{p}}(\omega, \omega') = \mathrm{res}_{\mathfrak{p}}\, f\, df'$; similarly, we have $j_{\mathfrak{p}}(\omega', \omega) = \mathrm{res}_{\mathfrak{p}}\, f'df$, and the sum of these numbers is

$$\mathrm{res}_{\mathfrak{p}}\, (f\, df' + f'df) = \mathrm{res}_{\mathfrak{p}}\, d(ff') = 0,$$

which proves our assertion.

Thus, it is natural to define $j_{\mathfrak{p}}(\omega, \omega')$ in the case where $\mathrm{res}_{\mathfrak{p}}\, \omega \neq 0$ (whence $\nu_{\mathfrak{p}}(\omega') \geqq 0$) by the formula $j_{\mathfrak{p}}(\omega, \omega') = -j_{\mathfrak{p}}(\omega', \omega)$.

It is clear that $j_{\mathfrak{p}}(\omega, \omega') = 0$ if $\mathfrak{p}$ is a pole of neither ω nor ω'. On the other hand, if $f \in R$, we have $j_{\mathfrak{p}}(df, \omega') = \mathrm{res}_{\mathfrak{p}}\, f\omega'$ if $\mathrm{res}_{\mathfrak{p}}\, \omega' = 0$ and $j_{\mathfrak{p}}(df, \omega') = \mathrm{res}_{\mathfrak{p}}\, f\omega' - f(\mathfrak{p})\, \mathrm{res}_{\mathfrak{p}}\, \omega'$ if $\mathrm{res}_{\mathfrak{p}}\, \omega' \neq 0$ (whence $\nu_{\mathfrak{p}}(f) \geqq 0$ if $j_{\mathfrak{p}}(df, \omega')$ is defined).

Now, let ω and ω' be differentials of R which are such that the residues of either one of them at the poles of the other are 0. Then $j_{\mathfrak{p}}(\omega, \omega')$ is defined for all places $\mathfrak{p}$ of R, and is $\neq 0$ for only a finite number of places. In this case, we shall define a number $j(\omega, \omega')$ by the formula $j(\omega, \omega') = \sum_{\mathfrak{p}} j_{\mathfrak{p}}(\omega, \omega')$, the summation being extended to all places of R. The following properties of the function $j(\omega, \omega')$ are obvious:

a) if $j(\omega, \omega')$ is defined, so is $j(\omega', \omega)$ and we have $j(\omega', \omega) = -j(\omega, \omega')$;

b) let ω' be a fixed differential of R; denote by P the set of poles of ω' and by P' the set of those poles of ω' at which the residue of ω' is $\neq 0$; then the differentials ω having no pole in the set P' and whose residues at all places of the set P are 0 form a vector space over the field of complex numbers; the function $\omega \to j(\omega, \omega')$ is defined and linear on this space;

c) if $f \in R$ is such that $j(df, \omega')$ is defined, then we have

$$j(df, \omega') = -\sum_{\mathfrak{p}} f(\mathfrak{p})\, \mathrm{res}_{\mathfrak{p}}\, \omega',$$

the sum being extended to all places $\mathfrak{p}$ which are not poles of f (this follows immediately from the fact that $\sum_{\mathfrak{p}} \mathrm{res}_{\mathfrak{p}}\, f\omega' = 0$; cf. Theorem 3, III, §5).

In what follows we shall denote by:

$\mathfrak{E}$ the space of differentials of the second kind of R;

$\mathfrak{F}$ the space of exact differentials of R;

P and Q two mutually disjoint finite sets of places of R;

$\mathfrak{E}_{P,Q}$ the set of differentials of R which have no pole in the set Q and whose residues at all places not in P are 0;

$\mathfrak{F}_Q$ the set of differentials of elements of R which admit the places of Q as zeros.

Then it is clear that $j(\omega, \omega')$ is defined whenever $\omega \in \mathfrak{E}_{P,Q}$, $\omega' \in \mathfrak{E}_{Q,P}$; we propose to study the properties of this function.

Consider first the case when P and Q are empty, whence $\mathfrak{E}_{P,Q} = \mathfrak{E}_{Q,P} = \mathfrak{E}$, $\mathfrak{F}_P = \mathfrak{F}_Q = \mathfrak{F}$. If $\omega \in \mathfrak{E}$, it is clear that $j(\omega, \omega') = 0$ whenever $\omega' \in \mathfrak{F}$. We shall prove that, conversely, if ω' is an element of $\mathfrak{E}$ such that $j(\omega, \omega') = 0$ for all $\omega \in \mathfrak{E}$, then $\omega' \in \mathfrak{F}$. Let g be the genus of R. Then we can find g distinct places $\mathfrak{p}_1, \cdots, \mathfrak{p}_g$ of R such that the divisor $\mathfrak{a} = \mathfrak{p}_1 \cdots \mathfrak{p}_g$ is non special (Lemma 4, VI, §8). Making use of Lemma 3, VI, §8, we see that ω' is congruent modulo $\mathfrak{F}$ to a differential which is multiple of $\mathfrak{a}^{-2}$; we may therefore assume without loss of generality that ω' is itself a multiple of $\mathfrak{a}^{-2}$. We shall then prove that $\omega' = 0$. Let ω be any differential of the second kind which is multiple of $\mathfrak{a}^{-2}$. Using the notation of Theorem 16, VI, §8, we denote by $a_i(\omega)$ and $b_i(\omega)$ the numbers such that $\nu_{\mathfrak{p}_i}(\omega - (a_i(\omega)x_i^{-2} + b_i(\omega))dx_i) > 0$ $(1 \leqq i \leqq g)$. Then we have

$$\nu_{\mathfrak{p}_i}(\omega + d(a_i(\omega)x_i^{-1} - b_i(\omega)x_i)) \geqq 1 \geqq -\nu_{\mathfrak{p}_i}(\omega') - 1,$$

from which it follows that $j_{\mathfrak{p}_i}(\omega, \omega')$ is equal to $\text{res}_{\mathfrak{p}_i} (-a_i(\omega)x_i^{-1} + b_i(\omega)x_i)\omega' = b_i(\omega)a_i(\omega') - b_i(\omega')a_i(\omega)$. Thus we have, for any $\omega \in \mathfrak{E}$ which is multiple of $\mathfrak{a}^{-2}$, $0 = j(\omega, \omega') = \sum_{i=1}^{g} (b_i(\omega)a_i(\omega') - b_i(\omega')a_i(\omega))$. Making use of Theorem 16, VI, §8, it follows immediately that $a_i(\omega') = b_i(\omega') = 0$, whence $\omega' = 0$.

Consider now the general case. If $\omega \in \mathfrak{E}_{P,Q}$ and if f is an element of R which has no pole on P, then we have $j(\omega, df) = \sum_{\mathfrak{p} \in P} f(\mathfrak{p}) \text{ res}_{\mathfrak{p}} \omega$; this is 0 if $df \in \mathfrak{F}_P$. Let conversely ω' be an element of $\mathfrak{E}_{Q,P}$ such that $j(\omega, \omega') = 0$ for all $\omega \in \mathfrak{E}_{P,Q}$. Take first $\omega = dg$, where g is an element of R which has no pole in Q. Then $0 = j(dg, \omega') = -\sum_{\mathfrak{q} \in Q} g(\mathfrak{q}) \text{ res}_{\mathfrak{q}} \omega'$. But we know that there exists an element g of R which takes arbitrarily preassigned values at all places of Q; it follows that $\text{res}_{\mathfrak{q}} \omega' = 0$ for all $\mathfrak{q} \in Q$, whence $\omega' \in \mathfrak{E}$. If ω is any differential of the second kind, then there exists an $h \in R$ such that $\omega - dh$ has no pole in Q (this follows for instance easily from Lemma 3, VI, §8 and Lemma 4, VI, §8). The quantity $j(\omega, \omega')$ is defined and equal to $j(dh, \omega') + j(\omega - dh, \omega') = 0$ (the first term is 0 because $dh \in \mathfrak{F}$ and $\omega' \in \mathfrak{E}$, the second because $\omega - dh \in \mathfrak{E}_{P,Q}$). Making use of what we have proved above, we see that $\omega' \in \mathfrak{F}$, whence $\omega' = df$, where f is an element of R which has no pole on P. If $\omega \in \mathfrak{E}_{P,Q}$, we have $0 = j(\omega, \omega') = \sum_{\mathfrak{p} \in P} f(\mathfrak{p}) \text{ res}_{\mathfrak{p}} \omega$. Making use of Theorem 3, III, §4, we see that we have

$$\sum_{\mathfrak{p} \in P} f(\mathfrak{p})\rho(\mathfrak{p}) = 0$$

whenever the numbers $\rho(\mathfrak{p})(\mathfrak{p} \in P)$ are such that $\sum_{\mathfrak{p} \in P} \rho(\mathfrak{p}) = 0$. It follows that the numbers $f(\mathfrak{p})$ are all equal to each other; if f_0 is their common value, we have $\omega' = d(f - f_0) \in \mathfrak{F}_P$.

The function j defines in a natural way a bilinear function on the product

$(\mathfrak{E}_{P,Q}/\mathfrak{F}_Q) \times (\mathfrak{E}_{Q,P}/\mathfrak{F}_P)$, and it follows from what we have just proved that this bilinear function is non degenerate. The space $\mathfrak{E}_{P,Q}/\mathfrak{F}_Q$ is of finite dimension in virtue of Theorem 15, VI, §8. Making use of the properties of non degenerate bilinear forms, we obtain the following result:

THEOREM 8. *Let P and Q be mutually disjoint finite subsets of the Riemann surface of the field R. Let λ be a linear function on the space of differentials of R which have no pole in the set Q and whose residues at all places not in P are 0; assume that $\lambda(df) = 0$ for every $f \in R$ which admits the places of Q as zeros. Then there exists a differential ω' of R which has no zero in the set P and whose residues at all places outside Q are 0 which is such that $\lambda(\omega) = j(\omega, \omega')$ for all ω of the domain of definition of λ, and the coset of ω' modulo the space of differentials of elements of R having the places of P as zeros is uniquely determined.*

§6. DEFINITION OF THE INTERSECTION NUMBERS

Let P and Q be mutually disjoint finite subsets of the Riemann surface S of the field R, and let c be any element of $H_1(S - P, Q)$. Define the space $\mathfrak{E}_{P,Q}$ as in §5; then $\int_c \omega$ is defined for every $\omega \in \mathfrak{E}_{P,Q}$ and the mapping $\omega \to \int_c \omega$ is a linear function on $\mathfrak{E}_{P,Q}$. Set $\partial c = \sum_{\mathfrak{q} \in Q} a(\mathfrak{q})\mathfrak{q}$; then, if the element $f \in R$ has no pole on Q, we have $\int_c df = \sum_{\mathfrak{q} \in Q} a(\mathfrak{q})f(\mathfrak{q})$ (Lemma 2, §4); this quantity is 0 if f admits the places of Q as zeros. Making use of Theorem 8, §5, we see that there exists a differential $\omega_c \in \mathfrak{E}_{Q,P}$ such that

$$\int_c \omega = j(\omega_c, \omega)$$

for all $\omega \in \mathfrak{E}_{P,Q}$. Furthermore, the space $\mathfrak{F}_P$ being defined as in §5, the coset of ω_c modulo $\mathfrak{F}_P$ is uniquely determined; we shall denote this coset by $\Omega(c)$.

Let generally Ω be any coset of the space of differentials modulo $\mathfrak{F}_P$. Then it is clear that, for any place $\mathfrak{p}$, the differentials belonging to Ω all have the same residue at $\mathfrak{p}$; this number will be called the *residue of* Ω *at* $\mathfrak{p}$ and denoted by $\text{res}_\mathfrak{p}\, \Omega$. We shall determine the residues of the coset $\Omega(c)$ introduced above. Since $\omega_c \in \mathfrak{E}_{Q,P}$, the residues of $\Omega(c)$ at all places not in Q are 0. If an element $f \in R$ has no pole on the set Q, then we have

$$\sum_{\mathfrak{q} \in Q} a(\mathfrak{q})f(\mathfrak{q}) = \int_c df = j(\omega_c, df) = \sum_{\mathfrak{q} \in Q} (\text{res}_\mathfrak{q}\, \omega_c)f(\mathfrak{q}).$$

Since we can find an element $f \in R$ taking arbitrarily preassigned values at the places of Q, we see that $\text{res}_\mathfrak{q}\, \Omega(c) = a(\mathfrak{q})$. Thus we have proved

THEOREM 9. *Let P and Q be mutually disjoint finite sets of places of the field R, and let c be an element of $H_1(S - P, Q)$. Then there exists a uniquely determined coset $\Omega(c)$ of the space of differentials having no pole on P and whose residues at all places not in Q are 0 modulo the space of differentials of elements of R having the elements of P as zeros such that the formula $\int_c \omega = j(\omega_c, \omega)$ holds for any $\omega_c \in \Omega(c)$ and for any differential ω having no pole on Q and whose residues at all places not* *in P are 0. If $\mathfrak{q} \in Q$, the residue at $\mathfrak{q}$ of a differential of the class $\Omega(c)$ is equal to the coefficient of $\mathfrak{q}$ in ∂c.*

It follows immediately that the logarithmic periods of any differential of the class $\Omega(c)$ are integral multiples of $2\pi(-1)^{1/2}$.

The mapping $c \rightarrow \Omega(c)$ is clearly a homomorphism of $H_1(S - P, Q)$ into the additive group of $\mathfrak{E}_{Q,P}/\mathfrak{F}_P$. Furthermore, we see easily that the following is true: let P_1 and Q_1 be mutually disjoint finite subsets of S such that $Q \subset Q_1$ and $P \supset P_1$, and let c_1 be the element of $H_1(S - P_1, Q_1)$ which corresponds to an element $c \in H_1(S - P, Q)$ by the identity mapping of the pair $(S - P, Q)$ into $(S - P_1, Q_1)$; then $\mathfrak{F}_P \subset \mathfrak{F}_{P_1}$ and $\Omega(c_1)$ is the coset modulo $\mathfrak{F}_{P_1}$ which contains $\Omega(c)$.

If P is any finite subset of S and γ a 1-chain on $S - P$, we shall set

$$\Omega(\gamma, P) = \Omega(c)$$

where c is the element of $H_1(S - P, |\partial\gamma|)$ which is represented by γ. Then, for a fixed P, the mapping $\gamma \rightarrow \Omega(\gamma, P)$ is easily seen to be a homomorphism of the group of 1-chains on $S - P$ into $\mathfrak{D}/\mathfrak{F}_P$, where $\mathfrak{D}$ is the space of all differentials. Moreover, we have Ω (subd $\gamma, P) = \Omega(\gamma, P)$.

If Ω is any coset of differentials modulo $\mathfrak{F}_P$ and if the elements of Ω belong to $\mathfrak{E}_{Q,P}$ (where Q is a finite set with no element in common with P), then $\int_{c'} \omega$ is defined for every $c' \in H_1(S - Q, P)$ and for every $\omega \in \Omega$, and the value of this integral depends only on c' and Ω; for, if f is an element of R which admits the places of P as zeros, then $\int_{c'} df = 0$ by Lemma 2, §4. The number $\int_{c'} \omega$ is called the *integral of Ω on c'* and is denoted by $\int_{c'} \Omega$.

Now, let c and c' be elements of $H_1(S - P, Q)$ and $H_1(S - Q, P)$ respectively. Then $\int_c \Omega(c')$ is defined. The number

$$i(c, c') = \frac{1}{2\pi(-1)^{1/2}} \int_c \Omega(c')$$

is called the *intersection number of c and c'*.

THEOREM 10. *Let P and Q be mutually disjoint finite subsets of the Riemann surface S of the field R, and let c and c' be elements of $H_1(S - P, Q)$ and $H_1(S - Q, P)$ respectively. Then the intersection number $i(c, c')$ is an integer, and we have $i(c', c) = -i(c, c')$. If c and c' can be represented by 1-chains γ and γ' respectively such that $|\gamma| \cap |\gamma'| = \emptyset$, then $i(c, c') = 0$.*

Let ω and ω' be differentials belonging to the cosets $\Omega(c)$ and $\Omega(c')$ respectively. Then we have $\int_c \Omega(c') = j(\omega, \omega')$, whence $i(c, c') = (2\pi(-1)^{1/2})^{-1} j(\omega, \omega')$. This proves the formula $i(c, c') = -i(c', c)$.

If γ and γ' are 1-chains such that

$$(1) \qquad |\gamma| \cap |\partial\gamma'| = \emptyset, \qquad |\gamma'| \cap |\partial\gamma| = \emptyset,$$

then we denote by $i(\gamma, \gamma')$ the number $i(c, c')$, where c is the homology class of γ in $H_1(S - |\partial\gamma'|, |\partial\gamma|)$ while c' is the homology class of γ' in $H_1(S - |\partial\gamma|,$

$|\partial\gamma'|$). Then, clearly, $i(\gamma', \gamma) = -i(\gamma, \gamma')$ and $i(\text{subd}\,\gamma, \gamma') = i(\gamma, \gamma')$. Moreover, if γ' is fixed, it is easily seen that the mapping $\gamma \to i(\gamma, \gamma')$ is a homomorphism of the additive group of 1-chains satisfying the conditions (1) into the additive group of complex numbers.

We shall first consider the following special case: γ and γ' are 1-simplices which lie on open discs V and V' respectively such that $V \cap V' = \emptyset$; under these conditions, we shall prove that $i(\gamma, \gamma') = 0$. We may of course assume that $S - (V \cup V')$ contains infinitely many points. We set $\partial\gamma = \mathfrak{b} - \mathfrak{a}$, $\partial\gamma' = \mathfrak{b}' - \mathfrak{a}'$ (with $\mathfrak{a}$, $\mathfrak{b}$ in V, $\mathfrak{a}'$, $\mathfrak{b}'$ in V'). We can find elements x and x' of R such that V is an x-disc, V' an x'-disc, and $x(\mathfrak{a}) = x'(\mathfrak{a}') = 0$. If $\xi \in x(V)$, we denote by $\mathfrak{p}(\xi)$ the point of V which is mapped on ξ by x, and we construct a 1-simplex γ_ξ on V such that $\partial\gamma_\xi = \mathfrak{p}(\xi) - \mathfrak{a}$. Since $H_1(V) = 0$ and $\mathfrak{a}'$ and $\mathfrak{b}'$ are not in V, it is clear that all simplices γ_ξ satisfying our conditions represent the same element of $H_1(S - (\{\mathfrak{a}'\} \cup \{\mathfrak{b}'\}, \{\mathfrak{a}\} \cup \{\mathfrak{p}(\xi)\})$ and therefore that the differential class $\Omega(\gamma_\xi, \{\mathfrak{a}'\} \cup \{\mathfrak{b}'\})$ depends only on ξ; we shall denote this class by Ω_ξ. Then $i(\gamma_\xi, \gamma') = -i(\gamma', \gamma_\xi)$ is equal to the product by $-(2\pi(-1)^{1/2})^{-1}$ of the integral on γ' of any differential of the class Ω_ξ.

If g is the genus of R, we can find g distinct points $\mathfrak{m}_1, \cdots, \mathfrak{m}_g$ of $S - (V \cup V')$ such that the divisor $\mathfrak{v}_0 = \mathfrak{m}_1 \cdots \mathfrak{m}_g$ is non special (Lemma 4, VI, §8). By the theorem of Riemann-Roch, we have $l(\mathfrak{v}_0^{-1}) = g - g + 1 + \delta(\mathfrak{v}_0) = 1$ (since $\mathfrak{v}_0$ is non special), and $l(\mathfrak{a}'\mathfrak{b}'\mathfrak{v}_0^{-1}) = g - 2 - g + 1 + \delta(\mathfrak{a}'^{-1}\mathfrak{b}'^{-1}\mathfrak{v}_0) = -1 + \delta(\mathfrak{a}'^{-1}\mathfrak{b}'^{-1}\mathfrak{v}_0)$. The first formula shows that the only multiples of $\mathfrak{v}_0^{-1}$ in R are the constants, whence $l(\mathfrak{a}'\mathfrak{b}'\mathfrak{v}_0^{-1}) = 0$, and the second formula then shows that $\delta(\mathfrak{a}'^{-1}\mathfrak{b}'^{-1}\mathfrak{v}_0) = 1$. Let ζ_0 be a basic element of the space of differentials which are $\equiv 0 \pmod{\mathfrak{a}'^{-1}\mathfrak{b}'^{-1}\mathfrak{v}_0}$, and let $\mathfrak{m}_0$ be a place contained in $S - (V \cup V')$ and which is not a zero of ζ_0. We set $\mathfrak{v} = \mathfrak{m}_0\mathfrak{v}_0$; then we assert that there exists in Ω_ξ a differential ω_ξ which is $\equiv 0 \pmod{\mathfrak{v}^{-2}\mathfrak{a}^{-1}(\mathfrak{p}(\xi))^{-1}}$, and that this differential is uniquely determined if $\xi \neq 0$. Let ω_ξ^* be any differential of the class Ω_ξ. For each place $\mathfrak{p}$ of R, we can find an element $f_\mathfrak{p} \in R$ such that $\nu_\mathfrak{p}(\omega_\xi^* - df_\mathfrak{p}) \geqq -1$ (Theorem 14, VI, §8). Let us try to find an $f \in R$ which satisfies the following conditions: if $\mathfrak{p}$ is different from $\mathfrak{a}'$, $\mathfrak{b}'$, $\mathfrak{m}_0, \mathfrak{m}_1, \cdots, \mathfrak{m}_g$, then $\nu_\mathfrak{p}(f - f_\mathfrak{p}) \geqq 0$; the numbers $\nu_{\mathfrak{a}'}(f)$, $\nu_{\mathfrak{b}'}(f)$ are > 0; we have $\nu_{\mathfrak{m}_j}(f - f_{\mathfrak{m}_j}) \geqq -1$ $(0 \leqq j \leqq g)$. Making use of Theorem 2, II, §5, we see that this is certainly possible provided there exists no differential of R which is $\equiv 0 \pmod{\mathfrak{a}'^{-1}\mathfrak{b}'^{-1}\mathfrak{v}}$. Such a differential would then be a multiple of $\mathfrak{a}'^{-1}\mathfrak{b}'^{-1}\mathfrak{v}_0$ and would therefore be a constant scalar multiple of ζ_0; it would also have $\mathfrak{m}_0$ as a zero, which is impossible since $\mathfrak{m}_0$ is not a zero of ζ_0. The existence of an element $f \in R$ satisfying our conditions is therefore established. The points $\mathfrak{a}'$ and $\mathfrak{b}'$ being zeros of f, $\omega_\xi^* - df = \omega_\xi$ is in the class Ω_ξ. Since neither $\mathfrak{a}'$ nor $\mathfrak{b}'$ is a pole of ω_ξ^*, we have $\nu_\mathfrak{p}(\omega_\xi) \geqq -1$ whenever $\mathfrak{p}$ is a place distinct from $\mathfrak{m}_0, \mathfrak{m}_1, \cdots, \mathfrak{m}_g$, while $\nu_{\mathfrak{m}_j}(\omega_\xi) \geqq -2$; if $\mathfrak{p}$ is distinct from $\mathfrak{a}$, $\mathfrak{p}(\xi)$, $\mathfrak{m}_0, \cdots, \mathfrak{m}_g$, then the residue of ω_ξ at $\mathfrak{p}$ is 0 (because $\omega_\xi \in \Omega_\xi$), and the inequality $\nu_\mathfrak{p}(\omega_\xi) \geqq -1$ implies $\nu_\mathfrak{p}(\omega_\xi) \geqq 0$. This proves that ω_ξ is multiple of $\mathfrak{v}^{-2}\mathfrak{a}^{-1}(\mathfrak{p}(\xi))^{-1}$. If ω_ξ' is any differential of the class Ω_ξ which is multiple of $\mathfrak{v}^{-2}\mathfrak{a}^{-1}(\mathfrak{p}(\xi))^{-1}$, then $\omega_\xi' - \omega_\xi$ is of the form $d\varphi$, where φ is an element of R which admits $\mathfrak{a}'$ and $\mathfrak{b}'$ as zeros.

If $\xi \neq 0$, i.e., $\mathfrak{p}(\xi) \neq \mathfrak{a}$, then $\mathfrak{a}$ and $\mathfrak{p}(\xi)$, which are poles of order $\leqq 1$ of $\omega_\xi' - \omega_\xi$, cannot be poles of $d\varphi$, which shows that φ is multiple of $\mathfrak{a}'\mathfrak{b}'\mathfrak{v}^{-1}$ (this argument assumes that $\mathfrak{a}' \neq \mathfrak{b}'$; if we had $\mathfrak{a}' = \mathfrak{b}'$, then γ' would be a cycle on V', and therefore homologous to 0 on V', and the equality $i(\gamma, \gamma') = 0$ would be obvious). We have $l(\mathfrak{a}'\mathfrak{b}'\mathfrak{v}^{-1}) = g - 1 - g + 1 + \delta(\mathfrak{a}'^{-1}\mathfrak{b}'^{-1}\mathfrak{v}) = 0$, whence $\varphi = 0$, and our assertion is established. The class Ω_0 obviously contains the zero differential; we shall set $\omega_0 = 0$.

Now, we propose to compute the value of the element ω_ξ/dx' of R at a point $\mathfrak{q}' \in V'$ distinct from $\mathfrak{a}'$ and $\mathfrak{b}'$. It is clear that $l(\mathfrak{q}'\mathfrak{a}'\mathfrak{b}'\mathfrak{v}^{-1}) = l(\mathfrak{q}'^2\mathfrak{a}'\mathfrak{b}'\mathfrak{v}^{-1}) = 0$, whence, by the Riemann-Roch theorem, $\delta(\mathfrak{q}'^{-1}\mathfrak{a}'^{-1}\mathfrak{b}'^{-1}\mathfrak{v}) = 1$ and $\delta(\mathfrak{q}'^{-2}\mathfrak{a}'^{-1}\mathfrak{b}'^{-1}\mathfrak{v}) = 2$. Let $\zeta_{\mathfrak{q}'}$ be a differential $\neq 0$ which is multiple of $\mathfrak{q}'^{-1}\mathfrak{a}'^{-1}\mathfrak{b}'^{-1}\mathfrak{v}$; since

$$\delta(\mathfrak{a}'^{-1}\mathfrak{b}'^{-1}\mathfrak{v}) = 0,$$

$\mathfrak{q}'$ is a pole of order 1 of $\zeta_{\mathfrak{q}'}$, whence $\mathrm{res}_{\mathfrak{q}'}\zeta_{\mathfrak{q}'} \neq 0$; multiplying $\zeta_{\mathfrak{q}'}$ by a suitable constant, we may assume that $\mathrm{res}_{\mathfrak{q}'}\, \zeta_{\mathfrak{q}'} = 1$. Take a differential which is multiple of $\mathfrak{q}'^{-2}\mathfrak{a}'^{-1}\mathfrak{b}'^{-1}\mathfrak{v}$ and which is linearly independent of $\zeta_{\mathfrak{q}'}$; subtracting from this differential a suitable constant multiple of $\zeta_{\mathfrak{q}'}$, we obtain a differential $\eta_{\mathfrak{q}'}$ which is multiple of $\mathfrak{q}'^{-2}\mathfrak{a}'^{-1}\mathfrak{b}'^{-1}\mathfrak{v}$ and which admits $\mathfrak{q}'$ as a pole of order 2 and residue 0; multiplying this differential by a suitable constant, we may assume that $\nu_{\mathfrak{q}'}(\eta_{\mathfrak{q}'} - (x' - x'(\mathfrak{q}'))^{-2}\, dx') \geqq 0$. It follows immediately from the definition of the class Ω_ξ that we have

$$j(\omega_\xi, \eta_{\mathfrak{q}'}) = \int_{\gamma_\xi} \eta_{\mathfrak{q}'} .$$

Now we prove

LEMMA 1. *Let ω and ω' be differentials of R and $\mathfrak{m}$ a place of R. If $\nu_\mathfrak{m}(\omega) + \nu_\mathfrak{m}(\omega') \geqq -1$, then the number $j_\mathfrak{m}(\omega, \omega')$ is defined and equal to* 0.

It is clear that $\mathfrak{m}$ cannot be a pole of both ω and ω', which proves that $j_\mathfrak{m}(\omega, \omega')$ is defined. Since $j_\mathfrak{m}(\omega, \omega') = -j_\mathfrak{m}(\omega', \omega)$, we may assume that $\nu_\mathfrak{m}(\omega) \geqq 0$. Let then y be an element of R such that $\nu_\mathfrak{m}(\omega - dy) \geqq -\nu_\mathfrak{m}(\omega') - 1$ and $\nu_\mathfrak{m}(y) > 0$; then we have $\nu_\mathfrak{m}(y) = \nu_\mathfrak{m}(dy) + 1$ (Lemma 1, VI, §8) and $\nu_\mathfrak{m}(dy) \geqq \min\{\nu_\mathfrak{m}(\omega), \nu_\mathfrak{m}(\omega - dy)\} \geqq -\nu_\mathfrak{m}(\omega') - 1$, whence $\nu_\mathfrak{m}(y\omega') \geqq 0$ and $j_\mathfrak{m}(\omega, \omega') = \mathrm{res}_\mathfrak{m}\, y\omega' = 0$.

It follows from Lemma 1 that $\mathfrak{q}'$ is the only place $\mathfrak{p}$ for which $j_\mathfrak{p}(\omega_\xi, \eta_{\mathfrak{q}'})$ may be $\neq 0$. Since $\nu_{\mathfrak{q}'}(\eta_{\mathfrak{q}'} + d((x' - x'(\mathfrak{q}'))^{-1}) \geqq 0$ and $j(\omega_\xi, \eta_{\mathfrak{q}'}) = -j(\eta_{\mathfrak{q}'}, \omega_\xi)$, we have

$$j(\omega_\xi, \eta_{\mathfrak{q}'}) = \mathrm{res}_{\mathfrak{q}'} \frac{\omega_\xi}{x' - x'(\mathfrak{q}')} = (\omega_\xi/dx')(\mathfrak{q}')$$

by the corollary to Theorem 6, VI, §3. Therefore, we have

$$(\omega_\xi/dx')(\mathfrak{q}') = \int_{\gamma_\xi} \eta_{\mathfrak{q}'} .$$

Leaving $\mathfrak{q}'$ fixed, this formula shows that $(\omega_\xi/dx')(\mathfrak{q}')$, considered as a function of $\mathfrak{p}(\xi)$, is a primitive of $\eta_{\mathfrak{q}'}$ in V; thus we see that, for any fixed $\mathfrak{q}'$, $(\omega_\xi/dx')(\mathfrak{q}')$ is a holomorphic function of ξ in $x(V)$ whose derivative is

$$\frac{d}{d\xi}\,((\omega_\xi/dx')(\mathfrak{q}')) = (\eta_{\mathfrak{q}'}/dx)(\mathfrak{p}(\xi)).$$

We shall now prove that $(\omega_\xi/dx')(\mathfrak{q}')$ may be expressed as a holomorphic function of the pair $(\xi, x'(\mathfrak{q}')) \in x(V) \times x'(V')$. To do this, we observe that $v_\xi = \omega_\xi/dx'$ is an element of R whose divisor of poles is of bounded degree; it follows that the number of points $\mathfrak{q}' \in V'$ at which v_ξ takes any given value is bounded independently of ξ. If $\mathfrak{q}'$ is any point of V', $v_\xi(\mathfrak{q}')$, which is a holomorphic function in $x(V)$, remains bounded when ξ remains in any closed disc $D \subset x(V)$. Making use of a theorem of P. Montel (*Leçons sur les familles normales de functions analytiques*, Paris, Gauthier-Villars, 1927, pp. 67 and 70), we conclude that the functions v_ξ, $\xi \in D$, form a "normal family", i.e., a relatively compact set of functions in the topology of the uniform convergence on every compact subset of V'. It follows that, if $\xi, \xi_1, \cdots, \xi_n, \cdots$ are points of $x(V)$ such that $\xi = \lim_{n\to\infty} \xi_n$, then v_{ξ_n} converges to v_ξ uniformly on every compact subset of V'. This means that $v_\xi(\mathfrak{q}')$ is a continuous function of the pair

$$(\xi, x'(\mathfrak{q}')) \in x(V) \times x'(V').$$

This function, being a holomorphic function of either one of its arguments when the other is fixed, is a holomorphic function of the pair $(\xi, x'(\mathfrak{q}'))$. Write $v_\xi(\mathfrak{q}') = \theta(\xi, x'(\mathfrak{q}'))$, and let $\Theta(\xi, x')$ be the holomorphic function of ξ and x' defined by the conditions

$$\partial\Theta/\partial x'(\xi, x') = \theta(\xi, x'), \qquad \Theta(\xi, 0) = 0.$$

Then it is clear that

$$\int_{\gamma'} \omega_\xi = \Theta(\xi, x'(\mathfrak{b}')).$$

This is the number which we want to prove to be 0. Now, $\partial\Theta/\partial\xi$ is, for ξ fixed, a primitive (with respect to x') of $\partial\theta/\partial\xi(\xi, x')$, whence

$$\partial^2\Theta/\partial\xi\partial x'(\xi, x') = (\eta_{\mathfrak{q}'}/dx)_{\mathfrak{p}(\xi)}.$$

We shall give another expression of this number. The degree of the divisor $(\mathfrak{p}(\xi))^{-1}\mathfrak{a}'\mathfrak{b}'\mathfrak{d}^{-1}$ is $-g$; it follows immediately from the theorem of Riemann-Roch that there exists a $u_\xi \neq 0$ in R which is multiple of this divisor. Since $l(\mathfrak{a}'\mathfrak{b}'\mathfrak{d}^{-1}) = 0$, u_ξ admits $\mathfrak{p}(\xi)$ as a pole of order 1; multiplying u_ξ by a suitable constant, we may assume that $(x - \xi)u_\xi$ takes the value 1 at $\mathfrak{p}(\xi)$. The sum of the residues of $u_\xi\eta_{\mathfrak{q}'}$ is 0 (Theorem 3, III, §5); we see immediately that the only residues of $u_\xi\eta_{\mathfrak{q}'}$ which may be $\neq 0$ are those at $\mathfrak{p}(\xi)$ and $\mathfrak{q}'$. We have

$$\operatorname{res}_{\mathfrak{p}(\xi)} u_\xi\eta_{\mathfrak{q}'} = \operatorname{res}_{\mathfrak{p}(\xi)} (x - \xi)^{-1}\eta_{\mathfrak{q}'} = (\eta_{\mathfrak{q}'}/dx)(\mathfrak{p}(\xi)).$$

Remembering that $\nu_{\mathfrak{q}'}(\eta_{\mathfrak{q}'} - (x' - x'(\mathfrak{q}'))^{-2}dx') \geqq 0$, we see easily that

$$\mathrm{res}_{\mathfrak{q}'}\, u_\xi \eta_{\mathfrak{q}'} = (du_\xi/dx')(\mathfrak{q}').$$

Thus we obtain the formula

$$\frac{\partial^2 \Theta}{\partial \xi\, \partial x'}(\xi, x'(\mathfrak{q}')) = -(du_\xi/dx')(\mathfrak{q}').$$

This formula has been established for the case where $\mathfrak{q}'$ is different from $\mathfrak{a}'$ and $\mathfrak{b}'$; but it is also true when $\mathfrak{q}'$ is $\mathfrak{a}'$ or $\mathfrak{b}'$ because both sides are continuous functions of $\mathfrak{q}'$. Now, we have $0 = u_\xi(\mathfrak{b}') - u_\xi(\mathfrak{a}') = \int_{\gamma'}(du_\xi/dx')\, dx'$ whence

$$\frac{\partial}{\partial \xi}(\Theta(\xi, x'(\mathfrak{b}')) - \Theta(\xi, 0)) = \frac{\partial}{\partial \xi}(\Theta(\xi, x'(\mathfrak{b}'))) = 0$$

and this proves that $\int_{\gamma'} \omega_\xi$ does not depend on ξ. Since $\omega_0 = 0$, we have $\int_{\gamma'} \omega_\xi = 0$ for all ξ, whence $i(\gamma, \gamma') = 0$.

Now, assume that γ and γ' are arbitrary 1-chains such that $|\gamma| \cap |\gamma'| = \emptyset$. Since $|\gamma|$ and $|\gamma'|$ are closed sets in the space S, which is compact and therefore normal, we can find open sets U and U' such that $|\gamma| \subset U$, $|\gamma'| \subset U'$, and $U \cap U' = \emptyset$. The sets U and U' can be covered by the open discs which are contained in them; it follows that we can find an integer $p \geqq 0$ which has the following properties: every simplex occurring in $\mathrm{subd}^p\, \gamma$ lies on some open disc contained in U, and every simplex occurring in $\mathrm{subd}^p\, \gamma'$ lies on some open disc contained in U'. If $\mathrm{subd}^p\, \gamma = \sum_{k=1}^{h} a_k \sigma_k$, the simplices σ_k $(1 \leqq k \leqq h)$ being those which occur in $\mathrm{subd}^p\, \gamma$, and similarly $\mathrm{subd}^p\, \gamma' = \sum_{k=1}^{h'} a'_k \sigma'_k$, then we have $i(\gamma, \gamma') = \sum_{k=1}^{h} \sum_{k'=1}^{h'} a_k a'_{k'} i(\sigma_k, \sigma'_{k'}) = 0$.

Finally, we have to prove that $i(\gamma, \gamma')$ is an integer whenever it is defined. Every point of $|\gamma|$ is then interior to some closed disc which does not meet $|\partial\gamma'|$; it follows that there exists an integer $p \geqq 0$ such that every simplex occurring in $\mathrm{subd}^p\, \gamma$ lies on the interior of some closed disc which does not meet $|\partial\gamma'|$. Set $\mathrm{subd}^p\, \gamma = \sum_{k=1}^{h} a_k \sigma_k$, where $\sigma_1, \cdots, \sigma_h$ are the simplices occurring in $\mathrm{subd}^p\, \gamma$. Then the set $|\partial\sigma_1| \cup \cdots \cup |\partial\sigma_h|$ is finite and does not meet $|\partial\gamma'|$; it follows that there exists a chain γ'_1 which is homologous to γ' on $S - |\partial\gamma|$ modulo $|\partial\gamma'|$ and which is on $S - \bigcup_{k=1}^{h} |\partial\sigma_k|$ (cf. Corollary to Lemma 4, §4). We have $i(\gamma, \gamma') = i(\gamma, \gamma'_1)$, $i(\mathrm{subd}^p\, \gamma, \gamma'_1) = \sum_{k=1}^{h} a_k i(\sigma_k, \gamma'_1)$. Thus we see that it is sufficient to prove that $i(\gamma, \gamma')$ is an integer in the case where $|\gamma|$ is contained in the interior of a closed disc D which does not meet $|\partial\gamma'|$. Let then ω be a differential of the class $\Omega(\gamma, |\partial\gamma'|)$. We have $i(\gamma, \gamma') = -(2\pi(-1)^{1/2})^{-1} \int_{c'} \omega$, where c' is the homology class of γ' in $H_1(S - |\partial\gamma|, |\partial\gamma'|)$. Let $\mathfrak{c}$ be an interior point of D; then c' contains a chain γ'_2 on $S - (|\partial\gamma| \cup \{\mathfrak{c}\})$ (Corollary to Lemma 4, §4). On the other hand, it is clear that, if V is the interior of D, then $S - V$ is a deformation retract of $S - \{\mathfrak{c}\}$. It follows that the homology class c'^* of γ'_2 in $H_1(S, |\partial\gamma'|)$ contains a chain γ'' on $S - V$. Let c'' be the homology class of γ'' in $H_1(S - |\partial\gamma|, |\partial\gamma'|)$. Then, since $|\gamma| \cap |\gamma''| = \emptyset$, we have $i(\gamma, \gamma'') = 0$, i.e., $\int_{c''} \omega = 0$. On the other hand, $c' - c''$ can be represented by a cycle $\gamma'_2 - \gamma''$

which is homologous to 0 on S. It follows that $\int_{c'-c''}\omega$ is a logarithmic period of ω. The residues of ω being integral numbers, in virtue of Theorem 9, we see that $\int_{c'}\omega = \int_{c'-c''}\omega$ is an integral multiple of $2\pi(-1)^{1/2}$. Theorem 10 is thereby proved.

§7. Geometric lemmas

Lemma 1. *Let D be a closed disc on the Riemann surface S of the field R, and let C be the circumference of D; let ζ be a 1-cycle on C whose homology class is the positive generator of $H_1(C)$. Let P be a finite subset of S which has no element on C, and let c be an element of $H_1(S, P)$. Denote by z the homology class of ζ in $H_1(S - P)$ and set $\partial c = \sum_{\mathfrak{p}\in P} a(\mathfrak{p})\mathfrak{p}$; then we have $i(z, c) = \sum_{\mathfrak{p}\in D\cap P} a(\mathfrak{p})$.*

Let γ be a 1-chain of the class c. Select a fixed point $\mathfrak{p}_0 \in D - C$, and, for each $\mathfrak{p} \in D \cap P$, let $\sigma(\mathfrak{p})$ be a 1-chain on $D - C$ such that $\partial\sigma(\mathfrak{p}) = \mathfrak{p}_0 - \mathfrak{p}$. Then we have, for $\mathfrak{p} \in D \cap P$, $i(\zeta, \sigma(\mathfrak{p})) = 0$, whence $i(\zeta, \gamma) = i(\zeta, \gamma + \sum_{\mathfrak{p}\in D\cap P} a(\mathfrak{p})\sigma(\mathfrak{p}))$. The only point of D occurring in $\partial(\gamma + \sum_{\mathfrak{p}\in D\cap P} a(\mathfrak{p})\sigma(\mathfrak{p}))$ is $\mathfrak{p}_0$, and $\mathfrak{p}_0$ occurs with the coefficient $\sum_{\mathfrak{p}\in D\cap P} a(\mathfrak{p})$. Let ω be a differential of the class $\Omega(\gamma + \sum_{\mathfrak{p}\in D\cap P} a(\mathfrak{p})\sigma(\mathfrak{p}))$ (cf. §6); then we have, by Theorem 9, §6, $\mathrm{res}_{\mathfrak{p}_0}\omega = \sum_{\mathfrak{p}\in D\cap P} a(\mathfrak{p})$ and $\mathrm{res}_{\mathfrak{q}}\,\omega = 0$ for all $\mathfrak{q} \neq \mathfrak{p}_0$ in D. It follows by Theorem 6, §4 that $\int_z \omega = 2\pi(-1)^{1/2} \sum_{\mathfrak{p}\in D\cap P} a(\mathfrak{p})$, and Lemma 1 follows directly from the definition of intersection numbers.

If $\mathfrak{a}$ and $\mathfrak{b}$ are distinct points of the circumference C of a disc D, then one and only one of the two components of $C - (\{\mathfrak{a}\} \cup \{\mathfrak{b}\})$ has the property that $\varepsilon(\mathfrak{a}, \mathfrak{c}, \mathfrak{b}) = 1$ for all points $\mathfrak{c}$ of this component (cf. end of §4); the adherence of this component is called the *positive arc* determined by $\mathfrak{a}$ and $\mathfrak{b}$ on C.

Lemma 2. *Let D be a closed disc on the Riemann surface S of the field R and let C be the circumference of D. Let γ be a 1-chain on D such that $|\partial\gamma| \subset C$; write $\partial\gamma = \sum_{\mathfrak{p}\in C} a(\mathfrak{p})\mathfrak{p}$. Let $\mathfrak{a}$ and $\mathfrak{b}$ be distinct points of C not on $|\gamma|$, and let Γ be the positive arc determined by $\mathfrak{a}$ and $\mathfrak{b}$ on C. Let λ be a 1-chain on D such that $\partial\lambda = \mathfrak{b} - \mathfrak{a}$, $|\lambda| \cap |\partial\gamma| = \emptyset$. Then we have $i(\gamma, \lambda) = \sum_{\mathfrak{p}\in\Gamma} a(\mathfrak{p})$.*

We can find an element x of R for which D is an x-disc. Submitting if necessary x to a homographic transformation, we may assume that $x(\mathfrak{a}) = -(-1)^{1/2}$, $x(\mathfrak{b}) = (-1)^{1/2}$. Let V be an open x-disc containing D; denote by U_1 the set of points $\mathfrak{p} \in V$ at which $x(\mathfrak{p})$ is not real $\geqq 0$ and by U_2 the set of points $\mathfrak{p} \in V$ at which $x(\mathfrak{p})$ is not real $\leqq 0$. Then it is easily seen that $x^{-1}\,dx$ admits a primitive f_i on U_i ($i = 1, 2$) such that $f_1(\mathfrak{a}) = (3/2)\pi(-1)^{1/2}$, $f_1(\mathfrak{b}) = (1/2)\pi(-1)^{1/2}$, $f_2(\mathfrak{a}) = -(1/2)\pi(-1)^{1/2}$, $f_2(\mathfrak{b}) = (1/2)\pi(-1)^{1/2}$. We shall see that $\Gamma \subset U_2$. Let $\mathfrak{c}$ be the point of C at which x takes the value 1. Let σ_1 be a 1-simplex on $C \cap U_2$ such that $\partial\sigma_1 = \mathfrak{c} - \mathfrak{a}$, σ_2 a 1-simplex on $C \cap U_2$ such that $\partial\sigma_2 = \mathfrak{b} - \mathfrak{c}$, and σ_3 a 1-simplex on $C \cap U_1$ such that $\partial\sigma_3 = \mathfrak{a} - \mathfrak{b}$. Then

$$\int_{\sigma_1+\sigma_2+\sigma_3} x^{-1}\,dx = (-1)^{1/2}\,(\pi/2 + \pi/2 + \pi) = 2\pi(-1)^{1/2},$$

whence $\varepsilon(\mathfrak{a}, \mathfrak{c}, \mathfrak{b}) = 1$ and $\Gamma \subset U_2$.

It is easily seen that there exists a real number α between 0 and 1 which has the following property: those points $\mathfrak{q} \in U$ for which $|x(\mathfrak{q}) - \alpha| \leqq |(-1)^{1/2} - \alpha|$ form a disc D' and $D' \cap C = \Gamma$. Then, if C' is the circumference of D', $\mathfrak{a}$ and $\mathfrak{b}$ belong to C' and $D \cap C'$ is one of the arcs determined by $\mathfrak{a}$ and $\mathfrak{b}$ on C'; moreover, we have $D \cap C' \subset U_1$. Let λ' be a 1-simplex on $D \cap C'$ such that $\partial\lambda' = \mathfrak{b} - \mathfrak{a}$. Then we assert that $i(\gamma, \lambda) = i(\gamma, \lambda')$. For each $\mathfrak{p} \in |\partial\gamma|$ we can find a 1-simplex $\mu(\mathfrak{p})$ such that a) $\mu(\mathfrak{p})$ lies on V; b) $|\mu(\mathfrak{p})| \cap D = \{\mathfrak{p}\}$; c) $\partial\mu(\mathfrak{p}) = \mathfrak{p}_1 - \mathfrak{p}$, where $\mathfrak{p}_1$ is a point of $V - D$. Then neither $|\lambda|$ nor $|\lambda'|$ has any point in common with any of the sets $|\mu(\mathfrak{p})|$, whence $i(\gamma, \lambda) = i(\gamma + \sum_{\mathfrak{p}\in|\partial\gamma|} a(\mathfrak{p})\,\mu(\mathfrak{p}), \lambda)$ and $i(\gamma, \lambda') = i(\gamma + \sum_{\mathfrak{p}\in|\partial\gamma|} a(\mathfrak{p})\,\mu(\mathfrak{p}), \lambda')$. The set $|\partial(\gamma + \sum_{\mathfrak{p}\in|\partial\gamma|} a(\mathfrak{p})\,\mu(\mathfrak{p}))|$ has no point in common with D. The chain $\lambda' - \lambda$ is a cycle on D, and therefore homologous to 0 on D, since $H_1(D) = 0$. It follows that $i(\gamma + \sum_{\mathfrak{p}\in|\partial\gamma|} a(\mathfrak{p})\,\mu(\mathfrak{p}), \lambda' - \lambda) = 0$, which proves our assertion. We can find a 1-simplex λ'' on C' such that $\partial\lambda'' = \mathfrak{a} - \mathfrak{b}$, $|\lambda''| \cap D = \{\mathfrak{a}\} \cup \{\mathfrak{b}\}$. Then $|\lambda''|$ has no point in common with $|\gamma|$, and $i(\gamma, \lambda') = i(-(\lambda' + \lambda''), \gamma)$. The chain $-(\lambda' + \lambda'')$ is a 1-cycle on C'; since $|\lambda'| \subset U_1$, $|\lambda''| \subset U_2$, we have $\int_{-\lambda'} x^{-1}\,dx = \pi(-1)^{1/2}$, $\int_{-\lambda''} x^{-1}\,dx = \pi(-1)^{1/2}$, whence $\int_{-(\lambda'+\lambda'')} x^{-1}\,dx = 2\pi(-1)^{1/2}$, which proves that $-(\lambda' + \lambda'')$ belongs to the positive generator of $H_1(C')$. The points of $|\partial\gamma|$ which are in D' are those which are in Γ; Lemma 2 follows therefore immediately from Lemma 1.

LEMMA 3. *Let D be a closed disc on the Riemann surface S of the field R and let E be a closed arcwise connected subset of R. Let V be an open subset of S containing D. Let n be an integer. Assume that a 1-cycle ζ on $S - E$ has the property that $i(\zeta, \lambda) \equiv 0 \pmod{n}$ for every 1-cycle λ on S. Then ζ may be written in the form $n\zeta_1 + \zeta_2 + \partial\tau$, where ζ_1 is a 1-cycle on S, ζ_2 a 1-cycle on $S - (D \cup E)$, and τ a 2-chain on $V \cup |\zeta|$. If φ is a 2-cycle on $S - E$, and if $D \cap E \neq \emptyset$, then φ is homologous on S to a 2-cycle on $S - (D \cup E)$.*

Let x be an element of R such that D is an x-disc, and let $\mathfrak{c}$ be the x-center of D. Let D' be a closed x-disc of x-center $\mathfrak{c}$, containing D in its interior and contained in V, and let C' be the circumference of D'. If $\mathfrak{q}_0 \in C'$, the set of points $\mathfrak{q} \in D'$ such that $x(\mathfrak{q}) = tx(\mathfrak{q}_0)$, t real, $0 \leqq t \leqq 1$, will be called the *radial set* of $\mathfrak{q}_0$.

We shall first prove the existence of a closed x-disc D'', of x-center $\mathfrak{c}$, with the following properties: D'' is in the interior of D' and D in the interior of D''; if F denotes the union of the radial sets of D' which contain points of the set $\overline{D' - D''} \cap E$ (where the bar means the adherence), then F has no point in common with $\overline{D' - D''} \cap (|\zeta| \cup |\varphi|)$. Assume for a moment that no such disc D'' exists. Let r' be the x-radius of D'. Then, for each integer $m > 0$, there exist points $\mathfrak{p}_m \in D' \cap E$ and $\mathfrak{q}_m \in D' \cap (|\zeta| \cup |\varphi|)$ which both lie on the same radial set of D' and which are such that $|x(\mathfrak{p}_m)| \geqq r' - m^{-1}$, $|x(\mathfrak{q}_m)| \geqq r' - m^{-1}$, whence $|x(\mathfrak{p}_m) - x(\mathfrak{q}_m)| \leqq m^{-1}$. Since $D' \cap E$ is compact, some subsequence of the sequence $(\mathfrak{p}_m)$ converges to a point $\mathfrak{p}_0 \in D' \cap E$. Then $\mathfrak{p}_0$ is also the limit of a subsequence of the sequence $(\mathfrak{q}_m)$, whence $\mathfrak{q}_0 \in |\zeta| \cup |\varphi|$, since $|\zeta| \cup |\varphi|$ is compact. But this is impossible since $|\zeta| \cup |\varphi| \subset S - E$.

The disc D'' having been selected so as to satisfy the prescribed conditions, it is clear that we can find an integer $p \geqq 0$ such that every simplex occurring in either one of the chains $\zeta^* = \text{subd}^p \zeta$ or $\varphi^* = \text{subd}^p \varphi$ and whose set of points meets D'' lies on D'. Let then γ_1 and ψ_1 be the chains formed with those simplices occurring in ζ^* and φ^* respectively and which lie on D', these simplices receiving in γ_1 and ψ_1 the same coefficients which they have in ζ^* and φ^* respectively. Then $\zeta^* - \gamma_1$ and $\varphi^* - \psi_1$ are chains on $S - (D \cup E)$.

If we map an arbitrary point $\mathfrak{p} \in \overline{D' - D''} - (\overline{D' - D''} \cap F)$ upon the point of C' which lies on the same radial set as $\mathfrak{p}$, we obtain a continuous mapping of $\overline{D' - D''} - (\overline{D' - D''} \cap F)$ which is clearly homotopic to the identity mapping and which coincides with the identity mapping on $C' - C' \cap F$. The cycles $\partial\gamma_1$, $\partial\psi_1$ lie on $\overline{D' - D''} - (\overline{D' - D''} \cap F)$; they are therefore homologous on this set to cycles on $C' - C' \cap F$. This means that there exist a 1-chain γ_2 and a 2-chain ψ_2 on $\overline{D' - D''} - (\overline{D' - D''} \cap F)$ such that $| \partial(\gamma_1 + \gamma_2) | \subset C'$, $| \partial(\psi_1 + \psi_2) | \subset C'$.

Write $\partial(\gamma_1 + \gamma_2) = \sum_{k=1}^h a_k \mathfrak{e}_k$, where $\mathfrak{e}_1, \cdots, \mathfrak{e}_h$ are points of C' and $a_k \neq 0$ $(1 \leqq k \leqq h)$. The points $\mathfrak{e}_k$ lie on $C' - C' \cap E$. We assert that, A being any connected component of $C' - C' \cap E$, we have $\sum_{\mathfrak{e}_k \in A} a_k \equiv 0 \pmod{n}$. This is clear if $C' \cap E$ either is empty or consists of a single point, since $\sum_{k=1}^h a_k = 0$. If $C' \cap E$ has at least two points, then the relative boundary of A on C' consists of two points $\mathfrak{a}$ and $\mathfrak{b}$ of E; we may assume that these points have been named in such a way that $\bar{A}$ is the positive arc determined by $\mathfrak{a}$ and $\mathfrak{b}$. It is clear that we can find a 1-chain λ_1 on F such that $\partial\lambda_1 = \mathfrak{b} - \mathfrak{a}$. Then we have $| \gamma_1 + \gamma_2 | \cap | \partial\lambda_1 | = \emptyset$, and $| \lambda_1 | \cap | \partial(\gamma_1 + \gamma_2) | = \emptyset$ and it follows from Lemma 2 that $\sum_{\mathfrak{e}_k \in A} a_k = i(\gamma_1 + \gamma_2, \lambda_1)$. Now, $| \gamma^* - \gamma_1 - \gamma_2 |$ does not meet F; $i(\gamma_1 + \gamma_2, \lambda_1)$ is therefore equal to $i(\gamma^*, \lambda_1) = i(\gamma, \lambda_1)$. Since E is arcwise connected, there exists a 1-chain λ_2 on E such that $\partial\lambda_2 = \mathfrak{a} - \mathfrak{b}$; $\lambda_1 + \lambda_2$ is a cycle and $i(\gamma, \lambda_1) = i(\gamma, \lambda_1 + \lambda_2)$ (because $| \gamma | \subset S - E$, whence $| \gamma | \cap | \lambda_2 | = \emptyset$). This number is by assumption $\equiv 0 \pmod{n}$.

For each component A of $C' - C' \cap E$ which meets $| \partial(\gamma_1 + \gamma_2) |$, select a point $\mathfrak{f}_A \in A$ and denote by b_A the integer $n^{-1} \sum_{\mathfrak{e}_k \in A} a_k$ (for $n \neq 0$; if $n = 0$, we set $b_A = 0$). Then it is clear that $n\sum_A b_A \mathfrak{f}_A - \partial(\gamma_1 + \gamma_2)$ bounds a 1-chain γ_3 on $C' - C' \cap E$; thus $\partial(\gamma_1 + \gamma_2 + \gamma_3) = n\sum_A b_A \mathfrak{f}_A$. It follows that $\sum_A b_A = 0$ and therefore that $\sum_A b_A \mathfrak{f}_A$ bounds a chain $-\gamma_4$ on C'. Write $\zeta^* = \gamma_1 + \gamma_2 + \gamma_3 + n\gamma_4 + (\zeta^* - \gamma_1 - \gamma_2 - \gamma_3 - n\gamma_4)$. The chain $\gamma_1 + \gamma_2 + \gamma_3 + n\gamma_4$ is a 1-cycle on D'; since $H_1(D') = \{0\}$, this cycle bounds a 2-chain τ_1 on D', whence $\zeta^* = \partial\tau_1 + (\zeta^* - \gamma_1 - \gamma_2 - \gamma_3 - n\gamma_4)$. Since ζ^* is a cycle, the same is true of $\zeta^* - \gamma_1 - \gamma_2 - \gamma_3 - n\gamma_4$. The chains $\zeta^* - \gamma_1 - \gamma_2 - \gamma_3$ lie on $S - (D \cup E)$; thus we see that $n\partial\gamma_4 = \partial(\zeta^* - \gamma_1 - \gamma_2 - \gamma_3)$ bounds a 1-chain on $S - (D \cup E)$. Now, it is easily seen that, for any space X, $H_0(X)$ never contains any element $\neq 0$ of finite order. We conclude that $\partial\gamma_4$ bounds a 1-chain γ_5 on $S - (D \cup E)$. We set $\zeta_1 = \gamma_5 - \gamma_4$, $\zeta_2 = \zeta^* - \gamma_1 - \gamma_2 - \gamma_3 - n\gamma_5$. Then ζ_1 is a 1-cycle, ζ_2 a 1-cycle on $S - (D \cup E)$, and we have $\zeta = (\zeta - \zeta^*) + \partial\tau_1 + n\zeta_1 + \zeta_2$. Since $\zeta^* = \text{subd}^p \zeta$, ζ^* is homologous to ζ on $| \zeta |$, and we have $\zeta - \zeta^* = \partial\tau_2$, where τ_2 is a 2-chain on $|\zeta|$. Since $|\tau_1| \subset D' \subset V$, we have $|\tau_1 + \tau_2| \subset |\zeta| \cup V$; the first assertion of Lemma 3 is thereby proved.

Assume from now on that $D \cap E \neq \emptyset$. We have $|\partial(\psi_1 + \psi_2)| \subset C' - C' \cap E$. If $C' \cap E$ is not empty, then $C' - C' \cap E$ is a union of mutually disjoint sets each of which is relatively open in $C' - C' \cap E$ and is homeomorphic to an open interval of the real axis. Any 1-cycle on $C' - C' \cap E$ lies on the union of a finite number of these sets, from which it follows immediately that $H_1(C' - C' \cap E) = \{0\}$ in this case, and therefore that $\partial(\psi_1 + \psi_2)$ bounds a 2-chain $-\psi_3$ on $C' - C' \cap E$. The same conclusion is still true if $C' \cap E$ is empty. For, let $\mathfrak{a}$ be a point of $D \cap E$, whence $\mathfrak{a} \notin |\psi_1 + \psi_2|$. Since $\mathfrak{a}$ is interior to D', C' is a deformation retract of $D' - \{\mathfrak{a}\}$. The chain $\partial(\psi_1 + \psi_2)$, which lies on C' and which bounds on $D' - \{\mathfrak{a}\}$, therefore bounds on $C' = C' - C' \cap E$.

Thus, in any case, $\partial(\psi_1 + \psi_2) = -\partial\psi_3$, where ψ_3 is a 2-chain on $C' - C' \cap E$. We write $\varphi^* = \psi_1 + \psi_2 + \psi_3 + (\varphi^* - \psi_1 - \psi_2 - \psi_3)$. The chain $\psi_1 + \psi_2 + \psi_3$ is a 2-cycle on D', and therefore homologous to 0 on D', since $H_2(D') = \{0\}$. On the other hand, we have $|\varphi^* - \psi_1| \subset S - (D \cup E)$, $|\psi_2| \subset S - (D \cup E)$, $|\psi_3| \subset S - (D \cup E)$. Since φ^* is homologous to φ on S, we see that φ is homologous to a 2-cycle on $S - (D \cup E)$. Lemma 3 is thereby proved.

LEMMA 4. *Let $\mathfrak{p}$ be a point of the Riemann surfaces S of the field R and let ζ be a 1-cycle on $S - \{\mathfrak{p}\}$ which is such that $i(\zeta, \lambda) \equiv 0 \pmod{n}$ for every 1-cycle λ on S, n being some fixed integer. Then ζ is homologous on $S - \{\mathfrak{p}\}$ to a cycle of the form $n\zeta_1$, ζ_1 being some cycle. If φ is a 2-chain on $S - \{\mathfrak{p}\}$, then φ is homologous to 0 on S.*

For each $\mathfrak{q} \in S$ select an open disc $V_\mathfrak{q}$ containing $\mathfrak{q}$ in such a way that the following conditions are satisfied: the set $V_\mathfrak{p}$ has no point in common with $|\zeta| \cup |\varphi|$; if $\mathfrak{q} \neq \mathfrak{p}$, then $\mathfrak{p} \notin V_\mathfrak{q}$. Let $D_\mathfrak{q}$ be a closed disc contained in $V_\mathfrak{q}$ and to which $\mathfrak{q}$ is interior, and let $U_\mathfrak{q}$ be the interior of $D_\mathfrak{q}$. The space S is compact and is covered by the sets $U_\mathfrak{q}$, $\mathfrak{q} \in S$; therefore, S may be covered by a finite number of sets $U_\mathfrak{q}$, say those relative to the points $\mathfrak{q}$ of some finite subset Q of S. Let a be the number of points in Q. We shall index these points by means of the integers from 1 to a. Since $\mathfrak{p} \notin V_\mathfrak{q}$ for $\mathfrak{q} \neq \mathfrak{p}$, we may assume that $\mathfrak{q}_1 = \mathfrak{p}$. Assume that $\mathfrak{q}_1, \cdots, \mathfrak{q}_i$ have already been defined, where i is some integer $<a$. If $D_{\mathfrak{q}_1} \cup \cdots \cup D_{\mathfrak{q}_i} = S$, then $\mathfrak{q}_{i+1}$ will be any one of the points of Q distinct from $\mathfrak{q}_1, \cdots, \mathfrak{q}_i$. If not, then, since S is connected, the set $D_{\mathfrak{q}_1} \cup \cdots \cup D_{\mathfrak{q}_i}$ has at least one boundary point $\mathfrak{r}_i$ on S, and $\mathfrak{q}_{i+1}$ will be a point of Q such that $\mathfrak{r}_i \in U_{\mathfrak{q}_{i+1}}$. This implies that $\mathfrak{q}_{i+1}$ is different from $\mathfrak{q}_1, \cdots, \mathfrak{q}_i$.

Set $E_i = D_{\mathfrak{q}_1} \cup \cdots \cup D_{\mathfrak{q}_i}$; then each E_i is closed and arcwise connected. This is true for $i = 1$, and, if it is true for some $i < a$, then it is true for $i + 1$ because $E_{i+1} = E_i \cup D_{\mathfrak{q}_{i+1}}$, and $D_{\mathfrak{q}_{i+1}}$ has the point $\mathfrak{r}_i$ in common with E_i (if $E_i \neq S$; if $E_i = S$, then $E_{i+1} = S$). Furthermore, we have $E_a = S$.

The cycles $\zeta_1 = \zeta$, $\varphi_1 = \varphi$ lie on $S - E_1$. Assume that, for some $i < a$, we have already proved that ζ is homologous on $S - \{\mathfrak{p}\}$ to a cycle of the form $n\theta_1 + \theta_2$, where θ_1 is a cycle on S and θ_2 a cycle on $S - E_i$. If λ is any cycle on S, then $i(\theta_2, \lambda) = i(\zeta, \lambda) - ni(\theta_1, \lambda) \equiv 0 \pmod{n}$. Making use of Lemma 3, we conclude that θ_2 is homologous on $|\theta_2| \cup V_{\mathfrak{q}_{i+1}}$ to a cycle of the form $n\theta_1' + \theta_2'$, where θ_1' is a cycle on S and θ_2' a cycle on $S - (D_{\mathfrak{q}_{i+1}} \cup E_i) = S - E_{i+1}$. Since $|\theta_2| \subset S - \{\mathfrak{p}\}$, $V_{\mathfrak{q}_{i+1}} \subset S - \{\mathfrak{p}\}$, we see that ζ is homologous on $S - \{\mathfrak{p}\}$ to $n(\theta_1 + \theta_1') + \theta_2'$, θ_2' being a cycle on $S - E_{i+1}$. Since $S - E_a = \emptyset$, the first assertion of Lemma 4 follows immediately from this inductive process.

Similarly, assume that, for some $i < a$, φ is homologous to a 2-cycle φ_i on $S - E_i$. We know that $D_{\mathfrak{q}_{i+1}} \cap E_i$ is not empty. Therefore, it follows from Lemma 3 that φ_i is homologous to a 2-cycle φ_{i+1} on $S - (D_{\mathfrak{q}_{i+1}} \cup E_i) = S - E_{i+1}$. Since $S - E_a$ is empty and φ is homologous to a 2-cycle on $S - E_a$, we see that φ is homologous to 0. Lemma 4 is now completely proved.

§8. The homology groups of the Riemann surface

Theorem 11. *Let P and Q be mutually disjoint finite subsets of the Riemann surface S of the field R. Denote by g the genus of R and by p and q the numbers of elements of P and Q respectively; set $p^* = \max\{p - 1, 0\}$, $q^* = \max\{q - 1, 0\}$. Then $H_1(S - P, Q)$ is a free abelian group on $2g + p^* + q^*$ generators. If $c \to \eta(c)$ is any homomorphism of this group into the additive group of integers, then there exists a unique element $c' \in H_1(S - Q, P)$ such that $\eta(c) = i(c, c')$ for every $c \in H_1(S - P, Q)$. If $c \to \pi(c)$ is any homomorphism of $H_1(S - P, Q)$ into the additive group of complex numbers, then there exists a differential ω having no pole in the set Q and whose residues at all places not in P are 0 such that $\pi(c) = \int_c \omega$ for all $c \in H_1(S - P, Q)$; the differential ω is determined by this condition to within an additive differential of the form df, where f is an element of R which admits the places of the set Q as zeros.*

For every point $\mathfrak{r} \in P \cup Q$, we select a closed disc $D_\mathfrak{r}$ containing $\mathfrak{r}$ in its interior but having no other point in common with $P \cup Q$. We denote by $C_\mathfrak{r}$ the circumference of $D_\mathfrak{r}$ and by $\zeta(\mathfrak{r})$ a cycle on $C_\mathfrak{r}$ which belongs to the positive generator of $H_1(C_\mathfrak{r})$. The set $S - Q$ is easily seen to be connected and therefore arcwise connected; if $\mathfrak{p}$ and $\mathfrak{p}'$ are distinct points of P, we denote by $\gamma(\mathfrak{p}, \mathfrak{p}')$ a 1-chain on $S - Q$ such that $\partial\gamma(\mathfrak{p}, \mathfrak{p}') = \mathfrak{p}' - \mathfrak{p}$. We shall first prove that $\sum_{\mathfrak{p} \in P} \zeta(\mathfrak{p})$ is homologous to 0 on $S - P$. We may of course assume that P is not empty; let $\mathfrak{p}$ be any point of P. If λ is any cycle on S, we have $i(\zeta(\mathfrak{p}), \lambda) = 0$ because $\zeta(\mathfrak{p})$ is homologous to 0 on $D_\mathfrak{p}$ and therefore on S. Making use of Lemma 4, §7 (where we take $n = 0$), we see that $\zeta(\mathfrak{p})$ is homologous to 0 on $S - \{\mathfrak{p}\}$. If $P \neq \{\mathfrak{p}\}$, then it follows from Lemma 4, §4, applied to the sets $U = S - \{\mathfrak{p}\}$ and $P - \{\mathfrak{p}\}$, that $\zeta(\mathfrak{p})$ is homologous on $S - P$ to a cycle of the form $\sum_{\mathfrak{p}' \in P, \mathfrak{p}' \neq \mathfrak{p}} a(\mathfrak{p}')\, \zeta(\mathfrak{p}')$. Since $\gamma(\mathfrak{p}, \mathfrak{p}')$ is a cycle modulo P, we have

$$i(\zeta(\mathfrak{p}) - \sum_{\mathfrak{p}' \in P, \mathfrak{p}' \neq \mathfrak{p}} a(\mathfrak{p}')\, \zeta(\mathfrak{p}'), \gamma(\mathfrak{p}, \mathfrak{p}')) = 0$$

for every $\mathfrak{p}' \in P - \{\mathfrak{p}\}$. Now, it follows from Lemma 1, §7 that

$$i(\zeta(\mathfrak{p}), \gamma(\mathfrak{p}, \mathfrak{p}')) = -1; \qquad i(\zeta(\mathfrak{p}'), \gamma(\mathfrak{p}, \mathfrak{p}')) = 1;$$

$$i(\zeta(\mathfrak{p}''), \gamma(\mathfrak{p}, \mathfrak{p}')) = 0 \text{ if } \mathfrak{p}'' \in P,\ \mathfrak{p}'' \neq \mathfrak{p}, \mathfrak{p}'.$$

Thus we have $-1 - a(\mathfrak{p}') = 0$ for all $\mathfrak{p}' \in P - \{\mathfrak{p}\}$, which proves that $\sum_{\mathfrak{p} \in P} \zeta(\mathfrak{p})$ is homologous to 0 on $S - P$.

Let n be an integer, and let c_0 be an element of $H_1(S - P, Q)$ such that $i(c_0, c') \equiv 0 \pmod{n}$ for every $c' \in H_1(S - Q, P)$. Then we shall prove that c_0 is of the form nc, $c \in H_1(S - P, Q)$. Set $\partial c_0 = \sum_{\mathfrak{q} \in Q} a(\mathfrak{q})\mathfrak{q}$, and take c' to be the homology class in $H_1(S - Q, P)$ of $\zeta(\mathfrak{q})$; then it follows from Lemma 1, §7 that $i(c_0, c') = -a(\mathfrak{q})$. Thus we have $a(\mathfrak{q}) = nb(\mathfrak{q})$, where $b(\mathfrak{q})$ is an integer, which we select to be 0 if $n = 0$. Since $\sum_{\mathfrak{q} \in Q} a(\mathfrak{q}) = 0$, we have $\sum_{\mathfrak{q} \in Q} b(\mathfrak{q}) = 0$. The space $S - P$ being arcwise connected, any two points of Q are homologous to each other on $S - P$. Since $\sum_{\mathfrak{q} \in Q} b(\mathfrak{q}) = 0$, the chain $\sum_{\mathfrak{q} \in Q} b(\mathfrak{q})\mathfrak{q}$ may be written as a linear combination with integral coefficients of chains of the form $\mathfrak{q}' - \mathfrak{q}''$, with $\mathfrak{q}'$ and $\mathfrak{q}''$ in Q. It follows that $\partial c_0 = n\partial c_1$, where c_1 is an element of $H_1(S - P, Q)$. Replacing c_0 by $c_0 - nc_1$, we see that we may assume that $\partial c_0 = 0$.

Let then γ be a cycle of the class c_0. Every 1-cycle λ on S is homologous on S to a cycle on $S - Q$ (Corollary to Lemma 4, §4). It follows that $i(\gamma, \lambda) \equiv 0 \pmod{n}$. Making use of Lemma 4, §7, we see that γ is homologous on S to a cycle of the form $n\gamma_2$, γ_2 being a cycle on S which may even be assumed to be on $S - P$ (Corollary to Lemma 4, §4). Let c_2 be the homology class of γ_2 in $H_1(S - P, Q)$. Then $c_0 - nc_2$ belongs to the kernel of the homomorphism of $H_1(S - P, Q)$ into $H_1(S, Q)$ which corresponds to the identity mapping of the pair $(S - P, Q)$ into (S, Q). Making use of Lemma 4, §4, we see that $c_0 - nc_2 = \sum_{\mathfrak{p} \in P} e(\mathfrak{p})z(\mathfrak{p})$, where $z(\mathfrak{p})$ is the homology class of $\zeta(\mathfrak{p})$ in $H_1(S - P, Q)$. If $\mathfrak{p}$ and $\mathfrak{p}'$ are distinct points of P, let $c'(\mathfrak{p}, \mathfrak{p}')$ be the homology class of $\gamma(\mathfrak{p}, \mathfrak{p}')$ in $H_1(S - Q, P)$. Making use of the formulas above, we see that $i(c_0 - nc_2, c'(\mathfrak{p}, \mathfrak{p}')) = e(\mathfrak{p}') - e(\mathfrak{p})$. This is by assumption $\equiv 0 \pmod{n}$. Thus, all numbers $e(\mathfrak{p})$ are congruent to each other modulo n. On the other hand, we have proved above that $\sum_{\mathfrak{p} \in P} z(\mathfrak{p}) = 0$. Thus we see that we may assume that all numbers $e(\mathfrak{p})$ are $\equiv 0 \pmod{n}$, which completes the proof of the fact that c_0 is of the form nc, $c \in H_1(S - P, Q)$. Taking in particular $n = 0$, we see that $i(c, c') = 0$ for every $c' \in H_1(S - Q, P)$ implies $c = 0$. From this we may conclude that $H_1(S - P, Q)$ has no element $\neq 0$ of finite order. For, if $uc = 0$ where u is an integer $\neq 0$, then we have, for $c' \in H_1(S - Q, P)$, $ui(c, c') = i(uc, c') = 0$, whence $i(c, c') = 0$ and $c = 0$.

Let the spaces $\mathfrak{E}_{P,Q}$, $\mathfrak{F}_P$ be defined as in §5, and, for $c \in H_1(S - P, Q)$, let the element $\Omega(c)$ of $\mathfrak{E}_{Q,P}/\mathfrak{F}_P$ be defined as in Theorem 9, §6. If $c' \in H_1(S - Q, P)$, then $i(c, c') = -(2\pi(-1)^{1/2})^{-1} \int_{c'} \Omega(c)$. Therefore, it follows immediately from what we have proved above that the homomorphism $c \to \Omega(c)$ of $H_1(S - P, Q)$ into $\mathfrak{E}_{Q,P}/\mathfrak{F}_P$ is an isomorphism. Let $\mathfrak{M}$ be the subspace of $\mathfrak{E}_{Q,P}/\mathfrak{F}_P$ which is spanned by the elements $\Omega(c)$, $c \in H_1(S - P, Q)$. We have observed in §5 that the function $j(\omega, \omega')$ defines in a natural way a non degenerate bilinear function on $(\mathfrak{E}_{Q,P}/\mathfrak{F}_P) \times (\mathfrak{E}_{P,Q}/\mathfrak{F}_Q)$; we shall still denote by j this bilinear function. Let Ω_0' be an element of $\mathfrak{E}_{P,Q}/\mathfrak{F}_Q$ such that $j(\Omega, \Omega_0') = 0$ for every $\Omega \in \mathfrak{M}$. If ω_0' is a differential belonging to the coset Ω_0', then we have $\int_c \omega_0' = 0$ for every $c \in H_1(S - P, Q)$. Since $\int_{\zeta(\mathfrak{p})} \omega_0' = 0$ for every $\mathfrak{p} \in P$, we have $\mathrm{res}_{\mathfrak{p}}\omega_0' = 0$ by Theorem 6, §4. Since $\omega_0' \in \mathfrak{E}_{P,Q}$, this means that ω_0' is of the second kind. Any homology class on

S may be represented by a cycle on $S - P$; therefore the integral of ω_0' on any homology class in $H_1(S)$ is 0, and it follows that ω_0' admits a primitive f on S (Theorem 7, §4). Since f is everywhere meromorphic, it is an element of R by Theorem 4, §2. Since $\omega_0' \in \mathfrak{E}_{P,Q}$, f has no pole on the set Q. If $\mathfrak{q}$ and $\mathfrak{q}'$ are distinct points of Q, let $\gamma(\mathfrak{q}, \mathfrak{q}')$ be a 1-chain on $S - P$ such that $\partial\gamma(\mathfrak{q}, \mathfrak{q}') = \mathfrak{q}' - \mathfrak{q}$. Then we have $0 = \int_{\gamma(\mathfrak{q},\mathfrak{q}')} df = f(\mathfrak{q}') - f(\mathfrak{q})$ by Lemma 2, §4. It follows that the numbers $f(\mathfrak{q})$, $\mathfrak{q} \in Q$, are all equal to each other. If f_0 is their common value, we have $\omega_0' = d(f - f_0) \in \mathfrak{F}_Q$, whence $\Omega_0' = 0$. The bilinear function j being non degenerate, we conclude that $\mathfrak{M}$ is the whole space $\mathfrak{E}_{Q,P}/\mathfrak{F}_P$. Set $m = 2g + p^* + q^*$; then it follows from Theorem 15, VI, §8 that $\mathfrak{E}_{Q,P}/\mathfrak{F}_P$ is of dimension m. Since $\mathfrak{M}$ is of dimension m, we can find m elements $d_1, \cdots, d_m$ of $H_1(S - P, Q)$ such that $\Omega(d_1), \cdots, \Omega(d_m)$ are linearly independent in $\mathfrak{E}_{Q,P}/\mathfrak{F}_P$. Similarly, we can find m' elements $d_1', \cdots, d_m'$ of $H_1(S - Q, P)$ such that $\Omega(d_1'), \cdots, \Omega(d_m')$ are linearly independent in $\mathfrak{E}_{P,Q}/\mathfrak{F}_Q$. Since j is non degenerate, the determinant of the matrix $(j(\Omega(d_i), \Omega(d_j')))_{1 \leqq i,j \leqq m}$ is $\neq 0$, from which it follows that the determinant D of the matrix $(i(d_i, d_j'))_{1 \leqq i,j \leqq m}$ is an integer $\neq 0$.

Let d be any element of $H_1(S - P, Q)$. Then $\Omega(d)$ is a linear combination $\sum_{i=1}^m \alpha_i \Omega(d_i)$ of $\Omega(d_1), \cdots, \Omega(d_m)$ with complex coefficients $\alpha_1, \cdots, \alpha_m$. We have $i(d, d_j') = -(2\pi(-1)^{1/2})^{-1} \int_{d_j'} \Omega(d) = -(2\pi(-1)^{1/2})^{-1} \sum_{i=1}^m \alpha_i \int_{d_j'} \Omega(d_i) = \sum_{i=1}^m \alpha_i i(d_i, d_j')$. Thus, we obtain a system of m linear equations in $\alpha_1, \cdots, \alpha_m$ with integral coefficients and right sides, and we know that the determinant D of this system is $\neq 0$. It follows that $\alpha_1, \cdots, \alpha_m$ are themselves rational numbers, which may be written with D as their common denominator. Therefore, for any $d \in H_1(S - P, Q)$, Dd is a linear combination of $d_1, \cdots, d_m$ with integral coefficients. Knowing that $H_1(S - P, Q)$ has no element $\neq 0$ of finite order, this implies (by well known theorems on abelian groups) that $H_1(S - P, Q)$ is a free abelian group on m generators. We shall assume from now on that $d_1, \cdots, d_m$ form a base of $H_1(S - P, Q)$, and, similarly, that $d_1', \cdots, d_m'$ form a base of $H_1(S - Q, P)$. Then we shall see that the number D is equal to ± 1. For, were this not the case, then D would have a prime divisor n, and the system of linear homogeneous congruences $\sum_{i=1}^m a_i i(d_i, d_j') \equiv 0 \pmod{n}$ $(1 \leqq j \leqq m)$ would have a non trivial solution $(a_1, \cdots, a_m)$. If $d = \sum_{i=1}^m a_i d_i$, then we would have (since $d_1', \cdots, d_m'$ form a base of $H_1(S - Q, P)$) $i(d, c') \equiv 0 \pmod{n}$ for every $c' \in H_1(S - Q, P)$. But this would imply that $d = nc$ with some c in $H_1(S - P, Q)$, which is impossible since $a_1, \cdots, a_m$ are not all $\equiv 0 \pmod{n}$ and $d_1, \cdots, d_m$ form a base of $H_1(S - P, Q)$.

Since $D = \pm 1$, it is well known that, the base $\{d_1, \cdots, d_m\}$ of $H_1(S - P, Q)$ being given, the base $(d_1', \cdots, d_m')$ of $H_1(S - Q, P)$ may be selected in such a way that $i(d_i, d_j') = 1$ if $i = j$ and 0 if $i \neq j$. This being the case, let there be given a homomorphism η of $H_1(S - P, Q)$ into the additive group of integers. If we set $c' = \sum_{i=1}^m \eta(d_i) d_i'$, we have $\eta(d_i) = i(d_i, c')$ $(1 \leqq i \leqq m)$, whence $\eta(c) = i(c, c')$ for every $c \in H_1(S - P, Q)$. Conversely, if c_1' is an element of $H_1(S - Q, P)$ such that $\eta(c) = i(c, c_1')$ for every $c \in H_1(S - P, Q)$, then

c'_1 can be written in the form $\sum_{i=1}^m a'_i d'_i$, and we have $a'_i = i(d_i, c'_1) = \eta(d_i)$ $(1 \leqq i \leqq m)$, whence $c'_1 = c'$.

Let π be a homomorphism of $H_1(S - P, Q)$ into the additive group of complex numbers. If ω is any differential of the coset $(2\pi(-1)^{1/2})^{-1} \sum_{i=1}^m \pi(d_i)\Omega(d'_i)$, then we have $\int_{d_i} \omega = \pi(d_i)(1 \leqq i \leqq m)$, whence $\int_c \omega = \pi(c)$ for every $c \in H_1(S - P, Q)$. This condition clearly determines ω to within an element of $\mathfrak{F}_Q$. Theorem 11 is thereby proved.

Let us consider the special case where P and Q are empty. Then $H_1(S - P, Q)$ and $H_1(S - Q, P)$ are both identical with $H_1(S)$. Thus we see that $H_1(S)$ is a free abelian group on $2g$ generators. This means that *the Riemann surface of a field of genus g is of topological genus g*. Moreover, if $(z_1^*, \cdots, z_{2g}^*)$ is any base of $H_1(S)$, then the determinant of the matrix $(i(z_i^*, z_j^*))$ is ± 1. This matrix being skew-symmetric, it follows that there exists an integral matrix L of determinant 1 such that

$$L \cdot (i(z_i^*, z_j^*)) \cdot {}^tL = \begin{pmatrix} 0 & I_g \\ -I_g & 0 \end{pmatrix} \tag{1}$$

(where ${}^tL$ denotes the transposed matrix of L, and I_g is the unit matrix of degree g). The matrix L allows us to pass from the base $(z_1^*, \cdots, z_{2g}^*)$ of $H_1(S)$ to a new base $(z_1, \cdots, z_{2g})$, and (1) implies that

$$i(z_i, z_j) = \begin{cases} 0 \text{ if } i \leqq g, j \neq i + g, \\ 1 \text{ if } i \leqq g, j = i + g, \\ 0 \text{ if } i > g, j \neq i - g, \\ -1 \text{ if } i > g, j = i - g. \end{cases} \tag{2}$$

Thus we have proved

THEOREM 12. *Let g be the genus of the field R. Then the first homology group $H_1(S)$ of the Riemann surface S of R is a free abelian group on 2g generators, and this group has a base formed by 2g elements $z_1, \cdots, z_{2g}$ whose mutual intersection numbers are given by the formulas* (2) *above.*

Any such base is called a *canonical base* of the group $H_1(S)$.

Let us now return to the consideration of the groups $H_1(S - P, Q)$ and $H_1(S - Q, P)$. Let $\{z_1, \cdots, z_{2g}\}$ be a canonical base of $H_1(S)$. In the homology class $z_k(1 \leqq k \leqq 2g)$, we can find a cycle ζ_k on $S - (P \cup Q)$ (Corollary to Lemma 4, §4). Denote by u_k and u'_k the homology classes of ζ_k in $H_1(S - P, Q)$ and $H_1(S - Q, P)$ respectively. If $\mathfrak{p} \in P$, let the cycle $\zeta(\mathfrak{p})$ be defined as in the proof of Theorem 11, and let $z(\mathfrak{p})$ be the homology class of $\zeta(\mathfrak{p})$ in $H_1(S - P, Q)$. We have seen that $\sum_{\mathfrak{p} \in P} z(\mathfrak{p}) = 0$. If $Q \neq \emptyset$, let $\mathfrak{q}_0$ be a point of Q; if $\mathfrak{q}$ is any point $\neq \mathfrak{q}_0$ in Q, let $c^*(\mathfrak{q})$ be an element of $H_1(S - P, Q)$ such that $\partial c^*(\mathfrak{q}) = \mathfrak{q} - \mathfrak{q}_0$. Set

$$c(\mathfrak{q}) = c^*(\mathfrak{q}) - \sum_{k=1}^{g} i(c^*(\mathfrak{q}), u'_{k+g})u_k + \sum_{k=1}^{g} i(c^*(\mathfrak{q}), u'_k)u_{k+g}.$$

Then we still have $\partial c(\mathfrak{q}) = \mathfrak{q} - \mathfrak{q}_0$, and furthermore $i(c(\mathfrak{q}), u'_k) = 0$ for $1 \leqq k \leqq 2g$.

If $P \neq \emptyset$, let $\mathfrak{p}_0$ be any point of P. Then we assert that the $2g + p^* + q^*$ elements

$$(3) \qquad u_k\ (1 \leqq k \leqq 2g); \qquad z(\mathfrak{p}) \text{ for } \mathfrak{p} \in P,\ \mathfrak{p} \neq \mathfrak{p}_0; \qquad c(\mathfrak{q}) \text{ for } \mathfrak{q} \in Q,\ \mathfrak{q} \neq \mathfrak{q}_0,$$

form a base of $H_1(S - P, Q)$. For, if we interchange the roles played by P and Q, then we find in a similar manner m elements u'_k $(1 \leqq k \leqq 2g)$, $z'(\mathfrak{q})$ $(\mathfrak{q} \in Q, \mathfrak{q} \neq \mathfrak{q}_0)$; $c'_1(\mathfrak{p})$ $(\mathfrak{p} \in P, \mathfrak{p} \neq \mathfrak{p}_0)$ of $H_1(S - Q, P)$ such that $i(u_k, c'_1(\mathfrak{p})) = 0$ $(1 \leqq k \leqq 2g$, $\mathfrak{p} \in P, \mathfrak{p} \neq \mathfrak{p}_0)$, $\partial c'_1(\mathfrak{p}) = \mathfrak{p} - \mathfrak{p}_0$. Set, for $\mathfrak{p} \in P$, $\mathfrak{p} \neq \mathfrak{p}_0$,

$$c'(\mathfrak{p}) = c'_1(\mathfrak{p}) + \sum_{\mathfrak{q} \in Q, \mathfrak{q} \neq \mathfrak{q}_0} i(c(\mathfrak{q}), c'_1(\mathfrak{p}))z'(\mathfrak{q});$$

then we still have $i(u_k, c'(\mathfrak{p})) = 0$ (because $i(u_k, z'(\mathfrak{q})) = 0$, since $z'(\mathfrak{q})$ may be represented by a cycle which is homologous to 0 on S) and $\partial c'(\mathfrak{p}) = \mathfrak{p} - \mathfrak{p}_0$. Furthermore, we have, by Lemma 1, §7, $i(c(\mathfrak{q}), z'(\mathfrak{q})) = -1$, $i(c(\mathfrak{q}), z'(\mathfrak{q}')) = 0$ (where $\mathfrak{q}$ and $\mathfrak{q}'$ are points of Q, distinct from $\mathfrak{q}_0$ and from each other). It follows that $i(c(\mathfrak{q}), c'(\mathfrak{p})) = 0$ for $\mathfrak{p} \in P$, $\mathfrak{p} \neq \mathfrak{p}_0$, $\mathfrak{q} \in Q$, $\mathfrak{q} \neq \mathfrak{q}_0$. We arrange the elements (3) of $H_1(S - P, Q)$ in a sequence $(d_1, \cdots, d_m)$ and the elements

$$(4) \quad u'_k\ (1 \leqq k \leqq 2g); \qquad z'(\mathfrak{q}), \text{ for } \mathfrak{q} \in Q,\ \mathfrak{q} \neq \mathfrak{q}_0; \qquad c'(\mathfrak{p}) \text{ for } \mathfrak{p} \in P,\ \mathfrak{p} \neq \mathfrak{p}_0$$

of $H_1(S - Q, P)$ in a sequence $(d'_1, \cdots, d'_m)$. Then it is easily seen that, for each index i $(1 \leqq i \leqq m)$, there exists only one index j_i for which $i(d_i, d'_{j_i}) \neq 0$, and that $i(d_i, d'_{j_i}) = \pm 1$. The determinant of the matrix $(i(d_i, d'_j))_{1 \leqq i,j \leqq m}$ is therefore ± 1, from which we conclude that $d_1, \cdots, d_m$ and $d'_1, \cdots, d'_m$ are bases of $H_1(S - P, Q)$ and $H_1(S - Q, P)$ respectively. These bases are said to constitute a pair of *associated canonical bases* for our two groups.

Denote by ω any differential of the set $\mathfrak{E}_{Q,P}$. Then the coset Ω of ω modulo $\mathfrak{F}_P$ may be written in the form

$$\Omega = \sum_{k=1}^{2g} \alpha_k \Omega(u_k) + \sum_{\mathfrak{p} \in P, \mathfrak{p} \neq \mathfrak{p}_0} \beta_{\mathfrak{p}} \Omega(z(\mathfrak{p})) + \sum_{\mathfrak{q} \in Q, \mathfrak{q} \neq \mathfrak{q}_0} \gamma_{\mathfrak{q}} \Omega(c'(\mathfrak{q}))$$

where the α_k, $\beta_{\mathfrak{p}}$, $\gamma_{\mathfrak{q}}$ are complex numbers. Similarly, if $\omega' \in \mathfrak{E}_{P,Q}$, then the coset Ω' of ω' modulo $\mathfrak{F}_Q$ may be written in the form

$$\Omega' = \sum_{k=1}^{2g} \alpha'_k \Omega(u'_k) + \sum_{\mathfrak{q} \in Q, \mathfrak{q} \neq \mathfrak{q}_0} \beta'_{\mathfrak{q}} \Omega(z'(\mathfrak{q})) + \sum_{\mathfrak{p} \in P, \mathfrak{p} \neq \mathfrak{p}_0} \gamma'_{\mathfrak{p}} \Omega(c'(\mathfrak{p}))$$

where the α'_k, $\beta'_{\mathfrak{q}}$, $\gamma'_{\mathfrak{p}}$ are complex numbers. It follows that

$$j(\omega, \omega') = j(\Omega, \Omega') = 2\pi(-1)^{1/2} \left(\sum_{k=1}^{g} \alpha_k \alpha'_{k+g} - \sum_{k=g+1}^{2g} \alpha_k \alpha'_{k-g} \right)$$
$$+ 2\pi(-1)^{1/2} \sum_{\mathfrak{p} \in P, \mathfrak{p} \neq \mathfrak{p}_0} \beta_{\mathfrak{p}} \gamma'_{\mathfrak{p}} - 2\pi(-1)^{1/2} \sum_{\mathfrak{q} \in Q, \mathfrak{q} \neq \mathfrak{q}_0} \gamma_{\mathfrak{q}} \beta'_{\mathfrak{q}}.$$

On the other hand, we have

$$2\pi(-1)^{1/2}\alpha_k = j(\Omega, \Omega(u'_{k+g})) = -\int_{u'_{k+g}} \Omega \qquad (1 \leqq k \leqq g),$$

$$2\pi(-1)^{1/2}\alpha_k = -j(\Omega, \Omega(u'_{k-g})) = \int_{u'_{k-g}} \Omega \qquad (g+1 \leqq k \leqq 2g),$$

and similarly

$$2\pi(-1)^{1/2}\alpha'_k = -\int_{u_{k+g}} \Omega' \qquad (1 \leqq k \leqq g),$$

$$2\pi(-1)^{1/2}\alpha'_k = \int_{u_{k-g}} \Omega' \qquad (g+1 \leqq k \leqq 2g).$$

Also

$$2\pi(-1)^{1/2}\beta_{\mathfrak{p}} = j(\Omega, \Omega(c'(\mathfrak{p}))) = -\int_{c'(\mathfrak{p})} \Omega,$$

$$2\pi(-1)^{1/2}\gamma_{\mathfrak{q}} = -j(\Omega, \Omega(z'(\mathfrak{q}))) = \int_{z'(\mathfrak{q})} \Omega = 2\pi(-1)^{1/2} \operatorname{res}_{\mathfrak{q}} \omega,$$

and similar formulas for $\beta'_{\mathfrak{q}}$, $\gamma'_{\mathfrak{p}}$. Thus we obtain the formula

$$(5) \qquad \begin{aligned} j(\omega, \omega') = \frac{-1}{2\pi(-1)^{1/2}} \sum_{k=1}^{g} \left(\int_{u'_{k+g}} \omega \int_{u_k} \omega' - \int_{u'_k} \omega \int_{u_{k+g}} \omega' \right) \\ - \sum_{\mathfrak{p} \in P, \mathfrak{p} \neq \mathfrak{p}_0} \left(\int_{c'(\mathfrak{p})} \omega \right) \operatorname{res}_{\mathfrak{p}} \omega' + \sum_{\mathfrak{q} \in Q, \mathfrak{q} \neq \mathfrak{q}_0} \left(\int_{c(\mathfrak{q})} \omega' \right) \operatorname{res}_{\mathfrak{q}} \omega. \end{aligned}$$

Let us apply this to the special case where ω and ω' are differentials of the first kind. Then, clearly, $j(\omega\ \omega') = \operatorname{res}_{\mathfrak{p}} \omega' = \operatorname{res}_{\mathfrak{q}} \omega = 0$ and we obtain

THEOREM 13. *Let $\{z_1, \cdots, z_{2g}\}$ be a canonical base of the first homology group of the Riemann surface of the field R, and let ω and ω' be differentials of the first kind of R. Then we have*

$$\sum_{k=1}^{g} \left(\int_{z_{k+g}} \omega \int_{z_k} \omega' - \int_{z_k} \omega \int_{z_{k+g}} \omega' \right) = 0.$$

These relations are called the *bilinear equalities* of Riemann.

To conclude this section, we prove

THEOREM 14. *The two-dimensional homology group of the Riemann surface of R is infinite cyclic.*

Let $\mathfrak{p}$ be any point of the Riemann surface S. Denote by D a closed disc containing $\mathfrak{p}$ in its interior and by C the circumference of D. Let γ be a 1-cycle on C belonging to the positive generator of $H_1(C)$. Since $H_1(D) = \{0\}$, γ bounds a 2-chain δ on D. On the other hand, it follows from Lemma 4, §7 that γ bounds a 2-chain δ' on $S - \{\mathfrak{p}\}$ (because, clearly, $i(\gamma, \lambda) = 0$ for every 1-cycle λ on S). The chain $\varphi = \delta - \delta'$ is a 2-cycle on S. Let m be an integer $\neq 0$; then $m\varphi$ is homologous to $m\delta$ modulo $S - \{\mathfrak{p}\}$, and $m\delta$ is not homologous to 0 modulo $S - \{\mathfrak{p}\}$ by Lemma 4, §4. It follows that the homology class of φ in $H_2(S)$ is of infinite order. Let φ' be any 2-cycle on S. Then φ' is homologous modulo $S - \{\mathfrak{p}\}$ to a chain of the form $k\delta$, where k is an integer (Lemma 4, §4), and therefore also to $k\varphi$. This means that $\varphi' - k\varphi$ is homologous on S to a 2-cycle on $S - \{\mathfrak{p}\}$. But it follows from Lemma 4, §7 that every 2-cycle on $S - \{\mathfrak{p}\}$ bounds on S, and therefore that φ' is homologous to $k\varphi$ on S. Theorem 14 is thereby proved.

§9. THE THEOREM OF ABEL

The group of divisors of the field R has beeu defined to be the free abelian group generated by the places of R. But the places of R are the same thing as the points of the Riemann surface S. Thus we see that the notions of divisor on R and of 0-chain on S are equivalent to each other. The only difference between divisors and 0-chains is a difference of notation, the group of divisors being written multiplicatively, while the group of 0-chains is written additively. We shall nevertheless make the convention to represent by the same symbol a 0-chain and the divisor which is associated with it.

Since S is arcwise connected, $H_0(S)$ is infinite cyclic, and a 0-chain $\sum_{k=1}^{h} a_k\mathfrak{p}_k$ bounds on S if and only if $\sum_{k=1}^{h} a_k = 0$. Thus we see that the 0-chains which bound are the divisors of degree 0.

THEOREM 15 (THEOREM OF ABEL). *Let $\mathfrak{d}$ be a divisor of the field R. Then a necessary and sufficient condition for $\mathfrak{d}$ to be the divisor of an element of R is that there should exist a 1-chain γ on the Riemann surface S of R such that $\partial\gamma = \mathfrak{d}$ and $\int_\gamma \eta = 0$ for every differential η of the first kind of R.*

The proof will be based on the following lemma.

LEMMA 1. *In order for a differential ω of R to be of the form $x^{-1}dx$, with some $x \in R$, it is necessary and sufficient that the following conditions be satisfied*: a) ω *has no pole of order* > 1; b) *every period of ω is an integral multiple of* $2\pi(-1)^{1/2}$.

We first prove that the conditions are necessary. For the condition a), this follows immediately from Lemma 1, VI, §8. Now, let V be an open disc containing no pole of $x^{-1}dx$, and let f be a primitive of $x^{-1}dx$ on V. Since x has no zero on V, $x^{-1}e^f$ is holomorphic on V; we have $d(x^{-1}e^f) = -x^{-2}e^f\,dx + x^{-1}e^f\,df = 0$. This shows that $e^f = Cx$, where C is a constant. If σ is any 1-simplex on V, and $\partial\sigma = \mathfrak{b} - \mathfrak{a}$, then we see that $\exp\,(\int_\sigma x^{-1}dx) = x(\mathfrak{b})\,(x(\mathfrak{a}))^{-1}$. If γ is a 1-chain such that every simplex occurring in γ lies on an open disc on which $x^{-1}\,dx$ has no pole, then we see that $\exp\,(\int_\gamma x^{-1}\,dx) = \prod_{k=1}^{h} (x(\mathfrak{p}_k))^{a_k}$ if $\partial\gamma = \sum_{k=1}^{h} a_k\mathfrak{p}_k$. Now, let ζ be a cycle such that $|\,\zeta\,|$ does not contain any pole of $x^{-1}\,dx$. Then there exists an integer $p \geqq 0$ such that every simplex occurring in $\text{subd}^p\zeta$ lies on an open disc on which $x^{-1}\,dx$ has no pole. It follows that $\exp\,(\int_\zeta x^{-1}\,dx) = 1$ and therefore that $\int_\zeta x^{-1}\,dx$ is an integral multiple of $2\pi(-1)^{1/2}$. This proves that condition b) is necessary.

Now, assume that conditions a) and b) are satisfied. Let P be the set of poles of ω and let $\mathfrak{p}_0$ be a fixed point of $S - P$. For any point $\mathfrak{p} \in S - P$, select a 1-chain $\gamma(\mathfrak{p})$ on $S - P$ such that $\partial\gamma(\mathfrak{p}) = \mathfrak{p} - \mathfrak{p}_0$, and set $g(\mathfrak{p}) = \exp(\int_{\gamma(\mathfrak{p})} \omega)$. We shall see that the function g is holomorphic on $S - P$. Let $\mathfrak{p}$ be a point of $S - P$, and V an open disc contained in $S - P$ and containing $\mathfrak{p}$. Then ω admits a primitive f on V. For each $\mathfrak{q} \in V$, let $\sigma(\mathfrak{q})$ be a 1-simplex on V such that $\partial\sigma(\mathfrak{q}) = \mathfrak{q} - \mathfrak{p}$. Then $\gamma(\mathfrak{q}) - (\gamma(\mathfrak{p}) + \sigma(\mathfrak{q}))$ is a cycle on $S - P$, whence $\int_{\gamma(\mathfrak{q})} \omega = \int_{\gamma(\mathfrak{p})} \omega + (f(\mathfrak{q}) - f(\mathfrak{p})) + 2k\pi\,(-1)^{1/2}$, where k is an integer, and therefore $g(\mathfrak{q}) = g(\mathfrak{p}) \cdot \exp(f(\mathfrak{q}) - f(\mathfrak{p}))$, which proves that g is holomorphic on V.

Now, let $\mathfrak{a}$ be a point of P, whence $\nu_{\mathfrak{a}}(\omega) = -1$. The logarithmic period of ω relative to $\mathfrak{a}$ being an integral multiple of $2\pi(-1)^{1/2}$, the residue of ω at $\mathfrak{a}$ is an integer m. Let y be an element of R such that $\nu_{\mathfrak{a}}(y) = 1$, $\nu_{\mathfrak{p}_0}(y) = 0$. Then we have $\nu_{\mathfrak{a}}(y^{-m}dy^m) = -1$, $\mathrm{res}_{\mathfrak{a}}(y^{-m}dy^m) = m$, which proves that $\omega - y^{-m}dy^m$ has no pole at $\mathfrak{a}$. We may assume that the chains $\gamma(\mathfrak{p})$ which were used in defining the function g are such that $|\gamma(\mathfrak{p})|$ never contains a zero or a pole of y unless $\mathfrak{p}$ is one. There exists a neighbourhood of $\mathfrak{a}$ in which y has no zero or pole except at $\mathfrak{a}$ itself; if $\mathfrak{p}$ is a point not equal to $\mathfrak{a}$ in this neighbourhood, then $(y^{-m}g)(\mathfrak{p}) = C \exp(\int_{\gamma(\mathfrak{p})}(\omega - y^{-m}dy^m))$ where C is a constant. Since $\omega - y^{-m}dy^m$ is holomorphic at $\mathfrak{a}$, we see that g may be extended to a function which is meromorphic at $\mathfrak{a}$. This being true for every $\mathfrak{a} \in P$, g may be extended to a meromorphic function on the whole of S. This function is an element of R by Theorem 4, §2, and it is clear that $\omega = g^{-1}\,dg$.

We are now in a position to prove the theorem of Abel. Let first $\mathfrak{d}$ be the divisor of an element x of R. Then the mapping $u \to (2\pi(-1)^{1/2})^{-1} \int_u x^{-1}\,dx$ is a homomorphism of $H_1(S - |\mathfrak{d}|)$ into the additive group of integers. By Theorem 11, §8, there exists an element $c \in H_1(S, |\mathfrak{d}|)$ such that $i(u, c) = (2\pi(-1)^{1/2})^{-1} \cdot \int_u x^{-1}\,dx$ for every $u \in H_1(S - |\mathfrak{d}|)$. Making use of Theorem 9, §6, we associate to c a coset $\Omega(c)$ of the space of differentials whose residues at all places not occurring in $\mathfrak{d}$ are 0 modulo the space of exact differentials; it is clear from the definition of $\Omega(c)$ that it contains $x^{-1}\,dx$. If $\mathfrak{p}$ is a place occurring in $\mathfrak{d}$, then the residue of $x^{-1}\,dx$ at $\mathfrak{p}$ is easily seen to be $\nu_{\mathfrak{p}}(x)$, while the residue of $\Omega(c)$ at $\mathfrak{p}$ is the coefficient with which $\mathfrak{p}$ occurs in ∂c (Theorem 9, §6); it follows that $\partial c = \mathfrak{d}$. Let η be a differential of the first kind; then $\int_c \eta = j(x^{-1}\,dx, \eta)$. Since $x^{-1}\,dx$ has no pole of order > 1 and η has no pole, $j(x^{-1}\,dx, \eta) = 0$ in virtue of Lemma 1, §6.

Conversely, let $\mathfrak{d}$ be a divisor of R such that there exists a $c \in H_1(S, |\mathfrak{d}|)$ such that $\partial c = \mathfrak{d}$ and $\int_c \eta = 0$ for every differential η of the first kind. Let ω be a differential of the class $\Omega(c)$. For each pole $\mathfrak{p}$ of ω let us select an $f_{\mathfrak{p}} \in R$ such that $\nu_{\mathfrak{p}}(\omega - df_{\mathfrak{p}}) \geqq -1$ (Theorem 14, VI, §8). If η is any differential of the first kind, then the definition of the function $j_{\mathfrak{p}}$ gives $j_{\mathfrak{p}}(\omega, \eta) = \mathrm{res}_{\mathfrak{p}} f_{\mathfrak{p}}\eta$. Thus we have $0 = \int_c \eta = j(\omega, \eta) = \sum_{\mathfrak{p}} \mathrm{res}_{\mathfrak{p}} f_{\mathfrak{p}}\eta$, the sum being extended to the poles of ω. If $\mathfrak{f}$ is the repartition of R which assigns $f_{\mathfrak{p}}$ to every pole $\mathfrak{p}$ of ω and 0 to the other places of R, then $\sum_{\mathfrak{p}} \mathrm{res}_{\mathfrak{p}} f_{\mathfrak{p}}\eta = \sum_{\mathfrak{p}} (f_{\mathfrak{p}}\eta)^{\mathfrak{p}}(1) = \eta(\mathfrak{f})$. Thus, it follows from Theorem 2, II, §5 that there exists an element f of R such that $\nu_{\mathfrak{p}}(f - f_{\mathfrak{p}}) \geqq 0$ for every pole $\mathfrak{p}$ of ω and $\nu_{\mathfrak{q}}(f) \geqq 0$ for every other place $\mathfrak{q}$. Then $\omega_1 = \omega - df$ is still a representative of the class $\Omega(c)$ and has no pole of order > 1. If u is an element of $H_1(S - |\mathfrak{d}|)$, then $\int_u \omega_1 = 2\pi(-1)^{1/2} i(c, u)$ is an integral multiple of $2\pi(-1)^{1/2}$. By Lemma 1, it follows that ω_1 is of the form $x^{-1}\, dx$, with some $x \in R$. Since $\partial c = \mathfrak{d}$, the residue of $x^{-1}\, dx$ at a place $\mathfrak{p}$ is the exponent with which $\mathfrak{p}$ occurs in $\mathfrak{d}$; on the other hand, this residue is easily seen to be equal to $\nu_{\mathfrak{p}}(x)$. This proves that $\mathfrak{d}$ is the divisor of x. The theorem of Abel is thereby proved.

Let g be the genus of R, and let $\{\eta_1, \cdots, \eta_g\}$ be a base of the space of differentials of the first kind of R. If $z \in H_1(S)$, we set $P_i(z) = \int_z \eta_i$ $(1 \leqq i \leqq g)$; then the mapping $z \to (P_1(z), \cdots, P_g(z))$ is a homomorphism of $H_1(S)$ into the additive group of the g-dimensional complex Cartesian space C^g; let $\mathfrak{P}$ be the image of $H_1(S)$ under this mapping.

Let $\mathfrak{a}$ be any divisor of degree 0 in R. Then we can find a 1-chain γ such that $\partial\gamma = \mathfrak{a}$. Set $B_i(\gamma) = \int_\gamma \eta_i$ $(1 \leqq i \leqq g)$, and denote by $A(\gamma)$ the residue class of the point $B(\gamma) = (B_1(\gamma), \cdots, B_g(\gamma)) \in C^g$ modulo $\mathfrak{P}$. Then $A(\gamma)$ depends only on $\mathfrak{a}$. For, if γ' is any other 1-chain such that $\partial\gamma' = \mathfrak{a}$, then $\gamma' - \gamma$ is a cycle. Since $B_i(\gamma') - B_i(\gamma) = \int_{\gamma'-\gamma} \eta_i$, we see that $B(\gamma') - B(\gamma) \in \mathfrak{P}$, which proves our assertion. We shall define $A(\mathfrak{a})$ to be $A(\gamma)$.

If $\mathfrak{a}_1, \mathfrak{a}_2$ are divisors of degree 0 and γ_1, γ_2 chains such that $\partial\gamma_1 = \mathfrak{a}_1$, $\partial\gamma_2 = \mathfrak{a}_2$, then $\partial(\gamma_1 + \gamma_2) = \mathfrak{a}_1\mathfrak{a}_2$, whence $A(\mathfrak{a}_1\mathfrak{a}_2) = A(\mathfrak{a}_1) + A(\mathfrak{a}_2)$. This shows that A is a homomorphism of the group of divisors of degree 0 into the group $C^g/\mathfrak{P}$. Let $\mathfrak{D}_0$ be the group of principal divisors (i.e. of divisors of elements of R). Then it follows immediately from the theorem of Abel that $A(\mathfrak{a}) = 0$ for all $\mathfrak{a} \in \mathfrak{D}_0$. Conversely, let $\mathfrak{a}$ be a divisor of degree 0 such that $A(\mathfrak{a}) = 0$. Let γ be a 1-chain such that $\partial\gamma = \mathfrak{a}$; then we have $B(\gamma) \in \mathfrak{P}$, and there exists a cycle ζ on S such that $B_i(\gamma) = \int_\zeta \eta_i$ $(1 \leqq i \leqq g)$. It follows that $\int_{\gamma-\zeta} \eta = 0$ for every differential η of the first kind; since $\partial(\gamma - \zeta) = \mathfrak{a}$, it follows from the theorem of Abel that $\mathfrak{a}$ is a principal divisor.

We see that $A(\mathfrak{a})$ depends only on the residue class $\bar{\mathfrak{a}}$ of $\mathfrak{a}$ modulo $\mathfrak{D}_0$; if we set $\bar{A}(\bar{\mathfrak{a}}) = A(\mathfrak{a})$, $\bar{A}$ is an isomorphism of $\mathfrak{D}/\mathfrak{D}_0$ (where $\mathfrak{D}$ is the group of all divisors of degree 0) with a subgroup of $C^g/\mathfrak{P}$. We shall prove that $\bar{A}(\mathfrak{D}/\mathfrak{D}_0) = C^g/\mathfrak{P}$.

Denote by Γ the subgroup of C^g containing $\mathfrak{P}$ which is such that $\Gamma/\mathfrak{P} = \bar{A}(\mathfrak{D}/\mathfrak{D}_0)$. In order to prove our assertion, it will clearly be sufficient to prove that Γ contains a neighbourhood of the origin in C^g, because this implies that $\Gamma = C^g$.

Select g distinct places $\mathfrak{v}_1, \cdots, \mathfrak{v}_g$ of R such that the divisor $\mathfrak{v}_1 \cdots \mathfrak{v}_g$ is non special (Lemma 4, VI, §8). Let x_i be a uniformizing variable at $\mathfrak{v}_i$ $(1 \leqq i \leqq g)$. We map every differential η of the first kind of R upon the point $((\eta/dx_1)(\mathfrak{v}_1), \cdots, (\eta/dx_g)(\mathfrak{v}_g)) \in C^g$. Since $\mathfrak{v}_1 \cdots \mathfrak{v}_g$ is non special, no differential $\eta \neq 0$ is mapped upon 0, and we obtain an isomorphism of the space of differentials of the first kind of R into C^g. The space of differentials of the first kind being of dimension g, our isomorphism is actually onto. Set $(\eta_i/dx_j)(\mathfrak{v}_j) = h_{ij}$ $(1 \leqq i, j \leqq g)$; then the matrix (h_{ij}) obviously has a determinant $\neq 0$.

For each j $(1 \leqq j \leqq g)$, let V_j be an open x_j-disc of x_j-center $\mathfrak{v}_j$, and let f_{ij} be a primitive of η_i on V_j such that $f_i(\mathfrak{v}_j) = 0$. If $\mathfrak{p}_j \in V_j$, we set $\mathfrak{C}_i(\mathfrak{p}_1, \cdots, \mathfrak{p}_g) =$

$\sum_{j=1}^{g} f_{ij}(\mathfrak{p}_j)$ $(1 \leqq i \leqq g)$ and we denote by $\mathfrak{C}(\mathfrak{p}_1, \cdots, \mathfrak{p}_g)$ the point $(\mathfrak{C}_1(\mathfrak{p}_1, \cdots, \mathfrak{p}_g), \cdots, \mathfrak{C}_g(\mathfrak{p}_1, \cdots, \mathfrak{p}_g)) \in C^g$. Let also $\sigma_j(\mathfrak{p}_j)$ be a 1-simplex on V_j such that $\partial\sigma_j(\mathfrak{p}_j) = \mathfrak{p}_j - \mathfrak{v}_j$, and set $\gamma(\mathfrak{p}_1, \cdots, \mathfrak{p}_g) = \sum_{j=1}^{g} \sigma_j(\mathfrak{p}_j)$. Then we have $\partial\gamma(\mathfrak{p}_1, \cdots, \mathfrak{p}_g) = (\mathfrak{p}_1 \cdots \mathfrak{p}_g)(\mathfrak{v}_1 \cdots \mathfrak{v}_g)^{-1}$, and, since $\mathfrak{C}_i(\mathfrak{p}_1, \cdots, \mathfrak{p}_g) = \int_{\gamma(\mathfrak{p}_1, \cdots, \mathfrak{p}_g)} \eta_i$, $\mathfrak{C}(\mathfrak{p}_1, \cdots, \mathfrak{p}_g) = B(\gamma(\mathfrak{p}_1, \cdots, \mathfrak{p}_g))$. Thus our assertion that Γ contains a neighbourhood of the origin in C^g will be proved if we show that $\mathfrak{C}$ maps $\prod_{j=1}^{g} V_j$ onto a neighbourhood of the origin in C^g. Now, $\mathfrak{C}_i(\mathfrak{p}_1, \cdots, \mathfrak{p}_g)$ may be written in the form $\mathfrak{C}_i^*(x_1(\mathfrak{p}_1), \cdots, x_g(\mathfrak{p}_g))$, where $\mathfrak{C}_i^*$ is a holomorphic function of its g arguments on the polycylinder $\prod_{j=1}^{g} x_j(V_j)$. Making use of the theorem on implicit functions, we see that our assertion will be proved if we show that the functional determinant of $\mathfrak{C}_1^*, \cdots, \mathfrak{C}_g^*$ is $\neq 0$ at the origin. Since f_{ij} is a primitive of η_i, we have $(df_{ij}/dx_j)(\mathfrak{v}_j) = (\eta_i/dx_j)(\mathfrak{v}_j) = h_{ij}$, whence $(\partial\mathfrak{C}_i^*/\partial x_j)(0, \cdots, 0) = h_{ij}$ and we know that the determinant of the matrix (h_{ij}) is $\neq 0$.

Thus we have proved that the group of classes of divisors of degree 0 of C^g is isomorphic with $C^g/\mathfrak{P}$. Now, we shall consider more closely the structure of the group $\mathfrak{P}$.

Using the same notation as above, we observe that the determinant of the matrix $((\eta_i/dx_j)(\mathfrak{p}_j))$ remains $\neq 0$ when $(\mathfrak{p}_1, \cdots, \mathfrak{p}_g)$ remains on some neighbourhood N of $(\mathfrak{v}_1, \cdots, \mathfrak{v}_g)$ in $\prod_{j=1}^{g} V_j$. This means that, if $(\mathfrak{p}_1, \cdots, \mathfrak{p}_g) \in N$, then no linear combination of $\eta_1, \cdots, \eta_g$ with constant coefficients not all 0 can have $\mathfrak{p}_1, \cdots, \mathfrak{p}_g$ as zeros, i.e. the divisor $\mathfrak{p}_1 \cdots \mathfrak{p}_g$ is non special. Making use of the Riemann-Roch theorem, we see that $l((\mathfrak{p}_1 \cdots \mathfrak{p}_g)^{-1}) = 1$, i.e. only the constants are multiples of $(\mathfrak{p}_1 \cdots \mathfrak{p}_g)^{-1}$, which proves that $(\mathfrak{v}_1 \cdots \mathfrak{v}_g)(\mathfrak{p}_1 \cdots \mathfrak{p}_g)^{-1}$ is not a principal divisor unless $\mathfrak{v}_1 \cdots \mathfrak{v}_g = \mathfrak{p}_1 \cdots \mathfrak{p}_g$. Since $\mathfrak{C}$ maps N upon a neighbourhood of the origin in C^g, we see that there is a neighbourhood of the origin in C^g which contains no point $\neq 0$ of $\mathfrak{P}$. This proves that $\mathfrak{P}$ is a discrete subgroup of C^g, and is therefore a free abelian group on a certain number $m \leqq 2g$ of generators. We shall see that m is equal to $2g$.

Let $\{z_1, \cdots, z_{2g}\}$ be a base of $H_1(S)$. Then the points $P(z_k) = (P_1(z_k), \cdots, P_g(z_k))$ $(1 \leqq k \leqq 2g)$ form a set of generators of $\mathfrak{P}$, and our assertion will be proved if we show that $\sum_{k=1}^{2g} a_k P(z_k) = 0$, where $a_1, \cdots, a_{2g}$ are integers, implies $a_k = 0$ $(1 \leqq k \leqq 2g)$. Set $z = \sum_{k=1}^{2g} a_k z_k$; then we have $\int_z \eta = 0$ for every differential η of the first kind. Our assertion will therefore follow from

LEMMA 2. *Let z be a homology class on the Riemann surface S of the field R. Assume that $\int_z \eta = 0$ for every differential of the first kind η. Then $z = 0$.*

Denote by $\mathfrak{D}$ the space of differentials of the second kind of R, by $\mathfrak{D}_1$ the space of differentials of the first kind, and by $\mathfrak{F}$ the space of exact differentials. Let $\Omega(z)$ be the element of $\mathfrak{D}/\mathfrak{F}$ which is associated to z by means of the result of Theorem 9, §6, and let ω be a differential belonging to the coset $\Omega(z)$. Then we have $j(\omega, \eta) = \int_z \eta = 0$ for every differential η of the first kind. On the other hand, we have $\mathfrak{D}_1 \cap \mathfrak{F} = \{0\}$, since, if $x \in R$, every pole of x is also a pole of dx; it follows that $(\mathfrak{D}_1 + \mathfrak{F})/\mathfrak{F}$ is of dimension g. We know that the function j defines a non degenerate bilinear function on $(\mathfrak{D}/\mathfrak{F}) \times (\mathfrak{D}/\mathfrak{F})$. It follows that the differentials $\theta \in \mathfrak{D}$ such that $j(\theta, \eta) = 0$ for every $\eta \in \mathfrak{D}_1$ form a subspace $\mathfrak{D}_2$ of $\mathfrak{D}$ such that $\dim (\mathfrak{D}_2/\mathfrak{F}) = \dim (\mathfrak{D}/\mathfrak{F}) - \dim \mathfrak{D}_1 = 2g - g = g$. But $\mathfrak{D}_2$ clearly contains $\mathfrak{D}_1$, whence $\mathfrak{D}_2 = \mathfrak{D}_1 + \mathfrak{F}$. It follows that the differential ω selected above may be assumed to lie in $\mathfrak{D}_1$. Since $\omega \in \Omega(z)$, the periods of ω are all integral multiples of $2\pi(-1)^{1/2}$; since ω is of the first kind, it follows from Lemma 1 that ω is of the form $x^{-1}dx$, $x \in R$. But then x can have no zero or pole, and is therefore a constant, whence $\omega = 0$ and $z = 0$.

Thus we have proved that $\mathfrak{P}$ is a discrete subgroup of C^g with $2g$ independent generators. This being the case, it is well known that $C^g/\mathfrak{P}$ is isomorphic with the $2g$-dimensional torus group. Thus we have proved

THEOREM 16. *If g is the genus of the field R, then the group of divisor classes of degree 0 of R is isomorphic with the $2g$-dimensional torus group.*

COROLLARY. *The notation being as in Theorem 16, the number of divisor classes of degree 0 which are of finite order and whose orders divide a given integer $n > 0$ is n^{2g}.*

§10. Fields of genus one

In this section, we shall assume that our field R is of genus 1.

Let $\mathfrak{D}$ be the group of divisors of degree 0 of R and let $\mathfrak{D}_0$ be the group of principal divisors. We shall see that the elements of $\mathfrak{D}/\mathfrak{D}_0$ are in a one-to-one correspondence with the points of the Riemann surface S of R. Let $\mathfrak{p}_0$ be any fixed point of S; if $\mathfrak{p} \in S$, denote by $h(\mathfrak{p})$ the coset of $\mathfrak{p}\mathfrak{p}_0^{-1}$ modulo $\mathfrak{D}_0$. We shall see that $h(\mathfrak{p}) \neq h(\mathfrak{q})$ if $\mathfrak{p} \neq \mathfrak{q}$. The divisor of a differential of R being of degree $2 \cdot 1 - 2 = 0$, no differential $\neq 0$ is multiple of $\mathfrak{p}$, and it follows from the theorem of Riemann-Roch that $l(\mathfrak{p}^{-1}) = 1$. The only elements of R which are $\equiv 0 \pmod{\mathfrak{p}^{-1}}$ are the constants, which proves that, if $\mathfrak{q} \neq \mathfrak{p}$, then $\mathfrak{q}\mathfrak{p}^{-1}$ is not a principal divisor and $h(\mathfrak{q}) \neq h(\mathfrak{p})$. Moreover, any class of divisors of degree 0 contains an element of the form $h(\mathfrak{p})$. For, let $\mathfrak{a}$ be any divisor of the class under consideration. We have, since $d(\mathfrak{a}) = 0$, $l(\mathfrak{a}^{-1}\mathfrak{p}_0^{-1}) \geqq 1$; let x be an element of R which is multiple of $\mathfrak{a}^{-1}\mathfrak{p}_0^{-1}$. Since the divisor $\mathfrak{d}(x)$ of x is of degree 0, we have $\mathfrak{d}(x) = \mathfrak{a}^{-1}\mathfrak{p}_0^{-1}\mathfrak{p}$, where $\mathfrak{p}$ is a place of R, which proves that the class of $\mathfrak{a}$ is $\mathfrak{p}\mathfrak{p}_0^{-1}$.

Denote by η a basic element of the space of differentials of the first kind of R, and by $\{z_1, z_2\}$ a base of $H_1(S)$. Set $\int_{z_1} \eta = \alpha_1$, $\int_{z_2} \eta = \alpha_2$. Then we know that α_1 and α_2 generate a discrete subgroup $\mathfrak{P}$ of the additive group C of complex numbers (cf. §9). Assign to every place $\mathfrak{p} \in S$ a 1-chain $\gamma(\mathfrak{p})$ on S such that $\partial\gamma(\mathfrak{p}) = \mathfrak{p} - \mathfrak{p}_0$, and let $A(\mathfrak{p})$ be the residue class of the number $\int_{\gamma(\mathfrak{p})} \eta$ modulo $\mathfrak{P}$ (we have seen in §9 that this residue class depends only on $\mathfrak{p}$). Then it follows from what was said just above and from the results of §9 that the mapping $\mathfrak{p} \to A(\mathfrak{p})$ maps S in a one-to-one way onto $C/\mathfrak{P}$. Now, the group $C/\mathfrak{P}$ carries a natural topology as a factor group of C. We shall see that the mapping $\mathfrak{p} \to A(\mathfrak{p})$ is continuous. Let $\mathfrak{p}$ be a point of S, and let V be an open disc containing $\mathfrak{p}$; denote by f the primitive of η on V which is such that $f(\mathfrak{p}) = 0$. If $\mathfrak{q} \in V$, let $\sigma(\mathfrak{q})$ be a

1-simplex on V such that $\delta\sigma(\mathfrak{q}) = \mathfrak{q} - \mathfrak{p}$. Then $\gamma(\mathfrak{p}) + \sigma(\mathfrak{q})$ differs from $\gamma(\mathfrak{q})$ by a cycle ζ, and we have $\int_\zeta \eta \in \mathfrak{P}$. It follows that $A(\mathfrak{q})$ is the sum of $A(\mathfrak{p})$ and of the residue class modulo $\mathfrak{P}$ of $\int_{\sigma(\mathfrak{q})} \eta = f(\mathfrak{q})$, which proves the continuity of A on V. Since S is compact, it follows that S is homeomorphic with $C/\mathfrak{P}$, i.e., with a 2-dimensional torus.

Now, let x be an element of R. Define a function x^* on C by the condition that $x^*(u) = x(\mathfrak{p}(u))$ $(u \in C)$, where $\mathfrak{p}(u)$ is the point of S which is mapped by A upon the residue class of u modulo $\mathfrak{P}$. We shall prove that x^* is a meromorphic function. Let u be any complex number; set $\mathfrak{p} = \mathfrak{p}(u)$, and define V and f as above. We can find a uniformizing variable y at $\mathfrak{p}$ such that V is a y-disc. If $\mathfrak{q} \in V$, we may write $f(\mathfrak{q}) = f^*(y(\mathfrak{q}))$, where f^* is a holomorphic function on $y(V)$. Since η has no zero on S and $df = \eta$, we see that the derivative of f^* is everywhere $\neq 0$ on V. We conclude that $y(\mathfrak{q})$ may be represented as a holomorphic function of $f(\mathfrak{q})$ for $\mathfrak{q}$ in a suitable neighbourhood of $\mathfrak{p}$. Since $f(\mathfrak{p}(v)) = v - u$ whenever $| v - u |$ is sufficiently small, we see that $y(\mathfrak{p}(v))$ is a holomorphic function of v on some neighbourhood of u. Since $x(\mathfrak{q})$ can be represented (for $\mathfrak{q} \in V$) as a meromorphic function of $y(\mathfrak{q})$, it follows that x^* is meromorphic at u.

Now, it is clear that x^* admits the periods α_1 and α_2. Therefore, x^* is an elliptic function with these periods (cf. II, §8). Conversely, let g^* be any elliptic function with the periods α_1 and α_2. Then, if $u \in C$, $g^*(u)$ depends only on the residue class of u modulo $\mathfrak{P}$. It follows that there exists a function g, defined on S, such that $g(\mathfrak{p}(u)) = g^*(u)$ for every $u \in C$. We see easily by the same kind of argument which was used above that g is everywhere meromorphic. Therefore, g is an element of R (Theorem 4, §2). Thus we have proved

THEOREM 17. *If the field R is of genus 1, there exist two complex numbers α_1 and α_2 such that R is isomorphic with the field of elliptic functions with the periods α_1 and α_2.*

Conversely, let α_1 and α_2 be complex numbers whose ratio is not real, and let R be the field of elliptic functions with the periods α_1 and α_2. Then we know (II, §8) that $R = C\langle f, f'\rangle$, where f is an elliptic function admitting as only poles the points of the group $\mathfrak{P}$ generated by α_1 and α_2 and such that $u^2f(u)$ takes the value 1 at $u = 0$, while f' is the derivative of f. Then it is easily seen that $\eta = f'^{-1} df$ is a differential of the first kind of f. The Riemann surface S of R may be identified with $C/\mathfrak{P}$. If $u \in C$, let $\gamma^*(u)$ be a 1-chain on C such that $\partial\gamma^*(u) = \{u\} - \{0\}$, and let $\gamma(u)$ be the image of $\gamma^*(u)$ in S under the natural mapping of C onto S. Since $df = f'du$, we see immediately that $\int_{\gamma(u)} \eta = u$. It follows that the group of periods of η is the group generated by α_1 and α_2. The differentials of the first kind of R which are $\neq 0$ are of the form $c\eta$, $c \in C$, $c \neq 0$, and the group of periods of $c\eta$ is generated by $c\alpha_1$ and $c\alpha_2$. We conclude that the field of elliptic functions with the periods β_1 and β_2 is isomorphic with the field of elliptic functions with the periods α_1 and α_2 if and only if there exists a complex number c such that the group generated by β_1 and β_2 is identical with the group generated by $c\alpha_1$ and $c\alpha_2$. This implies that $\beta_1 = c(k\alpha_1 + l\alpha_2)$, $\beta_2 = c(m\alpha_1 + n\alpha_2)$, where

k, l, m, n are integers such that $kn - lm = \pm 1$. We may also say that $\tau = \alpha_2/\alpha_1$ and $\tau' = \beta_2/\beta_1$ must be related by an equation of the form

$$\tau' = \frac{m + n\tau}{k + l\tau}$$

where k, l, m, n are integers such that $kn - lm = \pm 1$. This explains the relationship which exists between the theory of elliptic functions and that of modular functions.

§11. The Riemann surface as an analytic manifold

For the definition and elementary properties of analytic manifolds, we refer the reader to the author's book *Theory of Lie groups*, I, Chapters III and V.

We propose to define on the Riemann surface S of the field R a structure of analytic manifold. Let $\mathfrak{p}$ be any point of S and let x be a uniformizing variable at $\mathfrak{p}$. If $\mathfrak{q}$ is not a pole of S, we denote by $x_1(\mathfrak{q})$ and $x_2(\mathfrak{q})$ the real and the imaginary parts of $x(\mathfrak{q})$. Then x_1 and x_2 are real valued functions defined on a neighbourhood of $\mathfrak{p}$. Call $\mathfrak{A}_x(\mathfrak{p})$ the class of real valued functions defined on neighbourhoods of $\mathfrak{p}$ in S and which depend analytically on x_1 and x_2 around $\mathfrak{p}$. The class $\mathfrak{A}_x(\mathfrak{p})$ does not depend on the choice of x. For, let y be any uniformizing variable at $\mathfrak{p}$, and let y_1 and y_2 be the real and the imaginary part of y. On some neighbourhood of $\mathfrak{p}$, y may be expressed in the form $F(x)$, where F is a function which is defined and holomorphic on a neighbourhood of 0 in the complex plane. It follows immediately that y_1 and y_2 are in $\mathfrak{A}_x(\mathfrak{p})$, whence $\mathfrak{A}_y(\mathfrak{p}) \subset \mathfrak{A}_x(\mathfrak{p})$. We would see in the same way that $\mathfrak{A}_x(\mathfrak{p}) \subset \mathfrak{A}_y(\mathfrak{p})$, whence $\mathfrak{A}_x(\mathfrak{p}) = \mathfrak{A}_y(\mathfrak{p})$. The class $\mathfrak{A}_x(\mathfrak{p})$ will henceforth be denoted by $\mathfrak{A}(\mathfrak{p})$.

Let V be an open x-disc of x-center $\mathfrak{p}$. Then V contains a set V_1 which is mapped topologically by the mapping $\mathfrak{q} \to (x_1(\mathfrak{q}), x_2(\mathfrak{q}))$ onto a square of center origin in the 2-dimensional real Cartesian space. If $\mathfrak{q} \in V_1$, then $x - x(\mathfrak{q})$ is uniformizing at $\mathfrak{q}$. If we set $a_i = x_i(\mathfrak{q})$ $(i = 1, 2)$, then $\mathfrak{A}(\mathfrak{q})$ is the class of functions f defined on neighbourhoods of $\mathfrak{q}$ and which depend analytically on $x_1 - a_1$ and $x_2 - a_2$ around $\mathfrak{q}$. This last condition is equivalent to the condition that f should depend analytically on x_1 and x_2 around $\mathfrak{q}$. It follows that the assignment $\mathfrak{p} \to \mathfrak{A}(\mathfrak{p})$ defines a structure of analytic manifold of dimension 2 on S.

We shall say that a complex valued function f, defined on a neighbourhood of a point $\mathfrak{p} \in S$ is *analytic* at $\mathfrak{p}$ if its real and imaginary part are analytic at $\mathfrak{p}$; and we shall say that f is analytic on an open subset U of S on which it is defined if it is analytic at every point of U. This notion must be carefully distinguished from the notion of holomorphic function on U, as defined in §2: a holomorphic function on U is also analytic, but the converse is not true.

Let $L(\mathfrak{p})$ be the tangent vector space to S at a point $\mathfrak{p}$. In *Theory of Lie groups*, I, we have defined a differential $\omega_{\mathfrak{p}}$ at $\mathfrak{p}$ to be a real valued linear function on the space $L(\mathfrak{p})$, and we have shown that the differentials at $\mathfrak{p}$ form a vector space of dimension 2 over the field of real numbers. If x is a uniformizing variable at $\mathfrak{p}$, then the differentials at $\mathfrak{p}$ of the real and imaginary parts x_1 and x_2 of x form a base of the space of differentials at $\mathfrak{p}$. We shall generalize this notion by calling differential at $\mathfrak{p}$ any complex valued linear function on $L(\mathfrak{p})$. Thus the differentials at $\mathfrak{p}$ now form a vector space of dimension 2 over the field of complex numbers, which has a base consisting of the differentials at $\mathfrak{p}$ of the functions x_1 and x_2. If f is a complex valued function which is analytic at $\mathfrak{p}$, then we call differential of f at $\mathfrak{p}$ the differential $(df_1)_\mathfrak{p} + (-1)^{1/2} (df_2)_\mathfrak{p}$, where f_1 and f_2 are the real and imaginary parts of f.

Let U be an open subset of S. By a *differential form* of degree 1 on U we mean a mapping $\mathfrak{p} \to \omega_\mathfrak{p}$ which, to every $\mathfrak{p} \in U$, assigns an element $\omega_\mathfrak{p}$ of the space of differentials at $\mathfrak{p}$. This notion is distinct from, but related to, the notion of holomorphic differential which was defined in §2. Let ω be a holomorphic differential on U. Let $\mathfrak{p}$ be a point of U, and x a uniformizing variable at $\mathfrak{p}$. Then we may write $\omega = f\,dx$, where f is holomorphic at $\mathfrak{p}$. Denote by x_1 and x_2 the real and imaginary parts of x; then the expression $f(\mathfrak{p})\ ((dx_1)_\mathfrak{p} + (-1)^{1/2}(dx_2)_\mathfrak{p})$ represents an element $\omega_\mathfrak{p}$ of the space of differentials at $\mathfrak{p}$. This element does not depend on x. For, let y be any uniformizing variable at $\mathfrak{p}$. Then there is a neighbourhood of $\mathfrak{p}$ on which y may be expressed in the form $F(x)$, where F is a function which is defined and holomorphic on some neighbourhood of 0 in the complex plane. Let α be the value at 0 of the derivative of F; then we have $(dy/dx)\ (\mathfrak{p}) = \alpha$. If y_1 and y_2 are the real and imaginary parts of y, then we have $(dy_1)_\mathfrak{p} + (-1)^{1/2}(dy_2)_\mathfrak{p} = \alpha((dx_1)_\mathfrak{p} + (-1)^{1/2}\ (dx_2)_\mathfrak{p})$. If $\omega = g\ dy$, then we have $f = g(dy/dx)$, whence $f(\mathfrak{p})\ ((dx_1)_\mathfrak{p} + (-1)^{1/2}(dx_2)_\mathfrak{p}) = g(\mathfrak{p})\ \alpha((dx_1)_\mathfrak{p} + (-1)^{1/2}(dx_2)_\mathfrak{p}) = g(\mathfrak{p})((dy_1)_\mathfrak{p} + (-1)^{1/2}(dy_2)_\mathfrak{p})$, which proves our assertion. The mapping $\mathfrak{p} \to \omega_\mathfrak{p}$ is a differential form on U, which we shall denote by the same symbol ω as the given holomorphic differential. In particular, if h is any holomorphic function on U, then dh may be considered either as a holomorphic differential on U or as a differential form of degree 1 on U.

In *Theory of Lie groups*, I, we have defined the exterior product $\omega \square \omega'$ of two differential forms of degree 1 on an open set U; this notion may be extended without difficulty to the case of complex differential forms. It is noteworthy that $\omega \square \omega' = 0$ whenever ω and ω' are holomorphic (because, at every $\mathfrak{p} \in U$, we may write $\omega_\mathfrak{p} = \alpha(dx)_\mathfrak{p}$, $\omega'_\mathfrak{p} = \alpha'(dx)_\mathfrak{p}$, where α and α' are complex numbers, while x is a uniformizing variable at $\mathfrak{p}$; it follows that $\omega_\mathfrak{p} \square \omega'_\mathfrak{p} = \alpha\alpha'\ (dx)_\mathfrak{p} \square (dx)_\mathfrak{p} = 0$).

Denote by ω a differential form of degree 1 on an open set U, by $\mathfrak{p}$ a point of U, by x a uniformizing variable at $\mathfrak{p}$, and by x_1 and x_2 the real and imaginary parts of x. Then there exists a neighbourhood N of $\mathfrak{p}$ such that we have, for $\mathfrak{q} \in N$, $\omega_\mathfrak{q} = f_1(x_1(\mathfrak{q}), x_2(\mathfrak{q}))(dx_1)_\mathfrak{q} + f_2(x_1(\mathfrak{q}), x_2(\mathfrak{q}))(dx_2)_\mathfrak{q}$, where f_1 and f_2 are complex valued functions defined on a neighbourhood of the origin in the plane. If f_1 and f_2 are of class C_k at the origin (i.e., admit continuous partial derivatives up to order k on a neighbourhood of the origin), then ω is said to be *of class C_k at* $\mathfrak{p}$; if f_1 and f_2 are analytic functions of the real variables x_1 and x_2 on a neighbourhood of the origin, then ω is said to be *analytic* at $\mathfrak{p}$. It is easily seen that these definitions do not depend on the choice of the uniformizing variable x (for, if x is replaced by some other uniformizing variable at $\mathfrak{p}$, then f_1 and f_2 are replaced by functions which are linear combinations of f_1 and f_2 with coefficients which are themselves analytic functions of x_1 and x_2 on a neighbourhood of the origin). If ω is of class C_k (or analytic) at all points $\mathfrak{p}$ of U, then ω is said to be of class C_k (or analytic) on U.

In *Theory of Lie groups*, I, we have defined the differential $d\omega$ of a real analytic differential form ω. The definition may be extended without difficulty to the case of a complex differential form which is of class C_1. If f is a function of class C_1 and ω a differential form of class C_1, then we have $d(f\omega) = df \,\square\, \omega + f\, d\omega$. If ω is a differential form of class C_2, then $d\omega$ is of class C_1 and $dd\omega = 0$. If ω is a holomorphic differential on an open set U, then $d\omega = 0$. For, let $\mathfrak{p}$ be a point of U and let V be an open disc containing $\mathfrak{p}$ and contained in U. Then ω admits a primitive f on V, and we have, on V, $d\omega = ddf = 0$.

A 1-simplex σ on S is said to be *differentiable* if the mapping $t \to \sigma(1 - t, t)$ of the interval $[0, 1]$ into S may be extended to a mapping $t \to \sigma^*(t)$ of some open interval $]-\alpha, 1 + \beta[$ (with $\alpha > 0, \beta > 0$) into S which is everywhere of class C_1 (i.e., if f is any analytic function at $\sigma^*(t_0)$, where $-\alpha < t_0 < 1 + \beta$, then the function $f(\sigma^*(t))$ admits a continuous derivative on some neighbourhood of t_0). This being the case, the mapping $f \to (df(\sigma^*(t))/dt)_{t=t_0}$ of the class of analytic functions at $\sigma^*(t_0)$ into the field of real numbers is a tangent vector L_{t_0} at $\sigma^*(t_0)$, and this tangent vector depends only on σ and t_0 (not on the extension σ^*); L_{t_0} is called the *tangent vector to σ at the point of parameter* t_0. If ω is a holomorphic differential on an open set U containing $|\sigma|$ and if ω admits a primitive f on U, then we know that $\int_\sigma\omega = f(\sigma^*(1)) - f(\sigma^*(0))$ (Lemma 2, §4); this is equal to $\int_0^1 ((df(\sigma^*(t))/dt)\, dt = \int_0^1 \omega_{\sigma^*(t)}(L_t)\, dt$. It is therefore natural to set up the following definition: if ω is a differential form of class C_0 on some open set containing $|\sigma|$, then we call *integral of ω on σ* the number $\int_0^1 \omega_{\sigma^*(t)}(L_t)\, dt$.

A 1-chain γ on S is said to be differentiable if every simplex occurring in γ is differentiable. Let then $\gamma = \sum_{k=1}^h a_k\sigma_k$, where $\sigma_1, \cdots, \sigma_h$ are the simplices occurring in γ, and let ω be a differential form which is defined and of class C_0 on some open set containing $|\gamma|$. Then we define the integral of ω on γ to be the number $\int_\gamma \omega = \sum_{k=1}^h a_k \int_{\sigma_k} \omega$. It is easily seen that, if γ is a differentiable chain, then so is subd γ, and that $\int_{\text{subd}\,\gamma} \omega = \int_\gamma \omega$.

Now, we shall prove that S is orientable as an analytic manifold (cf. *Theory of Lie groups*, I), and we shall fix a definite orientation of S. Denote by $\mathfrak{D}_\mathfrak{p}$ the space of complex differentials at a point $\mathfrak{p}$ of S; generalizing what we did for the case of real differentials, we introduce the Grassman algebra $\mathfrak{G}_\mathfrak{p}$ over the space $\mathfrak{D}_\mathfrak{p}$. This is an algebra over the field of complex numbers, and the homogeneous elements of $\mathfrak{G}_\mathfrak{p}$ of degree 2 are the skew-symmetric complex valued bilinear functions on the tangent space $L(\mathfrak{p})$. They form a vector space $\mathfrak{D}_\mathfrak{p}^2$ of dimension 1 over the field of complex numbers, and to every pair (ω, ω') of elements of $\mathfrak{D}_\mathfrak{p}$ there corresponds an element $\omega \,\square\, \omega'$ of $\mathfrak{D}_\mathfrak{p}^2$, their exterior product. Those elements of $\mathfrak{D}_\mathfrak{p}^2$ which are real valued bilinear functions on $L(\mathfrak{p})$ will be called real; they are the homogeneous elements of degree 2 of the Grassman algebra over the space of real differentials. Now, let x be a uniformizing variable at $\mathfrak{p}$, and let x_1 and x_2 be the real and imaginary parts of x. Then we know that $(dx_1)_\mathfrak{p}$ and $(dx_2)_\mathfrak{p}$ form a base of $\mathfrak{D}_\mathfrak{p}$, and that $(dx_1)_\mathfrak{p} \square (dx_2)_\mathfrak{p}$ is a basic element of the space of real elements of $\mathfrak{D}_\mathfrak{p}^2$. Let $\bar{x}$ be the imaginary conjugate function of x. Since $(dx)_\mathfrak{p} = (dx_1)_\mathfrak{p} + (-1)^{1/2}(dx_2)_\mathfrak{p}$ and $(d\bar{x})_\mathfrak{p} = (dx_1)_\mathfrak{p} - (-1)^{1/2}(dx_2)_\mathfrak{p}$, $(dx)_\mathfrak{p}$ and $(d\bar{x})_\mathfrak{p}$ form a base of $\mathfrak{D}_\mathfrak{p}$, and we have $(dx)_\mathfrak{p} \square (d\bar{x})_\mathfrak{p} = -2(-1)^{1/2}(dx_1)_\mathfrak{p} \square (dx_2)_\mathfrak{p}$. This shows that $(-1)^{1/2}(dx)_\mathfrak{p} \square (d\bar{x})_\mathfrak{p}$ is a real basic element of $\mathfrak{D}_\mathfrak{p}^2$. Now, let y be any other uniformizing variable at $\mathfrak{p}$. Then we have $(dy)_\mathfrak{p} = \alpha(dx)_\mathfrak{p}$, where α is the value of dy/dx at $\mathfrak{p}$, whence $(-1)^{1/2}(dy)_\mathfrak{p} \square (d\bar{y})_\mathfrak{p} = \alpha\bar{\alpha}((-1)^{1/2}(dx)_\mathfrak{p} \square (d\bar{x})_\mathfrak{p})$, and the coefficient $\alpha\bar{\alpha}$ is > 0. This shows that we may define a real element φ of $\mathfrak{D}_\mathfrak{p}^2$ to be *positive* if $\varphi = \beta(-1)^{1/2}(dx)_\mathfrak{p} \square (d\bar{x})_\mathfrak{p}$ with $\beta > 0$: this definition does not depend on the choice of the uniformizing variable x.

Since $x - x(\mathfrak{q})$ is uniformizing at every point $\mathfrak{q}$ of some neighbourhood of $\mathfrak{p}$, we see immediately that the differential form of degree 2, $(-1)^{1/2}(dx) \square (d\bar{x})$, is positive at all points of some neighbourhood of $\mathfrak{p}$. This form being obviously continuous, we see that our definition of positivity for real elements of $\mathfrak{D}_\mathfrak{p}^2$ yields an orientation of S.

Having oriented the manifold S, we may integrate on S any continuous real differential form of degree 2. This definition can be extended without any difficulty to the case of continuous complex valued differential forms of degree 2.

Now, let φ be a differential form of degree 2 which is defined and continuous only on some open subset U of S. We wish to define the integral (when it exists) of φ on U. To do this, introduce the class F of continuous functions f on S which have the following properties: a) we have $|f(\mathfrak{p})| \leqq 1$ for all $\mathfrak{p} \in S$; b) there exists a compact subset K of U (depending on f) such that $f(\mathfrak{p}) = 0$ for all $\mathfrak{p} \notin K$. If $f \in F$, then $f\varphi$ may be extended to a continuous differential form (also denoted by $f\varphi$) on S, which is everywhere 0 outside U, and the integral $\int_S f\varphi$ is defined. If K is any compact subset of U, we denote by F_K the set of functions $f \in F$ which are equal to 1 on K. Then F_K is never empty (*Theory of Lie groups*, I, Lemma 1, Chapter V, §VII). Moreover, if K and K' are compact subsets of U, then so is $K \cup K'$, and $F_{K \cup K'} = F_K \cap F_{K'}$. Following the terminology of N. Bourbaki (*Eléments de Mathématique*, Livre III, *Topologie Générale*, Chapitre I), this implies that the sets F_K form a base of a filter Φ on the set F. We shall say that φ is *integrable* on U if the mapping $f \rightarrow \int_S f\varphi$ of F into the space of complex numbers tends to a limit I relatively to the filter Φ; I is then called the *integral of φ on U* and is denoted by $\int_U \varphi$. It is clear that, if $U = S$, then any continuous differential form φ of degree 2 on S is integrable in the sense of this definition, and that its integral is the already defined $\int_S \varphi$.

In order for φ to be integrable on U, it is sufficient that the numbers $\int_S f\varphi$ for all $f \in F$ form a bounded set. For, assume that this condition is satisfied. Since C is complete, in order to prove the existence of the limit I, it will be sufficient to prove that, for any $\varepsilon > 0$, there exists a compact subset K_ε of U such that $|\int_S f\varphi - \int_S f'\varphi| \leqq \varepsilon$ whenever f and f' belong to F_K. Assume for a moment that no such set exists. Then to every compact subset K of U we can associate a function $g_K \in F$ which is 0 in K and such that $| \int_S g_K \varphi | \geqq \varepsilon/2$ (we take $g_K = (1/2)(f - f')$, where f and f' are elements of F_K such that $| \int_S f\varphi - \int_S f'\varphi | > \varepsilon$). We define inductively a sequence (K_n) of compact subsets of U as follows: K_1 is the empty set; $K_1, \cdots, K_n$ being defined, K_{n+1} is a compact subset of U outside which $g_{K_1}, \cdots, g_{K_n}$ are all equal to 0. Let ζ_n be a number of absolute value 1 such that $\int_S \zeta_n g_{K_n} \varphi$ is real and > 0 (and therefore $> \varepsilon/2$). Then it is clear that $h_n = \zeta_1 g_{K_1} + \cdots + \zeta_n g_{K_n}$ is in F and that $\int_S h_n \varphi > \varepsilon n/2$, in contradiction with the assumption that $|\int_S f\varphi |$ is bounded for all $f \in F$.

If φ can be extended to a differential form φ_1 which is defined and continuous on some open set U_1 containing the adherence $\bar{U}$ of U, then φ is integrable on U. For we can find a continuous function h on S which is equal to 1 on $\bar{U}$ and to 0 outside U_1; then $h\varphi_1$ can be extended to a continous differential form φ_2 on S, and φ_2 coincides with φ on U.

§12. THE BILINEAR INEQUALITIES OF RIEMANN

Let $P = \{\mathfrak{p}_1, \cdots, \mathfrak{p}_h\}$ be a finite subset of the Riemann surface S of the field R, and let ω be a differential form of degree 1 which is defined and of class C_1 on the set $S - P$; then $d\omega$ is a continuous differential form of degree 2 on $S - P$.

For each i $(1 \leqq i \leqq h)$, let x_i be a uniformizing variable at $\mathfrak{p}_i$ and let D_i be a closed x_i-disc of x_i-center $\mathfrak{p}_i$; assume furthermore that $D_i \cap D_j = \emptyset$ if $i \neq j$. Denote by r_i the x_i-radius of D_i and by C_i its circumference. If θ is a real number, denote by $\mathfrak{q}_i(\theta)$ the point of C_i such that $x_i(\mathfrak{q}_i(\theta)) = r_i \exp(2\pi(-1)^{1/2}\theta)$. Then the mapping $(1 - \theta, \theta) \to \mathfrak{q}_i(\theta)$ (where θ varies from 0 to 1) is a differentiable 1-simplex γ_i on C_i. It is easily seen that γ_i is a cycle and belongs to the positive generator of $H_1(C_i)$. Denote by U the open set $S - \bigcup_{i=1}^h D_i$; then we wish to prove the formula

$$\int_U d\omega + \textstyle\sum_{i=1}^h \displaystyle\int_{\gamma_i} \omega = 0 \tag{1}$$

(which is a special case of the general Stokes formula).

Let $\bar{U}$ be the adherence of U. If $\mathfrak{q} \in U$, we select a uniformizing variable $y_\mathfrak{q}$ at $\mathfrak{q}$ and a closed neighbourhood $N(\mathfrak{q})$ of $\mathfrak{q}$ in U which is mapped topologically by $y_\mathfrak{q}$ upon a closed square of center origin in the plane. If $\mathfrak{q} \in \bar{U} - U$, then $\mathfrak{q}$ belongs to one and only one of the sets C_i. In that case, there is a neighbourhood of $\mathfrak{q}$ on which $x_i^{-1}dx_i$ admits a primitive $y_\mathfrak{q}$, which we normalize by the condition that $y_\mathfrak{q}(\mathfrak{q}) = 0$. It is easily seen that $\mathfrak{q}$ has a closed neighbourhood $N(\mathfrak{q})$ on which $y_\mathfrak{q}$ is defined and which is mapped topologically by $y_\mathfrak{q}$ onto a closed square of center origin in the plane. We may furthermore assume that $N(\mathfrak{q})$ does not meet any of the sets D_j for $j \neq i$.

Let $N'(\mathfrak{q})$ be a closed neighbourhood of $\mathfrak{q}$ contained in the interior of $N(\mathfrak{q})$. The compact set $\bar{U}$ is covered by the sets $N'(\mathfrak{q})$ for all $\mathfrak{q} \in \bar{U}$; it is therefore also covered by a finite number of these sets, say by those for which $\mathfrak{q} \in Q$, where Q is a finite subset of $\bar{U}$.

It is well known that we can assign to every $\mathfrak{q} \in Q$ a function $f_\mathfrak{q}$ of class C_1 on S which is everywhere 0 outside $N(\mathfrak{q})$ in such a way that $\sum_{\mathfrak{q} \in Q} f_\mathfrak{q}(\mathfrak{r}) = 1$ for every $\mathfrak{r} \in U$. Since ω coincides on U with $\sum_{\mathfrak{q} \in Q} f_{\mathfrak{q}}\, \omega$, we see that it is sufficient to prove formula (1) under the assumption that ω is everywhere 0 outside some set $N(\mathfrak{q})$.

Assume first that ω is everywhere 0 outside a set $N(\mathfrak{q})$ corresponding to a point $\mathfrak{q} \in U$. Then we have $\int_{\gamma_i} \omega = 0$ $(1 \leqq i \leqq h)$, and we have to prove that $\int_U d\omega = 0$. Let y_1 and y_2 be the real and imaginary parts of $y_{\mathfrak{q}}$. If $\mathfrak{r} \in N(\mathfrak{q})$, we may write

$$\omega_{\mathfrak{r}} = A_1(y_1(\mathfrak{r}), y_2(\mathfrak{r}))(dy_1)_{\mathfrak{r}} + A_2(y_1(\mathfrak{r}), y_2(\mathfrak{r}))(dy_2)_{\mathfrak{r}} \tag{2}$$

where A_1 and A_2 are continuously differentiable functions on the set $y_{\mathfrak{q}}(N(\mathfrak{q}))$ of the plane. We have

$$d\omega_{\mathfrak{r}} = \left(\frac{\partial A_2}{\partial y_1}(y_1(\mathfrak{r}), y_2(\mathfrak{r})) - \frac{\partial A_1}{\partial y_2}(y_1(\mathfrak{r}), y_2(\mathfrak{r}))\right)(dy_1)_{\mathfrak{r}} \,\square\, (dy_2)_{\mathfrak{r}}. \tag{3}$$

Since $d\omega_{\mathfrak{r}}$ is 0 outside $N(\mathfrak{q})$ and $dy_1 \,\square\, dy_2$ is a positive differential form of degree 2 on $N(\mathfrak{q})$, we have

$$\int_U d\omega = \int_{y_{\mathfrak{q}}(N(\mathfrak{q}))} \left(\frac{\partial A_2}{\partial y_1} - \frac{\partial A_1}{\partial y_2}\right) dy_1\, dy_2.$$

Let $2a$ be the side of the square $y_{\mathfrak{q}}(N(\mathfrak{q}))$. Then

$$\int_{y_{\mathfrak{q}}(N(\mathfrak{q}))} \frac{\partial A_2}{\partial y_1}\, dy_1\, dy_2 = \int_{-a}^{+a} \left(\int_{-a}^{+a} \frac{\partial A_2}{\partial y_1}\, dy_1\right) dy_2$$

$$= \int_{-a}^{+a} (A_2(a, y_2) - A_2(-a, y_2))\, dy_2 = 0$$

because A_1 and A_2 vanish on the boundary of $y_{\mathfrak{q}}(N(\mathfrak{q}))$, as follows from the fact that ω vanishes outside $N(\mathfrak{q})$. In the same way we would prove that $\int_{y_{\mathfrak{q}}(N(\mathfrak{q}))} (\partial A_1/\partial y_2)\, dy_1\, dy_2 = 0$, whence $\int_U d\omega = 0$.

Consider now the case where $\mathfrak{q} \in C_i$ for some i. Then we have to prove that $\int_U d\omega + \int_{\gamma_i} \omega = 0$. Let y_1 and y_2 be the real and imaginary parts of $y_{\mathfrak{q}}$. Then, if $\mathfrak{r} \in N(\mathfrak{q})$, $(dy_1)_{\mathfrak{r}}$ and $(dy_2)_{\mathfrak{r}}$ form a base of the space of differentials at $\mathfrak{r}$, because $dy_{\mathfrak{q}} = x_i^{-1} dx_i$, and it is easily seen that $dy_1 \,\square\, dy_2$ is positive at every point of $N(\mathfrak{q})$. If $\mathfrak{r} \in N(\mathfrak{q})$, we still express $\omega_{\mathfrak{r}}$ and $d\omega_{\mathfrak{r}}$ by the formulas (2) and (3) above. On the other hand, we have $x(\mathfrak{r}) = x(\mathfrak{q}) \exp(y(\mathfrak{r}))$ from which it follows that $U \cap N(\mathfrak{q})$ is the set of points $\mathfrak{r}$ of $N(\mathfrak{q})$ at which $y_1(\mathfrak{r}) > 0$. We assert that

$$\int_U d\omega = \int_{-a}^{+a} \left(\int_0^a \left(\frac{\partial A_2}{\partial y_1} - \frac{\partial A_1}{\partial y_2}\right) dy_1\right) dy_2, \tag{4}$$

where $2a$ is the side of the square $y_{\mathfrak{q}}(N(\mathfrak{q}))$. Set $B = \partial A_2/\partial y_1 - \partial A_1/\partial y_2$. Let f be any continuous function on S which is 0 outside a compact subset K of U. Then it is clear that $\int_U f\, d\omega = \int_{y_{\mathfrak{q}}(N(\mathfrak{q}))} f^*(y_1, y_2) B dy_1 dy_2$ where $f^*(y_1, y_2)$ is the value taken by f at the point $\mathfrak{r}$ of $N(\mathfrak{q})$ at which $y(\mathfrak{r}) = y_1 + (-1)^{1/2} y_2$. Since f is everywhere 0 outside U, we have $f^*(y_1, y_2) = 0$ if $y_1 \leqq 0$, whence

$$\int_U f\, d\omega = \int_{-a}^{+a} \left(\int_0^a f^*(y_1, y_2)\, B dy_1\right) dy_2.$$

Let ε be a positive number. If $f(\mathfrak{r}) = 1$ at every point $\mathfrak{r} \in N(\mathfrak{q})$ at which $y_1(\mathfrak{r}) \geqq \varepsilon$, and $|f(\mathfrak{r})| \leqq 1$ for all $\mathfrak{r} \in S$, then

$$\left| \int_U f\, d\omega - \int_{-a}^{+a} \left(\int_0^a B dy_1 \right) dy_2 \right| \leqq \int_{-a}^{+a} \left(\int_0^{\epsilon} |B|\, dy_1 \right) dy_2$$

and this quantity tends to 0 with ε, which proves formula (4). We have

$$\int_{-a}^{+a} \left(\int_0^a \frac{\partial A_2}{\partial y_1} dy_1 \right) dy_2 = -\int_{-a}^{+a} A_2(0, y_2)\, dy_2$$

because $A_2(a, y_2) = 0$ as follows from the fact that ω vanishes outside $N(\mathfrak{q})$. We shall prove that the right side of this formula is equal to $-\int_{\gamma_i} \omega$.

The point $\mathfrak{q}_i(\theta)$ being defined as above, we observe that, t being any real number, and $\mathfrak{A}(\mathfrak{q}_i(t))$ the class of analytic functions at $\mathfrak{q}_i(t)$ on S, the mapping $g \to (dg(\mathfrak{q}_i(\theta))/d\theta)_{\theta=t}$ is a tangent vector L_t to S at $\mathfrak{q}_i(t)$; if $0 \leqq t \leqq 1$, then L_t is the tangent vector to γ_i at the point of parameter t, whence $\int_{\gamma_i} \omega = \int_0^1 \omega_{\mathfrak{q}_i(t)}(L_t)\, dt$. On the other hand, we have $L_{t+1} = L_t$, whence also $\int_{\gamma_i} \omega = \int_b^{b+1} \omega_{\mathfrak{q}_i(t)}(L_t)\, dt$ for any real number b. Let t_0 be a real number such that $\mathfrak{q}_i(t_0) = \mathfrak{q}$. We know that we have, for $\mathfrak{r} \in N(\mathfrak{q})$, $x_i(\mathfrak{r}) = x_i(\mathfrak{q}) \exp(y_{\mathfrak{q}}(\mathfrak{r}))$. It follows that, if $\mathfrak{q}_i(t) \in N(\mathfrak{q})$, then $t - (1/2\pi)y_2(\mathfrak{q}_i(t)) - t_0$ is an integer. We see easily that $a < \pi$ and that $\mathfrak{q}_i(t) \in N(\mathfrak{q})$ for $|t - t_0| \leqq a/2\pi$. Thus we have

$$\int_{\gamma_i} \omega = \int_{t_0-1/2}^{t_0+1/2} \omega_{\mathfrak{q}_i(t)}(L_t)\, dt = \int_{t_0-a/2\pi}^{t_0+a/2\pi} \omega_{\mathfrak{q}_i(t)}(L_t)\, dt$$

because ω vanishes outside $N(\mathfrak{q})$. On the other hand, we have $(dy_{\mathfrak{q}})_{\mathfrak{q}_i(t)}(L_t) = (dy_{\mathfrak{q}}(\mathfrak{q}_i(\theta))/d\theta)_{\theta=t} = 2\pi(-1)^{1/2}$, because $y_{\mathfrak{q}}(\mathfrak{q}_i(\theta)) = 2\pi(-1)^{1/2}(\theta - t_0)$ if $|\theta - t_0| \leqq a/2\pi$. Thus $\omega_{\mathfrak{q}_i(t)}(L_t) = 2\pi A_2(0, y_2(\mathfrak{q}_i(t)))$, whence $\int_{\gamma_i} \omega = \int_{-a}^{+a} A_2(0, y_2) dy_2$. On the other hand, we have

$$\int_{-a}^{+a} \left(\int_0^a \frac{\partial A_1}{\partial y_2} dy_1 \right) dy_2 = \int_0^a \left(\int_{-a}^{+a} \frac{\partial A_1}{\partial y_2} dy_2 \right) dy_1$$

$$= \int_0^a (A_1(y_1, a) - A_1(y_1, -a))\, dy_1 = 0$$

because ω vanishes outside $N(\mathfrak{q})$. Formula (1) is now completely proved. We shall make use of this formula in the proof of

THEOREM 18 (BILINEAR INEQUALITIES OF RIEMANN). *Let ω be a differential of the first kind $\neq 0$ of the field R, and let $\{z_1, \cdots, z_{2g}\}$ be a canonical base of the first-dimensional homology group of the Riemann surface of R. Denote by π_k the period of ω relative to z_k ($1 \leqq k \leqq 2g$). Then we have*

$$(2\pi(-1)^{1/2})^{-1} \sum_{k=1}^{g} (\pi_{k+g}\, \bar{\pi}_k - \pi_k\, \bar{\pi}_{k+g}) > 0.$$

Making use of Theorem 11, §8, we see that there exists a differential η of the second kind of R such that $\int_{z_k} \eta = \bar{\pi}_k$ $(1 \leqq k \leqq 2g)$. It follows from formula (5), §8 that

$$\frac{1}{2\pi(-1)^{1/2}} \sum_{k=1}^{g} (\pi_{k+g}\,\bar{\pi}_k - \pi_k\,\bar{\pi}_{k+g}) = j(\eta, \omega).$$

Let $\mathfrak{p}_1, \cdots, \mathfrak{p}_h$ be the distinct poles of η. For each i $(1 \leqq i \leqq h)$ let x_i be a uniformizing variable at $\mathfrak{p}_i$. Denote by D_i a closed x_i-disc of x_i-center $\mathfrak{p}_i$ and x_i-radius r_i; assume that $D_i \cap D_j = \emptyset$ for $i \neq j$. Let C_i be the circumference of D_i and c_i the positive generator of $H_1(C_i)$. For each i we can find an element $f_i \in R$ such that $\eta - df_i$ has no pole at $\mathfrak{p}_i$ (Theorem 14, VI, §8). We may furthermore assume that r_i is so small that $\mathfrak{p}_i$ is the only pole of f_i in D_i. Then we have

$$j(\eta, \omega) = \textstyle\sum_{i=1}^{h} \operatorname{res}_{\mathfrak{p}_i} f_i\,\omega = \sum_{i=1}^{h} (2\pi(-1)^{1/2})^{-1} \displaystyle\int_{C_i} f_i\,\omega.$$

Now, we introduce the imaginary conjugate $\bar{\omega}$ of the form ω; $\bar{\omega}$ is defined by the formula $\bar{\omega}_\mathfrak{p}(L) = \overline{\omega_\mathfrak{p}(L)}$ whenever $\mathfrak{p} \in S$ and L is a tangent vector to S at $\mathfrak{p}$. If $\omega = u\,dx$, where x is a non constant element of R and $u \in R$, then it is easily seen that $\bar{\omega}_\mathfrak{p} = \overline{u(\mathfrak{p})}\,(d\bar{x})_\mathfrak{p}$, provided $\mathfrak{p}$ is not a zero or a pole of dx. It follows that $\bar{\omega}_\mathfrak{p}$ is everywhere analytic on S. If V is any open disc on S and λ a primitive of ω on V, then we have $(d\bar{\lambda})_\mathfrak{p} = \bar{\omega}_\mathfrak{p}$ for every $\mathfrak{p} \in V$.

We shall prove that the differential form $\eta - \bar{\omega}$, which is analytic on $U = S - \{\mathfrak{p}_1, \cdots, \mathfrak{p}_h\}$, is the differential of an analytic function on U. Let ζ be any differentiable 1-cycle on U. Then $\int_\zeta (\eta - \bar{\omega}) = \int_\zeta \eta - \int_\zeta \bar{\omega}$; the second term is obviously the imaginary conjugate of $\int_\zeta \omega$. If ζ belongs to the homology class $\sum_{k=1}^{2g} a_k z_k$ on S, then $\int_\zeta \eta = \sum_{k=1}^{2g} a_k \int_{z_k} \eta$, $\int_\zeta \omega = \sum_{k=1}^{2g} a_k \int_{z_k} \omega$, whence $\int_\zeta (\eta - \bar{\omega}) = 0$ since $\int_{z_k} \eta = \bar{\pi}_k$. This being said, let $\mathfrak{p}_0$ be any fixed point of U. Since U is connected, it is easily seen that, for each $\mathfrak{p} \in U$, there exists a differentiable chain $\gamma(\mathfrak{p})$ such that $\partial\gamma(\mathfrak{p}) = \mathfrak{p} - \mathfrak{p}_0$. In view of what we have proved a few lines above, the number $\varphi(\mathfrak{p}) = \int_{\gamma(\mathfrak{p})} (\eta - \bar{\omega})$ depends only on $\mathfrak{p}$, not on the choice of $\gamma(\mathfrak{p})$. We shall see that the function φ defined in this way is analytic on U and that $d\varphi = \eta - \bar{\omega}$. Let V be an open disc contained in U. Since η and ω admit primitives on V, there exists an analytic function φ_V on V such that $d\varphi_V = \eta - \bar{\omega}$ on V. Let $\mathfrak{p}$ and $\mathfrak{q}$ be points of V, and let τ be a differentiable 1-simplex on V such that $\partial\tau = \mathfrak{q} - \mathfrak{p}$. Then it follows immediately from the fact that $\int_\tau \bar{\omega}$ is the imaginary conjugate of $\int_\tau \omega$ that $\int_\tau (\eta - \bar{\omega}) = \varphi_V(\mathfrak{q}) - \varphi_V(\mathfrak{p})$. On the other hand, $\tau + \gamma(\mathfrak{p}) - \gamma(\mathfrak{q})$ is a differentiable cycle, whence $\int_\tau (\eta - \bar{\omega}) = \int_{\gamma(\mathfrak{q})} (\eta - \bar{\omega}) - \int_{\gamma(\mathfrak{p})} (\eta - \bar{\omega}) = \varphi(\mathfrak{q}) - \varphi(\mathfrak{p})$. Thus we see that $\varphi - \varphi_V$ is constant on V, which proves that φ is analytic on V and that $d\varphi = \eta - \bar{\omega}$.

Now we write

$$\begin{aligned} j(\eta, \omega) &= (2\pi(-1)^{1/2})^{-1} \sum_{i=1}^{h} \int_{\gamma_i} f_i\,\omega \\ &= (2\pi(-1)^{1/2})^{-1} \sum_{i=1}^{h} \int_{\gamma_i} (f_i - \varphi)\omega + (2\pi(-1)^{1/2})^{-1} \sum_{i=1}^{h} \int_{\gamma_i} \varphi\omega, \end{aligned}$$

where γ_i is defined as in the proof of formula (1). Denote by U' the complement of the set $\bigcup_{i=1}^h D_i$. Then we have, by formula (1), $\sum_{i=1}^h \int_{\gamma_i} \varphi\omega = -\int_{U'} d(\varphi\omega)$, Since $d\omega = 0$, $\omega \square \eta = 0$, we have $d(\varphi\omega) = d\varphi \square \omega = (\eta - \bar{\omega}) \square \omega = -\bar{\omega} \square \omega$. whence $(2\pi(-1)^{1/2})^{-1} \sum_{i=1}^h \int_{\gamma_i} \varphi\omega = (2\pi(-1)^{1/2})^{-1} \int_{U'} \bar{\omega} \square \omega$.

Consider now the quantity $\int_{\gamma_i} (f_i - \varphi)\omega$. Let V_i' be an open x_i-disc of x_i-center $\mathfrak{p}_i$ which contains D_i but which does not contain any pole of f_i other than $\mathfrak{p}_i$ or any of the points $\mathfrak{p}_j$ for $j \neq i$. Then $\eta - df_i$ is holomorphic on V_i' and has therefore a primitive g_i on this set. Let also λ_i be a primitive of ω on V_i'. Then $d\bar{\lambda}_i - dg_i = df_i - \eta + \eta - (\eta - \bar{\omega}) = df_i - d\varphi$. This means that $f_i - \varphi$ differs only by a constant from $\bar{\lambda}_i - g_i$ on $V_i' - \{\mathfrak{p}_i\}$, from which it follows that $| f_i - \varphi |$ remains bounded on $V_i' - \{\mathfrak{p}_i\}$. Let $x_{i,1}$ and $x_{i,2}$ be the real and imaginary parts of x_i; then we can write, for $\mathfrak{q} \in V_i'$, $\mathfrak{q} \neq \mathfrak{p}_i$, $((f_i - \varphi)\omega)_{\mathfrak{q}} = A_{i,1}(x_{i,1}(\mathfrak{q}), x_{i,2}(\mathfrak{q}))(dx_{i,1})_{\mathfrak{q}} + A_{i,2}(x_{i,1}(\mathfrak{q}), x_{i,2}(\mathfrak{q}))(dx_{i,2})_{\mathfrak{q}}$, where $A_{i,1}$ and $A_{i,2}$ are bounded analytic functions on $x_i(V_i' - \{\mathfrak{p}_i\})$. Thus we have

$$\int_{\gamma_i} (f_i - \varphi)\omega = r_i \int_0^{2\pi} (-A_{i,1}(r_i \cos\theta, r_i \sin\theta) \sin\theta$$
$$+ A_{i,2}(r_i \cos\theta, r_i \sin\theta) \cos\theta)\, d\theta$$

which proves that $\int_{\gamma_i} (f_i - \varphi)\omega$ tends to 0 when r_i tends to 0.

Let ε be a number >0 and <1. Denote by $D_{i,\varepsilon}$ the closed disc of x_i-center $\mathfrak{p}_i$ and x_i-radius $r_i\varepsilon$, and by U_ε' the complement of the set $\bigcup_{i=1}^h D_{i,\varepsilon}$. Then we have

$$j(\eta, \omega) = \lim_{\varepsilon \to 0} \int_{U_\varepsilon'} \frac{1}{2\pi(-1)^{1/2}} \bar{\omega} \square \omega.$$

Now, let $\mathfrak{p}$ be any point of U and let x be a uniformizing variable at $\mathfrak{p}$. If $\omega = u\, dx$, $u \in R$, we have $(2\pi(-1)^{1/2})^{-1}\bar{\omega}_{\mathfrak{p}} \square \omega_{\mathfrak{p}} = (2\pi(-1)^{1/2})^{-1} | u(\mathfrak{p}) |^2 (d\bar{x})_{\mathfrak{p}} \square (dx)_{\mathfrak{p}} = (2\pi)^{-1} | u(p) |^2(-1)^{1/2}(dx)_{\mathfrak{p}} \square (d\bar{x})_{\mathfrak{p}}$. Thus we see that $(2\pi(-1)^{1/2})^{-1} \bar{\omega} \square \omega$ is positive at $\mathfrak{p}$ provided $u(\mathfrak{p}) \neq 0$, i.e. provided $\mathfrak{p}$ is not a zero of ω. Since ω has only a finite number of zeros, we see that $(2\pi(-1)^{1/2})^{-1} \int_{U_\varepsilon'} \bar{\omega} \square \omega$ is a postive real number, and that this number increases when ε decreases. We conclude that $j(\eta, \omega)$ is a positive real number. Theorem 18 is thereby proved.

COROLLARY 1. *The notation being as in Theorem* 18, *the numbers* $\pi_1, \cdots, \pi_g$ *cannot all be* 0.

COROLLARY 2. *The notation being as in Theorem* 18, *let furthermore* $a_1, \cdots, a_g$ *be any* g *complex numbers. Then there exists a uniquely determined differential* ω *of the first kind such that* $\int_{z_k} \omega = a_k$ $(1 \leqq k \leqq g)$.

Let $\{\eta_1, \cdots, \eta_g\}$ be a base of the space of differentials of the first kind of R; set $\int_{z_k} \eta_l = \pi_{kl}(1 \leqq k, l \leqq g)$. Then it follows from Corollary 1 that the system of equations $\sum_{l=1}^g x_l \pi_{kl} = 0$ $(1 \leqq k \leqq g)$ has no non trivial solution. Therefore, the system of equations $\sum_{l=1}^g x_l \pi_{kl} = a_k$ $(1 \leqq k \leqq g)$ has a unique solution, which proves Corollary 2.

INDEX

Roman numbers refer to chapters, arabic numbers refer to sections.